LONDON MATHEMATICAL SOCIETY LECTURE NOTE SERIES

Managing Editor: Professor N.J. Hitchin, Mathematical Institute,
University of Oxford, 24–29 St Giles, Oxford OX1 3LB, United Kingdom

The titles below are available from booksellers, or, in case of difficulty, from Cambridge University Press.

46 *p*-adic Analysis: a short course on recent work, N. KOBLITZ
59 Applicable differential geometry, M. CRAMPIN & F.A.E. PIRANI
66 Several complex variables and complex manifolds II, M.J. FIELD
86 Topological topics, I.M. JAMES (ed)
88 FPF ring theory, C. FAITH & S. PAGE
90 Polytopes and symmetry, S.A. ROBERTSON
96 Diophantine equations over function fields, R.C. MASON
97 Varieties of constructive mathematics, D.S. BRIDGES & F. RICHMAN
99 Methods of differential geometry in algebraic topology, M. KAROUBI & C. LERUSTE
100 Stopping time techniques for analysts and probabilists, L. EGGHE
104 Elliptic structures on 3-manifolds, C.B. THOMAS
105 A local spectral theory for closed operators, I. ERDELYI & WANG SHENGWANG
107 Compactification of Siegel moduli schemes, C.-L. CHAI
109 Diophantine analysis, J. LOXTON & A. VAN DER POORTEN (eds)
113 Lectures on the asymptotic theory of ideals, D. REES
116 Representations of algebras, P.J. WEBB (ed)
119 Triangulated categories in the representation theory of finite-dimensional algebras, D. HAPPEL
121 Proceedings of *Groups - St Andrews 1985*, E. ROBERTSON & C. CAMPBELL (eds)
128 Descriptive set theory and the structure of sets of uniqueness, A.S. KECHRIS & A. LOUVEAU
130 Model theory and modules, M. PREST
131 Algebraic, extremal & metric combinatorics, M.-M. DEZA, P. FRANKL & I.G. ROSENBERG (eds)
138 Analysis at Urbana, II, E. BERKSON, T. PECK, & J. UHL (eds)
139 Advances in homotopy theory, S. SALAMON, B. STEER & W. SUTHERLAND (eds)
140 Geometric aspects of Banach spaces, E.M. PEINADOR & A. RODES (eds)
141 Surveys in combinatorics 1989, J. SIEMONS (ed)
144 Introduction to uniform spaces, I.M. JAMES
146 Cohen-Macaulay modules over Cohen-Macaulay rings, Y. YOSHINO
148 Helices and vector bundles, A.N. RUDAKOV *et al*
149 Solitons, nonlinear evolution equations and inverse scattering, M. ABLOWITZ & P. CLARKSON
150 Geometry of low-dimensional manifolds 1, S. DONALDSON & C.B. THOMAS (eds)
151 Geometry of low-dimensional manifolds 2, S. DONALDSON & C.B. THOMAS (eds)
152 Oligomorphic permutation groups, P. CAMERON
153 L-functions and arithmetic, J. COATES & M.J. TAYLOR (eds)
155 Classification theories of polarized varieties, TAKAO FUJITA
158 Geometry of Banach spaces, P.F.X. MÜLLER & W. SCHACHERMAYER (eds)
159 Groups St Andrews 1989 volume 1, C.M. CAMPBELL & E.F. ROBERTSON (eds)
160 Groups St Andrews 1989 volume 2, C.M. CAMPBELL & E.F. ROBERTSON (eds)
161 Lectures on block theory, BURKHARD KÜLSHAMMER
163 Topics in varieties of group representations, S.M. VOVSI
164 Quasi-symmetric designs, M.S. SHRIKANDE & S.S. SANE
166 Surveys in combinatorics, 1991, A.D. KEEDWELL (ed)
168 Representations of algebras, H. TACHIKAWA & S. BRENNER (eds)
169 Boolean function complexity, M.S. PATERSON (ed)
170 Manifolds with singularities and the Adams-Novikov spectral sequence, B. BOTVINNIK
171 Squares, A.R. RAJWADE
172 Algebraic varieties, GEORGE R. KEMPF
173 Discrete groups and geometry, W.J. HARVEY & C. MACLACHLAN (eds)
174 Lectures on mechanics, J.E. MARSDEN
175 Adams memorial symposium on algebraic topology 1, N. RAY & G. WALKER (eds)
176 Adams memorial symposium on algebraic topology 2, N. RAY & G. WALKER (eds)
177 Applications of categories in computer science, M. FOURMAN, P. JOHNSTONE & A. PITTS (eds)
178 Lower K- and L-theory, A. RANICKI
179 Complex projective geometry, G. ELLINGSRUD *et al*
180 Lectures on ergodic theory and Pesin theory on compact manifolds, M. POLLICOTT
181 Geometric group theory I, G.A. NIBLO & M.A. ROLLER (eds)
182 Geometric group theory II, G.A. NIBLO & M.A. ROLLER (eds)
183 Shintani zeta functions, A. YUKIE
184 Arithmetical functions, W. SCHWARZ & J. SPILKER
185 Representations of solvable groups, O. MANZ & T.R. WOLF
186 Complexity: knots, colourings and counting, D.J.A. WELSH
187 Surveys in combinatorics, 1993, K. WALKER (ed)
188 Local analysis for the odd order theorem, H. BENDER & G. GLAUBERMAN
189 Locally presentable and accessible categories, J. ADAMEK & J. ROSICKY
190 Polynomial invariants of finite groups, D.J. BENSON
191 Finite geometry and combinatorics, F. DE CLERCK *et al*
192 Symplectic geometry, D. SALAMON (ed)
194 Independent random variables and rearrangement invariant spaces, M. BRAVERMAN
195 Arithmetic of blowup algebras, WOLMER VASCONCELOS
196 Microlocal analysis for differential operators, A. GRIGIS & J. SJÖSTRAND
197 Two-dimensional homotopy and combinatorial group theory, C. HOG-ANGELONI *et al*

198 The algebraic characterization of geometric 4-manifolds, J.A. HILLMAN
199 Invariant potential theory in the unit ball of $\mathbf{C}^n$, MANFRED STOLL
200 The Grothendieck theory of dessins d'enfant, L. SCHNEPS (ed)
201 Singularities, JEAN-PAUL BRASSELET (ed)
202 The technique of pseudodifferential operators, H.O. CORDES
203 Hochschild cohomology of von Neumann algebras, A. SINCLAIR & R. SMITH
204 Combinatorial and geometric group theory, A.J. DUNCAN, N.D. GILBERT & J. HOWIE (eds)
205 Ergodic theory and its connections with harmonic analysis, K. PETERSEN & I. SALAMA (eds)
207 Groups of Lie type and their geometries, W.M. KANTOR & L. DI MARTINO (eds)
208 Vector bundles in algebraic geometry, N.J. HITCHIN, P. NEWSTEAD & W.M. OXBURY (eds)
209 Arithmetic of diagonal hypersurfaces over finite fields, F.Q. GOUVÊA & N. YUI
210 Hilbert C*-modules, E.C. LANCE
211 Groups 93 Galway / St Andrews I, C.M. CAMPBELL *et al* (eds)
212 Groups 93 Galway / St Andrews II, C.M. CAMPBELL *et al* (eds)
214 Generalised Euler-Jacobi inversion formula and asymptotics beyond all orders, V. KOWALENKO *et al*
215 Number theory 1992–93, S. DAVID (ed)
216 Stochastic partial differential equations, A. ETHERIDGE (ed)
217 Quadratic forms with applications to algebraic geometry and topology, A. PFISTER
218 Surveys in combinatorics, 1995, PETER ROWLINSON (ed)
220 Algebraic set theory, A. JOYAL & I. MOERDIJK
221 Harmonic approximation, S.J. GARDINER
222 Advances in linear logic, J.-Y. GIRARD, Y. LAFONT & L. REGNIER (eds)
223 Analytic semigroups and semilinear initial boundary value problems, KAZUAKI TAIRA
224 Computability, enumerability, unsolvability, S.B. COOPER, T.A. SLAMAN & S.S. WAINER (eds)
225 A mathematical introduction to string theory, S. ALBEVERIO, J. JOST, S. PAYCHA, S. SCARLATTI
226 Novikov conjectures, index theorems and rigidity I, S. FERRY, A. RANICKI & J. ROSENBERG (eds)
227 Novikov conjectures, index theorems and rigidity II, S. FERRY, A. RANICKI & J. ROSENBERG (eds)
228 Ergodic theory of $\mathbf{Z}^d$ actions, M. POLLICOTT & K. SCHMIDT (eds)
229 Ergodicity for infinite dimensional systems, G. DA PRATO & J. ZABCZYK
230 Prolegomena to a middlebrow arithmetic of curves of genus 2, J.W.S. CASSELS & E.V. FLYNN
231 Semigroup theory and its applications, K.H. HOFMANN & M.W. MISLOVE (eds)
232 The descriptive set theory of Polish group actions, H. BECKER & A.S. KECHRIS
233 Finite fields and applications, S. COHEN & H. NIEDERREITER (eds)
234 Introduction to subfactors, V. JONES & V.S. SUNDER
235 Number theory 1993–94, S. DAVID (ed)
236 The James forest, H. FETTER & B. GAMBOA DE BUEN
237 Sieve methods, exponential sums, and their applications in number theory, G.R.H. GREAVES *et al*
238 Representation theory and algebraic geometry, A. MARTSINKOVSKY & G. TODOROV (eds)
239 Clifford algebras and spinors, P. LOUNESTO
240 Stable groups, FRANK O. WAGNER
241 Surveys in combinatorics, 1997, R.A. BAILEY (ed)
242 Geometric Galois actions I, L. SCHNEPS & P. LOCHAK (eds)
243 Geometric Galois actions II, L. SCHNEPS & P. LOCHAK (eds)
244 Model theory of groups and automorphism groups, D. EVANS (ed)
245 Geometry, combinatorial designs and related structures, J.W.P. HIRSCHFELD *et al*
246 *p*-Automorphisms of finite *p*-groups, E.I. KHUKHRO
247 Analytic number theory, Y. MOTOHASHI (ed)
248 Tame topology and o-minimal structures, LOU VAN DEN DRIES
249 The atlas of finite groups: ten years on, ROBERT CURTIS & ROBERT WILSON (eds)
250 Characters and blocks of finite groups, G. NAVARRO
251 Gröbner bases and applications, B. BUCHBERGER & F. WINKLER (eds)
252 Geometry and cohomology in group theory, P. KROPHOLLER, G. NIBLO, R. STÖHR (eds)
253 The *q*-Schur algebra, S. DONKIN
254 Galois representations in arithmetic algebraic geometry, A.J. SCHOLL & R.L. TAYLOR (eds)
255 Symmetries and integrability of difference equations, P.A. CLARKSON & F.W. NIJHOFF (eds)
256 Aspects of Galois theory, HELMUT VÖLKLEIN *et al*
257 An introduction to noncommutative differential geometry and its physical applications 2ed, J. MADORE
258 Sets and proofs, S.B. COOPER & J. TRUSS (eds)
259 Models and computability, S.B. COOPER & J. TRUSS (eds)
260 Groups St Andrews 1997 in Bath, I, C.M. CAMPBELL *et al*
261 Groups St Andrews 1997 in Bath, II, C.M. CAMPBELL *et al*
263 Singularity theory, BILL BRUCE & DAVID MOND (eds)
264 New trends in algebraic geometry, K. HULEK, F. CATANESE, C. PETERS & M. REID (eds)
265 Elliptic curves in cryptography, I. BLAKE, G. SEROUSSI & N. SMART
267 Surveys in combinatorics, 1999, J.D. LAMB & D.A. PREECE (eds)
268 Spectral asymptotics in the semi-classical limit, M. DIMASSI & J. SJÖSTRAND
269 Ergodic theory and topological dynamics, M.B. BEKKA & M. MAYER
270 Analysis on Lie Groups, N.T. VAROPOULOS & S. MUSTAPHA
271 Singular perturbations of differential operators, S. ALBEVERIO & P. KURASOV
272 Character theory for the odd order function, T. PETERFALVI
273 Spectral theory and geometry, E.B. DAVIES & Y. SAFAROV (eds)
274 The Mandelbrot set, theme and variations, TAN LEI (ed)
276 Singularities of plane curves, E. CASAS-ALVERO
278 Global attractors in abstract parabolic problems, J.W. CHOLEWA & T. DLOTKO
279 Topics in symbolic dynamics and applications, F. BLANCHARD, A. MAASS & A. NOGUEIRA (eds)
280 Characers and Automorphism Groups of Compact Riemann Surfaces, T. BREUER
281 Explicit birational geometry of 3-folds, ALESSIO CORTI & MILES REID (eds)

London Mathematical Society Lecture Note Series. 282

Auslander-Buchweitz Approximations of Equivariant Modules

Mitsuyasu Hashimoto
Nagoya University

CAMBRIDGE
UNIVERSITY PRESS

PUBLISHED BY THE PRESS SYNDICATE OF THE UNIVERSITY OF CAMBRIDGE
The Pitt Building, Trumpington Street, Cambridge, United Kingdom

CAMBRIDGE UNIVERSITY PRESS
The Edinburgh Building, Cambridge, CB2 2RU, UK
40 West 20th Street, New York, NY 10011–4211, USA
10 Stamford Road, Oakleigh, VIC 3166, Australia
Ruiz de Alarcón 13, 28014 Madrid, Spain

http://www.cambridge.org

First published 2000

Printed in the United Kingdom at the University Press, Cambridge

A catalogue record for this book is available from the British Library

ISBN 0 521 79696 2 paperback

To Tomoko

Contents

Introduction

Let R be a commutative ring, and G an affine flat group scheme over R. We say that A is a (commutative) G-algebra if A is a G-module and is a (commutative) R-algebra, and the product $A \otimes A \to A$ is G-linear. We say that M is a (G, A)-module (or G-equivariant A-module) if M is an A-module and is a G-module, and the A-action $A \otimes M \to M$ is G-linear. A (G, A)-linear map simply means a G-linear A-linear map. Thus, we get an abelian category ${}_{G,A}\mathbb{M}$ with enough injectives. The main purpose of these notes is to discuss homological aspects of (G, A)-modules, from the viewpoint of commutative ring theory of R and A.

In particular, we study various (weak) Auslander–Buchweitz contexts which appear there. The theory of Cohen–Macaulay approximations over Cohen–Macaulay local rings by Auslander and Buchweitz [10] contributes greatly to the new developments in commutative ring theory [148]. On the other hand, their theory of approximations is given in rather general form as a theory of abelian categories [10, 11], and its applications are appearing in so many topics of algebras. (Weak) Auslander–Buchweitz contexts (I.1.12) are one of its formulations.

Auslander and Reiten [11] proved that, in the category of finite modules over a finite dimensional algebra over a field, Auslander–Buchweitz contexts and basic cotilting modules are in one-to-one correspondence. Miyachi [112] proved that we have an Auslander–Buchweitz context from a cotilting module in a rather general situation. Cohen–Macaulay approximations over Cohen–Macaulay local rings with canonical modules are a special case.

Moreover, as an application of the Auslander–Reiten correspondence above, C. M. Ringel proved the existence of Δ-good approximations over a quasi-hereditary algebra [131]. From Ringel's theorem, the existence of Δ-good approximations of finite dimensional representations of reductive groups immediately follows, using Schur algebras. Quasi-hereditary algebras were originally introduced by Cline–Parshall–Scott [36, 134] to study the representation categories of reductive groups. S. Donkin studied tilting modules of reductive groups [50], and this direction is developing. Note that a tilting module of a reductive group, which is one of the important consequences of Ringel's approximations, is the cotilting module corresponding to Ringel's Auslander–Buchweitz context over a Schur algebra by the Auslander–Reiten correspondence. It is also a tilting module in the sense of [113], and thus the name 'tilting' is used for both concepts.

Our goal is to construct (weak) Auslander–Buchweitz contexts in the category of A-finite (G, A)-modules when G is a split reductive group over a commutative noetherian ring R, and A is a commutative noetherian G-algebra. These approximations generalize Cohen–Macaulay approximations in commutative ring theory and Ringel's approximations in the representation theory of algebraic groups simultaneously.

Some applications are also included. As an application of the construction of Ext_A-groups as equivariant modules, we prove a new criterion of Cohen–Macaulay, Gorenstein, local complete intersection, and regular properties of noetherian G-algebras (Matijevic–Roberts type theorem). The case where G is a split torus is known as a criterion for multi-graded rings by Goto–Watanabe [60]. Homological theory for graded modules over graded rings by Goto–Watanabe [59, 60] is a strong motivation for the study of homological theory of equivariant modules here.

There are applications of the Auslander–Buchweitz theory of equivariant modules, too. The first one is an application to the problem of resolutions of determinantal rings. We prove that there is a (non-minimal) equivariant resolution of a determinantal ring with the same length as the minimal free resolution, and with each term a direct summand of a finite direct sum of tensor products of exterior powers. For the maximal minor case, Buchsbaum [28] constructed such a resolution explicitly.

The second one is an application to invariant theory. Let G be a reductive group over a perfect field of positive characteristic, V a finite dimensional G-module, and assume that the symmetric algebra $S := \mathrm{Sym}\, V$ admits a good filtration as a G-module. Then S^G is strongly F-regular [76]. This result yields a similar result over an arbitrary base ring, using (u-)good modules over an arbitrary base discussed here (in preparation, and we omit it).

Other related results on integral representations, most of which are necessary for Auslander–Buchweitz theory, are included. Highest weight theory for coalgebras over an arbitrary base, including the theory of Schur algebras and tilting modules, is also new here. Note that representations and Schur algebras over a Dedekind ring are treated in [48, 51], and quasi-hereditary algebras over an arbitrary base are treated in [38]. Our approach is slightly different, and is a natural extension of the original approach for the field base due to Donkin in [48].

In Chapter I, we introduce basic notions from each algebraic topic used later. Namely, from homological algebra, commutative ring theory, theory of schemes and sheaves, Hopf algebra theory, and representation theory of algebraic groups and algebras. We have devoted considerable space to this chapter. This is partly to make the notes accessible to a wider range of readers interested in algebra, including graduate students, and partly to lay a foundation for the homological theory of Hopf algebras over an arbitrary base ring. We only assume that readers are familiar with the elements of each theory. In general, the results in the chapter are preliminary results and/or well-known, and most of the known results are listed without proofs but with references as far as possible.

Some generalities are included only for record. For example, (I.5.1) and (I.5.2) will not be used later very much. Similarly for the summary on exact categories and their derived categories in (I.1.3)–(I.1.6), and the reader may skip these subsections in the first reading, provided he or she is familiar with the homological algebra of the usual abelian categories (the differences from the usual homological algebra will be used only in (I.3.6)).

A G-module is nothing but an $R[G]$-comodule. A (G, A)-module is nothing but an $(R[G], A)$-Hopf module. For future reference, we have included a long section on elementary homological algebras of comodules and Hopf modules (I.3).

In chapter II, we construct Ext_A and Tor^A functors of (G, A)-modules. When G is the split torus $\mathbb{G}_m^n$, then to say that A is a G-algebra is the same as to say that A is $\mathbb{Z}^n$-graded. Similarly, a (G, A)-module is nothing but a graded A-module. Homological algebra of $\mathbb{Z}^n$-graded rings and modules was studied by Goto–Watanabe [59, 60] from the viewpoint of commutative ring theory. In their study, the facts that if M and N are graded A-modules then $\mathrm{Tor}_i^A(M, N)$ is again graded in a natural way, and that if A is noetherian and M is A-finite then $\mathrm{Ext}_A^i(M, N)$ is also graded in a natural way, are used effectively. The purpose here is to generalize these facts to more general G. Moreover, some results on rings and modules are generalized to those of quasi-coherent sheaves over schemes in (II.2.3).

In (II.2.4), we give an application of the results in chapter II. Namely, we generalize a result on homological properties (Cohen–Macaulay, Gorenstein, l.c.i., and regular) for graded rings by Matijevic–Roberts and others. Moreover, in (II.2.2) we give a sufficient condition for the category of G-modules to be a full subcategory of the module category of the hyperalgebra of G. We also discuss the projectivity of the coordinate ring in (II.2.2) and give a new criterion.

In chapter III, we generalize Ringel's approximation for quasi-hereditary algebras, which is an important example of an Auslander–Buchweitz context, to those for some general coalgebras over an arbitrary noetherian commutative ring R. We discuss a variation (or a coalgebra version) of the theory by Cline–Parshall–Scott and Donkin (over a field), before we consider a general base ring.

All the results in this chapter are about G-modules (or C-comodules), and the commutative algebra A (on which G or C acts or coacts) does not appear. However, as a G-module is an R-module, ring theoretic properties of R and module theoretic properties of modules over R play important roles here. Although projectivity of (non-finite) modules is usually less important than flatness in algebraic geometry, the projectivity of the coalgebra C (as an R-module) is very important here, as in Seshadri's important paper [135].

We review the theory of Cline–Parshall–Scott and Donkin, and this part (III.1) is reasonably self-contained. Everything is done in the coalgebra language here, for later generalization to the case where the base ring is arbitrary.

The notion of good modules (or modules with good filtrations) is divided into two, u-good modules and good modules. U-goodness, which is an abbreviation for universal goodness, means goodness which is stable under base change. A good module is not u-good in general. However, we prove that an R-finite R-projective good module is u-good.

In (III.4.4), we construct an Auslander–Buchweitz context which generalizes both Ringel's approximation and Cohen–Macaulay approximation simultaneously. A generalization of Sharp's theorem by Avramov–Foxby (Theorem I.4.10.19) is important here.

In chapter IV, we construct an Auslander–Buchweitz context in the category of (G, A)-modules. We only consider positively graded A and graded (G, A)-modules M, with each homogeneous component of A or M a G-submodule. Such a situation is realized on replacing G by another reductive group, see (IV.1.1). As in chapter III, we have two (weak) Auslander–Buchweitz contexts corresponding to regular and Cohen–Macaulay properties. The results are new, even if we assume that the base ring R is a field. In (IV.2), as an example and an application, we study resolutions of determinantal rings.

Determinantal rings have always been interesting examples in commutative ring theory. Subsection (IV.2.1) is a survey of the problem of resolutions of determinantal rings. Applications of the equivariant method have been successful in studying determinantal rings. A. Lascoux and Pragacz–Weyman determined the equivariant minimal free resolution of $A = S/I_t$ as a graded S-module in characteristic zero. In particular, the Betti numbers $\dim_k \operatorname{Tor}_i^S(k, A)$ were determined with the determination of the irreducible decomposition of $\operatorname{Tor}_i^S(k, A)$ as representations. After Lascoux's work, the equivariant method has been used to study syzygies of determinantal rings over an arbitrary base ring (see [77]).

As an application of the construction of an Auslander–Buchweitz context in the category of (G, S)-modules, we show that there is an equivariant finite free resolution of S/I_t whose length is equal to $\operatorname{proj.dim}_S S/I_t$ and each term of which consists of $S \otimes T$, with T tilting. This gives a partial answer to the question posed by D. A. Buchsbaum and J. Weyman, and generalizes the generalized Koszul complex by Buchsbaum.

Acknowledgement. These notes are widely extended versions of the informal notes ([74], in Japanese) for the author's lectures at Tokyo Metropolitan University in 1996. Many of the results proved here have been announced in [75]. I am grateful to my distinguished friend Professor Kazuhiko Kurano for providing the opportunity to give the lecture, and to write the notes. He also kindly helped me in error-corrections of the draft. Special thanks are also due to Professor Luchezar L. Avramov, Professor David A. Buchsbaum, Professor Yukio Doi, Professor David Eisenbud, Professor Shiro Goto, Professor Masaharu Kaneda, Professor Jun-ichi Miyachi, Professor Masayoshi

Miyanishi, Professor Shigeru Mukai, Professor Claus Michael Ringel, Professor Mitsuhiro Takeuchi, Professor Takayoshi Wakamatsu, Professor Kei-ichi Watanabe, and Professor Yasuo Yoshinobu for helpful discussions and valuable advice.

Conventions and terminology

An equation of the form $A := B$ reads 'A is defined to be B.'

For a set X, $\#X$ stands for the cardinality of X. The symbols $\mathbb{N}$ and $\mathbb{N}_0$ respectively stand for the set of positive and non-negative integers. Semigroups and rings are always required to have unit elements, and semigroup homomorphisms and ring homomorphisms are always required to preserve unit elements. Subsemigroups and subrings are required to have the unit elements in common. Unit elements are required to act as identities for semigroup actions on sets and ring actions on additive groups. For a semigroup G, we denote by $G^\times$ the group of invertible elements of G. For a ring A, this notation applies to the multiplicative semigroup A, and hence $A^\times$ is the unit group of A.

Throughout these notes, the symbol R always stands for a commutative ring. The symbols $\otimes$ and Hom stand for $\otimes_R$ and Hom_R, respectively. If the ring R in question happens to be a field, then we sometimes let $R = k$, and use k. In this case, $\otimes$ and Hom stand for $\otimes_k$ and Hom_k, respectively.

The word 'scheme' always means a separated scheme. For R-schemes X and Y, $X(Y)$ stands for the set of R-morphisms from Y to X. For a scheme X and an X-scheme Y, we say that Y is algebraic over X if Y is of finite type over X. We say that Y is a *variety* over X if Y is algebraic over X and Y is an integral scheme. If Y is a closed subscheme of X and is integral, then we say that Y is a *subvariety* of X. A *geometric point* of an R-scheme X is an algebraically closed field K which is an R-algebra, together with an R-morphism $\mathrm{Spec}\, K \to X$. If R is not specified, then we assume $R = \mathbb{Z}$. A geometric fiber of an R-morphism $Y \to X$ is the fiber $Y \times_X \mathrm{Spec}\, K$ for some geometric point $\mathrm{Spec}\, K \to X$. For an R-scheme X and a commutative R-algebra R', the base change $\mathrm{Spec}\, R' \times_{\mathrm{Spec}\, R} X$ is sometimes denoted by $R' \otimes X$.

For a scheme X, an *$\mathcal{O}_X$-module* means an $\mathcal{O}_X$-module sheaf. If we want to mean a presheaf, then we call it an $\mathcal{O}_X$-module presheaf. The abelian category of $\mathcal{O}_X$-modules is denoted by ${}_X\mathbb{M}$. The full subcategory of quasi-coherent (resp. coherent) $\mathcal{O}_X$-modules is denoted by $\mathrm{Qco}(X)$ (resp. $\mathrm{Coh}(X)$).

For a commutative ring A, $\mathrm{Spec}\, A$ denotes the prime spectrum of A (and the set of prime ideals of A). For an A-module M and $\mathfrak{p} \in \mathrm{Spec}\, A$, $M_\mathfrak{p}$ denotes the localization of M with respect to the multiplicatively closed subset $A \setminus \mathfrak{p}$ of A. The residue field $A_\mathfrak{p}/\mathfrak{p}A_\mathfrak{p}$ of the local ring $A_\mathfrak{p}$ is denoted by $\kappa(\mathfrak{p})$. For an A-module M, $M(\mathfrak{p})$ stands for the $\kappa(\mathfrak{p})$-vector space $\kappa(\mathfrak{p}) \otimes_A M$. For an A-scheme X, the fiber $\kappa(\mathfrak{p}) \otimes_A X$ is denoted by $X(\mathfrak{p})$. The symbol $\operatorname{Max} A$ stands for the set of maximal ideals of A. For an A-module M, $\operatorname{supp} M$ stands for the support $\{\mathfrak{p} \in \operatorname{Spec} A \mid M_\mathfrak{p} \neq 0\}$ of M. A minimal element of $\operatorname{supp} M$ (with respect to the incidence relation) is called a *minimal prime* of M, and the set of minimal primes of M is denoted by $\operatorname{Min} M$. The set of associated primes of M is denoted by $\operatorname{Ass} M$. A finite A-module means a finitely generated A-module. For an A-module M, the corresponding quasi-coherent sheaf over $\operatorname{Spec} A$ is denoted by $\tilde{M}$.

Let A be a local ring with the unique maximal ideal $\mathfrak{m}$. We express this situation by saying that $(A, \mathfrak{m})$ is a local ring.

For a scheme X, $x \in X$ and an abelian presheaf $\mathcal{M}$ over X, $\mathcal{M}_x$ denotes the stalk of $\mathcal{M}$ at x. The maximal ideal of $\mathcal{O}_{X,x}$ is denoted by $\mathfrak{m}_{X,x}$ or sometimes by $\mathfrak{m}_x$, and the residue field $\mathcal{O}_{X,x}/\mathfrak{m}_x$ is denoted by $\kappa(x)$.

If $f : X \to \operatorname{Spec} R$ is an R-scheme, M is an R-module, and $\mathcal{N} \in {}_X\mathbb{M}$, then we denote $f^*\tilde{M} \otimes_{\mathcal{O}_X} \mathcal{N}$ by $M \otimes \mathcal{N}$.

For an affine R-scheme $X = \operatorname{Spec} A$, we sometimes denote its coordinate ring $A = \Gamma(X, \mathcal{O}_X)$ by $R[X]$.

For a ring A, an A-module means a left A-module, unless otherwise specified. The category of A-modules is denoted by ${}_A\mathbb{M}$. The opposite ring of A is denoted by A^{op}. The category of right A-modules is denoted by $\mathbb{M}_A$ or ${}_{A^{\mathrm{op}}}\mathbb{M}$. For an A-module M, we denote the injective hull (injective envelope) of M by $E_A(M)$.

Some remarks on terminology in category theory are collected in Remark I.1.12.15.

These notes consist of four chapters I – IV, and each chapter consists of several sections, and each section is divided into several subsections. Theorem, paragraph and equation numbering are unified, and any number is of the form `sec.subsec.thm(par,eqn)num`.

Cross references in these notes are in the form

`chap.sec.subsec.thm(par,eqn),`

where `chap` denotes the chapter number (in a capital Roman numeral), and `sec`, `subsec`, `thm`, `par`, `eqn` respectively denote the section, subsection, theorem, paragraph, and the equation number (in Arabic). However, for references from the same chapter, the chapter number is omitted. A reference with only two Arabic numerals (preceded by a chapter number if any) such as (IV.2.2) shows the `[chap.]sec.subsec` number, and refers to the corresponding whole subsection. At the beginning (or sometimes in the middle) of each subsection or paragraph, any general assumption effective throughout the subsection or paragraph is shown, and usually such assumptions are not shown again in each theorem (proposition or lemma) therein.

Chapter I

Background Materials

1 From homological algebra

1.1 Yoneda's lemma

(1.1.1) We denote the category of sets by Set, the category of groups by Grp, and the category of abelian groups by Ab or by ${}_{\mathbb{Z}}\mathbb{M}$. If $\mathcal{C}$ is a category, then we denote the set of objects of $\mathcal{C}$ by $\operatorname{ob}(\mathcal{C})$. However, by abuse of notation, we sometimes denote it also by $\mathcal{C}$. For $M, N \in \mathcal{C}$, the set of morphisms from M to N in $\mathcal{C}$ is denoted by $\mathcal{C}(M, N)$ or $\operatorname{Hom}_{\mathcal{C}}(M, N)$. The opposite category of $\mathcal{C}$ is denoted by $\mathcal{C}^{\mathrm{op}}$.

(1.1.2) For categories $\mathcal{C}$, $\mathcal{C}'$, a functor $F : \mathcal{C} \to \mathcal{C}'$ and a full subcategory $\mathcal{S}$ of $\mathcal{C}$, we denote the full subcategory of $\mathcal{C}'$ consisting of objects isomorphic to some $F(S)$ for $S \in \mathcal{S}$ by $F(\mathcal{S})$. For a category $\mathcal{C}$, we denote both a single null object of $\mathcal{C}$ and the set of all null objects by the same symbol 0, which will not cause confusion. For $C \in \mathcal{C}$, the category of objects over C is denoted by $\mathcal{C}/C$. An object of $\mathcal{C}/C$ is a morphism $\varphi : D \to C$ with the codomain C, and a morphism from $\varphi : D \to C$ to $\psi : E \to C$ is a morphism $f : D \to E$ of $\mathcal{C}$ such that $\psi f = \varphi$. The composition of morphisms of $\mathcal{C}/C$ is the same as that of $\mathcal{C}$.

(1.1.3) Let $\mathcal{U}$ be a *universe* (see, e.g., [9]) and $\mathcal{C}$ a category. We say that $\mathcal{C}$ is a $\mathcal{U}$-category if $\mathcal{C}(C, C') \in \mathcal{U}$ for objects C, C' of $\mathcal{C}$. We fix once for all a universe $\mathcal{U}$, and we only consider this universe $\mathcal{U}$ unless otherwise specified. An element of $\mathcal{U}$ is sometimes referred as a *small set.* Unless otherwise specified, a category means a $\mathcal{U}$-category. However, when we form a new category from some $\mathcal{U}$-categories, exceptions (see (1.1.5) and (1.4.7)) may apply. If we need to emphasize a category is a $\mathcal{U}$-category, we use the expression 'category with small hom sets.'

The category Set means the category of elements of $\mathcal{U}$, and the category Grp means the category of small groups (i.e., groups in $\mathcal{U}$), and so on, and these concrete categories are also $\mathcal{U}$-categories.

We say that a category $\mathcal{C}$ is *small* if $\mathrm{ob}(\mathcal{C}) \in \mathcal{U}$. We say that a category $\mathcal{C}$ is *svelte* (or *skeletally small*) if there exists a small full subcategory $\mathcal{D}$ of $\mathcal{C}$ such that the inclusion $\mathcal{D} \hookrightarrow \mathcal{C}$ is an equivalence.

(1.1.4) Let $\mathcal{C}$ be a category, and R a commutative ring. We say that $\mathcal{C}$ is an *R-category* if $\mathcal{C}(M,N)$ is an R-module for $M,N \in \mathcal{C}$, and the composition $\mathcal{C}(M,N) \times \mathcal{C}(L,M) \to \mathcal{C}(L,N)$ is R-bilinear. A $\mathbb{Z}$-category is also called an $\underline{\mathrm{Ab}}$-category, or a *preadditive category.* A preadditive category with finite direct products (in particular, with a terminal object) is called an *additive category.* A finite direct product in an additive category is naturally isomorphic to the coproduct, and in particular, the category has a null object.

Let $\mathcal{A}$ and $\mathcal{B}$ be R-categories, and $F : \mathcal{A} \to \mathcal{B}$ a functor. We say that F is an *R-linear functor* if the canonical map

$$F : \mathrm{Hom}_{\mathcal{A}}(a,b) \to \mathrm{Hom}_{\mathcal{B}}(Fa, Fb)$$

is R-linear for any $a, b \in \mathcal{A}$. A $\mathbb{Z}$-linear functor is called an *additive functor.*

(1.1.5) For categories $\mathcal{A}$ and $\mathcal{B}$, we denote the set of functors from $\mathcal{A}$ to $\mathcal{B}$ by $\mathrm{Func}(\mathcal{A},\mathcal{B})$. For $F, G \in \mathrm{Func}(\mathcal{A},\mathcal{B})$, we denote the set of natural transformations from F to G by $\mathrm{Nat}(F,G)$. Note that $\mathrm{Func}(\mathcal{A},\mathcal{B})$ is a (not necessarily small) category with $\mathrm{Nat}(F,G)$ its hom set.

For $A \in \mathcal{A}$, when we define $y(A) := \mathrm{Hom}_{\mathcal{A}}(?, A)$, we get a functor $y : \mathcal{A} \to \mathrm{Func}(\mathcal{A}^{\mathrm{op}}, \underline{\mathrm{Set}})$. The following is well-known as Yoneda's lemma.

Lemma 1.1.6 *Let $\mathcal{A}$ be a category, and $T : \mathcal{A}^{\mathrm{op}} \to \underline{\mathrm{Set}}$ a functor. Then we have that the natural map $Y : T \to \mathrm{Nat}(y(?), T)$ given by*

$$(Y(t))(\varphi) = (T(\varphi))(t) \quad (\text{for } A, B \in \mathrm{ob}(\mathcal{A}),\ t \in T(A),\ \varphi \in \mathcal{A}(B,A))$$

is an isomorphism, whose inverse is given by

$$\mathrm{Nat}(y(A), T) \to T(A) \qquad (f \mapsto f_A(\mathrm{id}_A)).$$

In particular, considering the case $T = y(C)$ for some $C \in \mathcal{A}$, the map

$$y : y(C)(A) = \mathcal{A}(A,C) \to \mathrm{Nat}(y(A), y(C)) \qquad (t \mapsto y(t) \text{ for } t \in \mathcal{A}(A,C))$$

is an isomorphism. Hence, the functor $y : \mathcal{A} \to \mathrm{Func}(\mathcal{A}^{\mathrm{op}}, \underline{\mathrm{Set}})$ is fully faithful.

(1.1.7) As the functor y is a full embedding, we sometimes identify $\mathcal{A}$ with the full subcategory $y(\mathcal{A})$ of $\mathrm{Func}(\mathcal{A}^{\mathrm{op}}, \underline{\mathrm{Set}})$, and $y(A)$ is denoted simply by A for $A \in \mathcal{A}$. For $F \in \mathrm{Func}(\mathcal{A}^{\mathrm{op}}, \underline{\mathrm{Set}})$, we have $A \in \mathcal{A}$ such that $F \cong y(A)$ is unique up to isomorphisms, if it exists. If such an A exists, then we say that F is *representable*, and we say that F is represented by A. Thus, $F \in \mathrm{Func}(\mathcal{A}^{\mathrm{op}}, \underline{\mathrm{Set}})$ is representable if and only if $F \in y(\mathcal{A})$.

Similarly, a covariant functor $G \in \mathrm{Func}(\mathcal{A}, \underline{\mathrm{Set}})$ is said to be representable if G is isomorphic to $\mathcal{A}(A, ?)$ for some $A \in \mathrm{ob}(\mathcal{A})$.

1.2 Adjoint functors and limits

(1.2.1) Let $\mathcal{A}$ and $\mathcal{B}$ be categories, and $F : \mathcal{A} \to \mathcal{B}$ a functor with the right adjoint $G : \mathcal{B} \to \mathcal{A}$. Namely, let us assume that $G : \mathcal{B} \to \mathcal{A}$ is a functor, and there exists some isomorphism

$$\Phi_{A,B} : \mathcal{B}(FA, B) \to \mathcal{A}(A, GB)$$

which is natural with respect to A and B. When we define

$$u_A := \Phi_{A,FA}(1_{FA}) : A \to (GF)A$$

for $A \in \mathcal{A}$, then u is a natural map from $\mathrm{Id}_{\mathcal{A}}$ to GF. We call u the *unit* of adjunction. Similarly,

$$\varepsilon_B := \Phi^{-1}_{GB,B}(1_{GB}) : (FG)B \to B$$

is natural with respect to B, and we call $\varepsilon : FG \to \mathrm{Id}_{\mathcal{B}}$ the *counit* of adjunction. They satisfy the relation

$$(1.2.2) \qquad (\varepsilon F) \circ (Fu) = 1_F, \qquad (G\varepsilon) \circ (uG) = 1_G.$$

Conversely, if two functors $F : \mathcal{A} \to \mathcal{B}$ and $G : \mathcal{B} \to \mathcal{A}$ and natural transformations $u : \mathrm{Id}_{\mathcal{A}} \to GF$ and $\varepsilon : FG \to \mathrm{Id}_{\mathcal{B}}$ are given and the relation (1.2.2) is satisfied, then an equivalence Φ is defined by $\Phi_{A,B}(f) := G(f) \circ u_A$ for $A \in \mathcal{A}$, $B \in \mathcal{B}$ and $f \in \mathcal{B}(FA, B)$, and G is right adjoint to F. The natural maps u and ε determined by this adjunction agree with the original ones, respectively.

Note that a functor right adjoint to F is unique up to equivalence. This follows easily from Lemma 1.1.6.

Let $\mathcal{A}$ and $\mathcal{B}$ be categories. It is easy to see that $\mathrm{Func}(\mathcal{A}, \mathcal{B})$ is a category with small hom sets, if $\mathcal{A}$ is svelte. An object of $\mathrm{Func}(\mathcal{A}^{\mathrm{op}}, \mathcal{B})$ is sometimes referred as a ($\mathcal{B}$-valued) *presheaf* over $\mathcal{A}$.

Let $B \in \mathcal{B}$. The functor $c(B) \in \mathrm{Func}(\mathcal{A}, \mathcal{B})$ defined by $c(B)(A) = B$ and $c(B)(f) = \mathrm{id}_B$ for any $A \in \mathrm{ob}(\mathcal{A})$ and any $f \in \mathrm{Mor}(\mathcal{A})$ is called the *constant functor* with the constant value B. Thus, we obtain a functor

$c(?) : \mathcal{B} \to \mathrm{Func}(\mathcal{A}, \mathcal{B})$. Let $F \in \mathrm{Func}(\mathcal{A}, \mathcal{B})$. If $\mathrm{Nat}(F, c(?)) : \mathcal{B} \to \underline{\mathrm{Set}}$ is representable, then the object in $\mathcal{B}$ which represents $\mathrm{Nat}(F, c(?))$ is called the *inductive limit* of F, and is denoted by $\varinjlim F$. We say that the category $\mathcal{B}$ has inductive limits if for any svelte category $\mathcal{A}$ and $F \in \mathrm{Func}(\mathcal{A}, \mathcal{B})$, the functor $\mathrm{Nat}(F, c(?))$ is representable. If $\mathcal{B}$ has inductive limits, then we can make $\varinjlim : \mathrm{Func}(\mathcal{A}, \mathcal{B}) \to \mathcal{B}$ be a functor so that $\varinjlim$ is a left adjoint functor of c via the isomorphism $\mathrm{Nat}(F, c(?)) \cong \mathcal{B}(\varinjlim F, ?)$.

Dually, we define $\varprojlim F$, the *projective limit* of F, to be the object of $\mathcal{B}$ which represents $\mathrm{Nat}(c(?), F)$. We say that $\mathcal{B}$ has projective limits if for any svelte category $\mathcal{A}$ and $F \in \mathrm{Func}(\mathcal{A}, \mathcal{B})$, the functor $\mathrm{Nat}(c(?), F)$ is representable. If $\mathcal{B}$ has projective limits, then $\varprojlim : \mathrm{Func}(\mathcal{A}, \mathcal{B}) \to \mathcal{B}$ is a functor which is right adjoint to c.

Note that $\underline{\mathrm{Set}}$ and $\underline{\mathrm{Ab}}$ have both inductive limits and projective limits.

Definition 1.2.3 Let I be a category. We say that I is *filtered* if

1 For any $i, j \in I$, there exists some $k \in I$ such that $I(i, k) \neq \emptyset \neq I(j, k)$.

2 For any $i, j \in I$ and $f, g \in I(i, j)$, there exists some $k \in I$ and some $h \in I(j, k)$ such that $hf = hg$.

(1.2.4) An element of $\mathrm{Func}(I, \mathcal{B})$ is said to be a *filtered inductive system* (resp. *filtered projective system*) if I (resp. I^{op}) is filtered.

We say that $\varinjlim F$ is a *filtered inductive limit* if F is a filtered inductive system (and if it exists).

Let J be a full subcategory of a filtered category I. We say that J is a *final subcategory* of I if for any $i \in I$, there exists some $j \in J$ such that $I(i, j) \neq \emptyset$. We also say that I is *cofinal* with J. If this is the case, the restriction $\mathrm{Nat}_I(F, c(?)) \to \mathrm{Nat}_J(F|_J, c(?))$ is an isomorphism for $F \in \mathrm{Func}(I, \mathcal{B})$. In particular, if $\varinjlim F|_J$ exists, then it agrees with $\varinjlim F$.

Let P be a preordered set. Then P is a small category by letting $P(a, b)$ be a singleton if $a \leq b$ and the empty set if $a \not\leq b$. An ordered set P is filtered as a small category if and only if P is a directed set.

(1.2.5) Let $M : I \to \mathcal{A}$ be a functor such that I is svelte and $\varinjlim M$ exists. Then there is a natural map $\eta_M : M \to c(\varinjlim M)$ corresponding to $\mathrm{id} : \varinjlim M \to \varinjlim M$ by the isomorphism $\mathcal{A}(\varinjlim M, \varinjlim M) \cong \mathrm{Nat}(M, c(\varinjlim M))$. We say that a functor $F : \mathcal{A} \to \mathcal{B}$ preserves inductive limits if for any $M : I \to \mathcal{A}$ such that I is svelte and $\varinjlim M$ exists, the inductive limit $\varinjlim(F \circ M)$ also exists, and the map $\varinjlim(F \circ M) \to F(\varinjlim M)$ which corresponds to the natural map

$$F \circ M \xrightarrow{F\eta_M} F \circ (c(\varinjlim M)) = c(F(\varinjlim M))$$

is an isomorphism.

Lemma 1.2.6 *Let $\mathcal{A}$ and $\mathcal{B}$ be categories, $F : \mathcal{A} \to \mathcal{B}$ a functor with the right adjoint G. Then the following hold:*

1 *F preserves inductive limits. In particular, if $\mathcal{A}$ and $\mathcal{B}$ are abelian, then F is right exact (as biproducts and cokernels are inductive limits). Moreover, the isomorphism Φ is an isomorphism of abelian groups.*

1* *G preserves projective limits.*

2 *If $\mathcal{A}$ and $\mathcal{B}$ are abelian and F is right exact, then G preserves injective objects.*

3 *F is faithful if and only if u is a monomorphism (i.e., for any A, u_A is a monomorphism).*

4 *F is fully faithful if and only if u is an isomorphism. In particular, if F is fully faithful and $\mathcal{B}$ has projective limits, then $\mathcal{A}$ also has projective limits. In fact, we have $G \varprojlim Ff = \varprojlim f$.*

4* *If G is fully faithful and $\mathcal{A}$ has inductive limits, then $\mathcal{B}$ has inductive limits. In fact, $F \varinjlim Gf = \varinjlim f$.*

Proof. We prove only **3** and **4**. Note that u is a monomorphism (resp. an isomorphism) if and only if $u_* : \mathcal{A}(A, A') \to \mathcal{A}(A, GFA')$ given by $u_*(f) = u \circ f$ is injective (resp. bijective) for any $A, A' \in \mathcal{A}$. We have for $f \in \mathcal{A}(A, A')$ by the naturality of Φ and the naturality of u,

$$\begin{aligned}\Phi_{A,FA'}(F(f)) = \Phi_{A,FA'}(\mathcal{B}(FA, F(f))(1_{FA})) &= \mathcal{A}(A, (GF)(f))(\Phi_{A,FA}(1_{FA})) \\ &= \mathcal{A}(A, (GF)(f))(u_A) = (GF)(f) \circ u_A = u_{A'} \circ f = u_*(f).\end{aligned}$$

As Φ is an isomorphism, u_* is injective (resp. bijective) if and only if

$$F : \mathcal{A}(A, A') \to \mathcal{B}(FA, FA')$$

has the same property. Hence, **3** and the first part of **4** follow.

If $u : \mathrm{Id}_{\mathcal{A}} \to GF$ is an isomorphism and $\mathcal{B}$ has projective limits, then for $f \in \mathrm{Func}(I, \mathcal{A})$, we have $G \varprojlim Ff \cong \varprojlim GFf \cong \varprojlim f$, and $\mathcal{A}$ also has projective limits. □

1.3 Exact categories

Let $\mathcal{A}$ be an additive category.

Definition 1.3.1 We say that $\mathcal{A}$ is an *exact category* if two classes of morphisms $\mathcal{E}_m$ and $\mathcal{E}_e$ of $\mathcal{A}$ are specified and the following conditions are satisfied:

E1 If $i, j \in \mathcal{E}_m$ and ij is defined, then $ij \in \mathcal{E}_m$.

E1* If $p, q \in \mathcal{E}_e$ and pq is defined, then $pq \in \mathcal{E}_e$.

E2 Any split monomorphism which has a cokernel belongs to $\mathcal{E}_m$.

E3 If $i \in \mathcal{E}_m$, then i has a cokernel, $\operatorname{coker} i \in \mathcal{E}_e$, and i is a kernel of $\operatorname{coker} i$.

E3* If $p \in \mathcal{E}_e$, then p has a kernel, $\ker p \in \mathcal{E}_m$, and p is a cokernel of $\ker p$.

E4 Let $p : B \to C$ and $g : C' \to C$ be morphisms of $\mathcal{A}$, and assume that $p \in \mathcal{E}_e$. Then there exists a pull-back of p and g, and the base change p' of p belongs to $\mathcal{E}_e$.

E4* Let $i : A \to B$ and $f : A \to A'$ be morphisms of $\mathcal{A}$, and assume that $i \in \mathcal{E}_m$. Then there exists a push-out of i and f, and the cobase change i' of i belongs to $\mathcal{E}_m$.

E5 If $i : A \to B$ and $f : B \to B'$ are morphisms of $\mathcal{A}$, $fi \in \mathcal{E}_m$, and i has a cokernel, then $i \in \mathcal{E}_m$.

E5* If $p : B \to C$ and $g : B' \to B$ are morphisms of $\mathcal{A}$, $pg \in \mathcal{E}_e$, and p has a kernel, then $p \in \mathcal{E}_e$.

(1.3.2) A sequence of morphisms

$$0 \to A \xrightarrow{i} B \xrightarrow{p} C \to 0 \tag{1.3.3}$$

of an exact category $\mathcal{A}$ is called a *short exact sequence* if i is a kernel of p, p is a cokernel of i, and $i \in \mathcal{E}_m$ (or equivalently, $p \in \mathcal{E}_e$). We also say that (i, p) is a short exact sequence. Let us denote the set (not necessarily small) of short exact sequences in $\mathcal{A}$ by $\mathcal{E}$. Then $\mathcal{E}_m$ is the set of morphisms of $\mathcal{A}$ such that there exists some morphism p with $(i, p) \in \mathcal{E}$. Similarly, $\mathcal{E}_e$ is also determined by $\mathcal{E}$. Hence, we also say that $(\mathcal{A}, \mathcal{E})$ is an exact category. Moreover, $\mathcal{E}_e$ is the set of morphisms which are cokernels of some morphisms of $\mathcal{E}_m$. Similarly, $\mathcal{E}_e$ is determined by $\mathcal{E}_m$. Hence, any one of $\mathcal{E}_e$, $\mathcal{E}_m$ and $\mathcal{E}$ determines the others.

(1.3.4) We say that a morphism f of an exact category $\mathcal{A}$ is *admissible* if there is a factorization $f = ip$ such that $i \in \mathcal{E}_m$ and $p \in \mathcal{E}_e$. For any epi-mono decomposition $f = i'p'$ of an admissible morphism $f = ip$, there exists some isomorphism α such that $i' = i\alpha$ and $p' = \alpha^{-1}p$. An admissible mono(resp. epi-)morphism is nothing but a morphism of $\mathcal{E}_m$ (resp. $\mathcal{E}_e$).

We say that a sequence of morphisms

$$A \xrightarrow{f} B \xrightarrow{g} C$$

in $\mathcal{A}$ is exact if f and g are admissible with epi-mono decompositions $f = ip$ and $g = jq$, respectively, such that $(i, q) \in \mathcal{E}$. We also say that $[f, g]$ is exact. A complex

$$\cdots \to A^{i-1} \xrightarrow{\partial^{i-1}} A^i \xrightarrow{\partial^i} A^{i+1} \to \cdots$$

in $\mathcal{A}$ is called exact if $[\partial^{i-1}, \partial^i]$ is exact for any $i \in \mathbb{Z}$.

(1.3.5) An additive functor between exact categories is called an *exact functor* if it preserves short exact sequences. An additive functor between exact categories F is called *half exact* if $[Fi, Fp]$ is exact for any short exact sequence (i, p). Similarly, left and right exact functors are also defined.

Definition 1.3.6 An additive category $\mathcal{A}$ is called *semisaturated* (resp. *Karoubian* or sometimes *saturated*) if any split epimorphism has a kernel (resp. any projector (i.e., idempotent endomorphism) has an image).

(1.3.7) An abelian category is Karoubian. Any Karoubian additive category is semisaturated. If $\mathcal{A}$ is semisaturated (resp. Karoubian), then so is $\mathcal{A}^{\mathrm{op}}$.

Definition 1.3.8 Let $\mathcal{A}$ be an exact category, and $\mathcal{B}$ a full subcategory of $\mathcal{A}$. We say that $\mathcal{B}$ is closed under extensions (resp. monocokernels, epikernels) in $\mathcal{A}$ if $\mathcal{B}$ contains some null object of $\mathcal{A}$, and $A, C \in \mathcal{B}$ (resp. $A, B \in \mathcal{B}$, $B, C \in \mathcal{B}$) implies $A, B, C \in \mathcal{B}$ for any exact sequence (1.3.3) in $\mathcal{A}$. If $B \in \mathcal{B}$ implies $A \in \mathcal{B}$ (resp. $C \in \mathcal{B}$) and $\mathcal{B}$ is non-empty, then we say that $\mathcal{B}$ is closed under subobjects (resp. quotients). We say that $\mathcal{B}$ is closed under subquotients if $\mathcal{B}$ is closed under both subobjects and quotient objects. If $A \in \mathcal{B}$ (resp. $D \in \mathcal{B}$) for any exact sequence

$$0 \to A \to B \to C \to D \to 0$$

in $\mathcal{A}$ with $B, C \in \mathcal{B}$, then we say that $\mathcal{B}$ is closed under kernels (resp. cokernels). If $\mathcal{B}$ is closed under kernels, cokernels and extensions, then we say that $\mathcal{B}$ is a *thick subcategory* of $\mathcal{A}$. A thick subcategory closed under subquotients is said to be *very thick*.

(1.3.9) If $\mathcal{B}$ is closed under extensions, then it is closed under isomorphisms and finite direct sums, and $\mathcal{B}$ itself is an additive category so that the inclusion $\mathcal{B} \hookrightarrow \mathcal{A}$ is additive. A thick subcategory of an abelian category is abelian, and the inclusion functor is fully faithful and exact.

Exercise 1.3.10 Prove the following.

1 Let $\mathcal{A}$ be an abelian category. Letting $\mathcal{E}_m$ be its set of monomorphisms and $\mathcal{E}_e$ its set of epimorphisms, $\mathcal{A}$ is an exact category. In this case, $\mathcal{E}$ is the set of (usual) short exact sequences.

2 If $(\mathcal{A}, \mathcal{E})$ is an exact category, then $(\mathcal{A}^{\mathrm{op}}, \mathcal{E}^{\mathrm{op}})$ is also an exact category with $\mathcal{E}_m^{\mathrm{op}}$ and $\mathcal{E}_e^{\mathrm{op}}$ its set of admissible epimorphisms and monomorphisms, respectively.

3 Let $\mathcal{B}'$ be a full subcategory of an exact category $\mathcal{B}$ closed under extensions. Defining a short exact sequence in $\mathcal{B}'$ to be a short exact sequence in $\mathcal{B}$ consisting of objects of $\mathcal{B}'$, $\mathcal{B}'$ is an exact category, and the inclusion $\mathcal{B}' \hookrightarrow \mathcal{B}$ is exact.

3′ Let $F : \mathcal{B} \to \mathcal{C}$ be a half exact functor between exact categories. Then $\operatorname{Ker} F := \{B \in \mathcal{B} \mid FB \cong 0\}$ is closed under extensions in $\mathcal{B}$.

4 If $(\mathcal{A}, \mathcal{E}_\lambda)$ is a family of exact categories, then $(\mathcal{A}, \bigcap_\lambda \mathcal{E}_\lambda)$ is an exact category.

4′ Let $F : (\mathcal{A}, \mathcal{E}) \to (\mathcal{B}, \mathcal{E}')$ be an exact functor between exact categories. If $\mathcal{E}'' \subset \mathcal{E}'$ and $(\mathcal{B}, \mathcal{E}'')$ is also an exact category, then defining

$$\mathcal{E}''' := \{E \in \mathcal{E} \mid F(E) \in \mathcal{E}''\},$$

$(\mathcal{A}, \mathcal{E}''')$ is also an exact category. In particular, if $\mathcal{A}$ and $\mathcal{B}$ are abelian categories (with the structures of exact categories as in **1**) and when we denote the set of exact sequences E in $\mathcal{A}$ such that $F(E)$ is split exact by $\mathcal{E}(F)$, then $(\mathcal{A}, \mathcal{E}(F))$ is an exact category.

Any exact category $\mathcal{B}'$ is always produced as in **3** above, with $\mathcal{B}$ abelian (but not necessarily a $\mathcal{U}$-category), as follows.

Theorem 1.3.11 *Let $\mathcal{A}$ be an exact R-category. Then the category $\mathcal{B} :=$ $\mathrm{Sex}_R(\mathcal{A}^{\mathrm{op}}, {}_R\mathbb{M})$ of contravariant left exact R-functors from $\mathcal{A}$ to the category of R-modules ${}_R\mathbb{M}$ is abelian (the notation* Sex *is due to Gabriel, and explained thus:* **s**inister **ex**act*). The Yoneda embedding $y : \mathcal{A} \to \mathcal{B}$ is an R-equivalence from $\mathcal{A}$ to a full subcategory $\mathcal{B}'$ of $\mathcal{B}$ closed under extensions. Moreover, for a sequence of morphisms (i, p) in $\mathcal{B}$, (i, p) is a short exact sequence if and only if (yi, yp) is a short exact sequence. If moreover $\mathcal{A}$ is semisaturated, then $\mathcal{B}'$ is closed under epikernels in $\mathcal{B}$.*

This theorem was proved by Quillen [126]. For the proof, see [141]. The Yoneda embedding y is also called the *Gabriel–Quillen embedding* in this case.

Corollary 1.3.12 *The five lemma and the 3×3 lemma are true in exact categories.*

Corollary 1.3.13 *The canonical functor from the category of Karoubian (resp. semisaturated) exact categories to the category of exact categories has* *a left adjoint. Namely, if $\mathcal{B}$ is an exact category, then there exists a (unique) Karoubian exact category $\mathcal{B}^s$ (resp. semisaturated exact category $\mathcal{B}^{ss}$) and an exact functor $f : \mathcal{B} \to \mathcal{B}^s$ (resp. $f' : \mathcal{B} \to \mathcal{B}^{ss}$) such that for any Karoubian (resp. semisaturated) exact category $\mathcal{A}$ and any exact functor $g : \mathcal{B} \to \mathcal{A}$, there exists a unique exact functor $h : \mathcal{B}^s \to \mathcal{A}$ (resp. $h' : \mathcal{B}^{ss} \to \mathcal{A}$) such that hf (resp. $h'f'$) and g are equivalent.*

For an exact category $\mathcal{B}$, we call $\mathcal{B}^s$ (resp. $\mathcal{B}^{ss}$) the *saturation* (resp. *semisaturation*) of $\mathcal{B}$.

1.4 Derived categories and derived functors

(1.4.1) Let $\mathcal{A}$ be an additive category. We only treat cohomological chain complexes here. We say that $\mathbb{F} = (F^i, d^i)$ is a chain complex in $\mathcal{A}$ if $(F^i)_{i\in\mathbb{Z}}$ is a collection of objects of $\mathcal{A}$, $d^i \in \mathcal{A}(F^i, F^{i+1})$, and $d^{i+1} \circ d^i = 0$ $(i \in \mathbb{Z})$. A homological complex can be viewed as a cohomological complex with the identification $F^i := F_{-i}$.

A chain complex $\mathbb{F}$ is said to be *bounded below* (resp. *bounded above*) if $F^i = 0$ for $i \ll 0$ (resp. $i \gg 0$). It is said to be *bounded* if it is both bounded below and bounded above. The length of a complex $\mathbb{F}$ is defined to be $\sup\{i - j \mid F_i \neq 0,\ F_j \neq 0\}$, so the length of $\mathbb{F}$ is not ∞ if and only if $\mathbb{F}$ is bounded. As a convention, we define the length of the zero complex to be $-\infty$. The category $C(\mathcal{A})$ of chain complexes and chain maps in $\mathcal{A}$ is an additive category. If $\mathcal{A}$ is an R-category, then so is $C(\mathcal{A})$. The full subcategory of complexes bounded below (resp. bounded above, bounded) is denoted by $C^+(\mathcal{A})$ (resp. $C^-(\mathcal{A})$, $C^b(\mathcal{A})$).

(1.4.2) Let $\mathbb{F}$ and $\mathbb{G}$ be complexes in $\mathcal{A}$. Then we have the following chain complexes obtained from $\mathbb{F}$ and $\mathbb{G}$.

$\mathbb{F}[n]$: Defining $F[n]^i := F^{n+i}$ and $d^i_{\mathbb{F}[n]} := (-1)^n d^{n+i}_{\mathbb{F}}$, we have a complex $\mathbb{F}[n]$ in $\mathcal{A}$.

$\operatorname{Hom}^\bullet_{\mathcal{A}}(\mathbb{F}, \mathbb{G})$: We define $\operatorname{Hom}^\bullet_{\mathcal{A}}(\mathbb{F}, \mathbb{G})^i := \prod_{n\in\mathbb{Z}} \operatorname{Hom}_{\mathcal{A}}(F^n, G^{n+i})$, and

$$d^i_{\operatorname{Hom}}((f^n)_n) := (d^{n+i}_{\mathbb{G}} f^n - (-1)^i f^{n+1} d^n_{\mathbb{F}})_n.$$

Then $\operatorname{Hom}^\bullet_{\mathcal{A}}(\mathbb{F}, \mathbb{G})$ is a complex of abelian groups. If $\mathcal{A}$ is an R-category, then it is a complex of R-modules.

$\mathbb{F} \otimes^\bullet_R \mathbb{G}$: Let $\mathcal{A}$ be the category of R-modules ${}_R\mathbb{M}$. Then defining $(\mathbb{F} \otimes^\bullet_R \mathbb{G})^i := \bigoplus_{m+n=i} F^m \otimes G^n$ and $d^i(f^m \otimes g^n) = d^m f^m \otimes g^n + (-1)^m f^m \otimes d^n g^n$, we get a complex $\mathbb{F} \otimes^\bullet_R \mathbb{G}$ in $\mathcal{A}$. A natural map $H^m(\mathbb{F}) \otimes H^n(\mathbb{G}) \to H^{m+n}(\mathbb{F} \otimes^\bullet_R \mathbb{G})$ is induced.

The following is checked easily.

Exercise 1.4.3 The composition

$$\mathrm{Hom}^\bullet_{\mathcal{A}}(\mathbb{G},\mathbb{H}) \otimes^\bullet_{\mathbb{Z}} \mathrm{Hom}^\bullet_{\mathcal{A}}(\mathbb{F},\mathbb{G}) \to \mathrm{Hom}^\bullet_{\mathcal{A}}(\mathbb{F},\mathbb{H}) \tag{1.4.4}$$

is a chain map.

Note that an n-cocycle in $\mathrm{Hom}^\bullet_{\mathcal{A}}(\mathbb{F},\mathbb{G})$ is nothing but a chain map from $\mathbb{F}$ to $\mathbb{G}[n]$. An n-coboundary is a null homotopic chain map from $\mathbb{F}$ to $\mathbb{G}[n]$.

Letting chain complexes in $\mathcal{A}$ be the objects and $H^0(\mathrm{Hom}^\bullet_{\mathcal{A}}(\mathbb{F},\mathbb{G}))$ the hom set from $\mathbb{F}$ to $\mathbb{G}$, we get a category $K(\mathcal{A})$. The composition is that of chain maps. It is well-defined, because it agrees with the map of cohomology induced by (1.4.4). The full subcategory of $K(\mathcal{A})$ consisting of complexes bounded below (resp. bounded above, bounded) is denoted by $K^+(\mathcal{A})$ (resp. $K^-(\mathcal{A})$, $K^b(\mathcal{A})$).

(1.4.5) If $\mathcal{A}$ is an exact category, then $K(\mathcal{A})$ has the structure of a triangulated category [120]. The translation functor T is given by $T(\mathbb{F}) := \mathbb{F}[1]$. For basics on triangulated categories, see [143].

Let ? be either b, $+$, $-$, or $\emptyset$. We denote the full subcategory of $K^?(\mathcal{A})$ consisting of all exact sequences in $K^?(\mathcal{A})$ by $E^?(\mathcal{A})$.

Proposition 1.4.6 (Neeman) *With the notation above, $E^?(\mathcal{A})$ is a triangulated subcategory of $K^?(\mathcal{A})$. If $\mathcal{A}$ is Karoubian, then $E^?(\mathcal{A})$ is épaisse. If $? = +,-,b$ and $\mathcal{A}$ is semisaturated, then $E^?(\mathcal{A})$ is épaisse.*

(1.4.7) The quotient $K^?(\mathcal{A})/E^?(\mathcal{A})^e$ (for definition, see [143]) is denoted by $D^?(\mathcal{A})$, and it is called the *derived category* of $\mathcal{A}$, where $E^?(\mathcal{A})^e$ denotes the épaisse closure of $E^?(\mathcal{A})$. By Rickard's criterion [130], $E^?(\mathcal{A})^e$ is the set of direct summands of exact sequences in $K^?(\mathcal{A})$. Note that $D^?(\mathcal{A})$ may not be a category with small hom sets any more. As an easy criterion, note that $D^?(\mathcal{A})$ is a category with small hom sets if $\mathcal{A}$ is svelte (follows from [143, p. 298]).

Let $f : \mathbb{F} \to \mathbb{G}$ be a morphism of $C(\mathcal{A})$. We say that f is a *quasi-isomorphism* if the mapping cone $C(f)$ is a direct summand of an exact sequence. If $f - f'$ is null homotopic, then $C(f)$ and $C(f')$ are isomorphic in $C(\mathcal{A})$. So the notion of quasi-isomorphism is also defined for morphisms of $K(\mathcal{A})$. Note that $D^?(\mathcal{A})$ is obtained by localizing quasi-isomorphisms of $K^?(\mathcal{A})$.

Definition 1.4.8 Let $\mathcal{A}$ be an exact category, and $A \in \mathrm{ob}(\mathcal{A})$. We say that an object I of $\mathcal{A}$ is injective if any admissible monomorphism $I \to B$ splits. We say that $\mathcal{A}$ has enough injectives if for any $A \in \mathrm{ob}(\mathcal{A})$, there exists some injective object I of $\mathcal{A}$ and an admissible monomorphism $A \to I$.

Dually, projective objects in $\mathcal{A}$, and the notion of enough projectives, are also defined. If $\mathcal{A}$ has enough injectives, then $D^+(\mathcal{A})$ is a category with small hom sets.

For $\mathbb{F}, \mathbb{G} \in \mathrm{ob}(C(\mathcal{A}))$, we denote $\mathrm{Hom}_{D(\mathcal{A})}(\mathbb{F}, \mathbb{G}[n])$ by $\mathrm{Ext}^n_{\mathcal{A}}(\mathbb{F}, \mathbb{G})$, and call it the nth hyperextension group.

Let $\mathcal{A}$ and $\mathcal{B}$ be Karoubian exact categories, and $F : \mathcal{A} \to \mathcal{B}$ an additive functor. The induced functor $C^?(\mathcal{A}) \to C^?(\mathcal{B})$ induces $F' : K^?(\mathcal{A}) \to K^?(\mathcal{B})$. The right (resp. left) derived functor $D^?(\mathcal{A}) \to D^?(\mathcal{B})$ [143] of F' is simply called the right (resp. left) derived functor of F, and is denoted by $\underline{R}^? F$ (resp. $\underline{L}^? F$). If no confusion is possible, then we sometimes denote $\underline{R}^? F$ (resp. $\underline{L}^? F$) simply by $\underline{R}F$ (resp. $\underline{L}F$).

Theorem 1.4.9 *Let $\mathcal{A}$ and $\mathcal{A}'$ be semisaturated exact categories, and $\mathcal{B}$ a full subcategory of $\mathcal{A}$ closed under monocokernels and extensions. Let $F : \mathcal{A} \to \mathcal{A}'$ be an additive functor. Assume that for any object A of $\mathcal{A}$, there exist some object B of $\mathcal{B}$ and an admissible mono $A \to B$. Moreover, we assume that the restriction of F to $\mathcal{B}$ is exact (note that $\mathcal{B}$ is an exact category by* Exercise 1.3.10, **3**). *Then there exists some derived functor $\underline{R}^+F : D^+\mathcal{A} \to D^+\mathcal{A}'$ of F. More precisely, for $\mathbb{F} \in K^+(\mathcal{A})$, there exist some $\mathbb{G} \in K^+(\mathcal{B})$ and a quasi-isomorphism $\mathbb{F} \to \mathbb{G}$ such that $\underline{R}^+F(\mathbb{F}) \cong F'\mathbb{G}$ (independent of the choice of $\mathbb{G}$).*

Using the Gabriel–Quillen embedding $\mathcal{A} \to \mathrm{Sex}(\mathcal{A}, \underline{\mathrm{Ab}})^{\mathrm{op}}$, the proof is easily reduced to the case where $\mathcal{A}$ is abelian. This case is proved in [69, Corollary I.5.3].

Assume that the hypothesis of the theorem is satisfied and $\mathcal{A}'$ is abelian. Then for $i \in \mathbb{Z}$, we define the functor R^iF to be the cohomology $H^i(\underline{R}^+F)$. Similarly, if $\underline{L}F$ exists, then we define L_iF to be $H^{-i}(\underline{L}F)$.

For $A \in \mathrm{ob}(\mathcal{A})$, A is identified with the complex

$$\cdots \to 0 \to 0 \to A \to 0 \to 0 \to \cdots$$

concentrated in degree zero, and we obtain a canonical composite functor $\mathcal{A} \to C^b(\mathcal{A}) \to K^b(\mathcal{A}) \to D^b(\mathcal{A})$. Thus, R^iFA and L_iFA are defined for $i \in \mathbb{Z}$.

The easiest case where the assumption of the theorem is satisfied, is the case where $\mathcal{A}$ has enough injectives. In this case, when we define $\mathcal{B}$ to be the full subcategory of injective objects of $\mathcal{A}$, then the assumption of the theorem is satisfied for any additive functor $F : \mathcal{A} \to \mathcal{A}'$. In particular, if $\mathcal{A}$ is abelian with enough injectives, then $R^iF(A)$ is defined, and agrees with the usual derived functor.

(1.4.10) We will use the following classical theorem of [69] later.

Theorem 1.4.11 *Let $\mathcal{A}$ and $\mathcal{A}'$ be abelian categories, $\mathcal{B}$ a full subcategory of $\mathcal{A}$, and $F : \mathcal{A} \to \mathcal{A}'$ an additive functor. Assume that $\mathcal{A}$ has enough injectives so that the derived functor $\underline{R}^+F$ exists. If there exists $n \in \mathbb{N}$ such that $R^iFA = 0$ for any $A \in \mathcal{A}$ and $i \geq n$, then the derived functor $\underline{R}F : D(\mathcal{A}) \to D(\mathcal{A}')$ exists. For $\mathbb{F} \in C(\mathcal{A})$, there is a quasi-isomorphism $\mathbb{F} \to \mathbb{J}$ with each term J_i of $\mathbb{J}$ F-acyclic (i.e., $R^jFJ_i = 0$ for $j > 0$), and $\underline{R}F(\mathbb{F})$ agrees with the image of $F(\mathbb{J})$ in $D(\mathcal{A}')$. In particular, the restriction of $\underline{R}F$ to $D^+(\mathcal{A})$ is a functor to $D^+(\mathcal{A}')$, and agrees with $\underline{R}^+F$.*

This theorem is a consequence of the following well-known existence theorem [69, Theorem I.5.1], [143, Theorem 2.2.2].

Theorem 1.4.12 *Let $\mathcal{A}$ and $\mathcal{A}'$ be abelian categories. Let $\mathcal{B}$ be a localizing subcategory of $K(\mathcal{A})$, and $\mathcal{L}$ a triangulated subcategory of $\mathcal{B}$. Let $F : \mathcal{B} \to K^?(\mathcal{A}')$ be a triangulated functor. Assume that for any exact sequence $\mathbb{E}$ in $\mathcal{L}$, $F(\mathbb{E})$ is exact. Assume further that for any object $\mathbb{B}$ in $\mathcal{B}$, there is a quasi-isomorphism $\mathbb{B} \to \mathbb{L}$ such that $\mathbb{L} \in \mathcal{L}$. Then the derived functor $\underline{R}F : D(\mathcal{B}) \to D^?(\mathcal{A}')$ exists, and for any object $\mathbb{L}$ in $\mathcal{L}$, we have the canonical map $F(\mathbb{L}) \to \underline{R}F(\mathbb{L})$ is an isomorphism of $D^?(\mathcal{A}')$.*

Let $\mathcal{A}$ be an abelian category, and $\mathbb{I} \in C(\mathcal{A})$. We say that $\mathbb{I}$ is *K-injective* if $\operatorname{Hom}^\bullet_{\mathcal{A}}(\mathbb{F}, \mathbb{I})$ is exact for any exact $\mathcal{F} \in C(\mathcal{A})$. A *$K$-injective resolution* of $\mathbb{G} \in C(\mathcal{A})$ is a quasi-isomorphism $\mathbb{G} \to \mathbb{I}$ with $\mathbb{I}$ K-injective.

If a K-injective complex is exact, then it is zero in $K(\mathcal{A})$ (i.e., null homotopic). Combining this observation and the theorem above, we have another existence theorem for unbounded derived functors. The following theorem (the statement is slightly different, but almost the same proof works) is due to Spaltenstein [137].

Theorem 1.4.13 *Let $\mathcal{A}$ be a Grothendieck category. Then any object $\mathbb{F} \in C(\mathcal{A})$ admits a K-injective resolution. For any additive functor $F : \mathcal{A} \to \mathcal{A}'$, the unbounded derived functor $\underline{R}F : D(\mathcal{A}) \to D(\mathcal{A}')$ exists.*

For Grothendieck categories, see (1.7). As a consequence of the theorem, some derived functors for sheaves over ringed spaces are defined for unbounded derived categories, see [137].

1.5 Extensions and Ext groups

(1.5.1) Let $\mathcal{A}$ be an exact category. We defined the hyperextension group $\operatorname{Ext}^n_{\mathcal{A}}(\mathbb{F}, \mathbb{G})$ for $\mathbb{F}, \mathbb{G} \in D(\mathcal{A})$ in the last subsection. Hence, for $A, C \in \operatorname{ob}(\mathcal{A})$, $\operatorname{Ext}^n_{\mathcal{A}}(C, A)$ is defined. For $n < 0$, we have $\operatorname{Ext}^n_{\mathcal{A}}(C, A) = 0$, and $\operatorname{Ext}^0_{\mathcal{A}}(C, A) \cong \operatorname{Hom}_{\mathcal{A}}(C, A)$ [143]. In general, $\operatorname{Ext}^n_{\mathcal{A}}(C, A)$ is an abelian group which is not necessarily small (even for the case $n = 1$), see [56]. We survey another construction of Ext^n here. In the sequel, the construction of Ext^1 (only for the case where $\mathcal{A}$ is abelian) is important.

(1.5.2) Let $n \geq 1$. For $A, C \in \mathcal{A}$, we define $E^n_{C,A}$ to be the set of exact sequences

$$(f_0, \dots, f_n) : 0 \to A \xrightarrow{f_n} B_n \xrightarrow{f_{n-1}} \cdots \to B_1 \xrightarrow{f_0} C \to 0$$

in $\mathcal{A}$. If $n = 1$, then an element of $E^n_{C,A}$ is a short exact sequence. We define an equivalence relation $\equiv$ in $E^n_{C,A}$ to be the symmetric transitive closure of $\sim$, where we say that $S \sim S'$ for $S, S' \in E^n_{C,A}$ if there exists a commutative diagram

$$\begin{array}{cccccccccc} S: & 0 \to & A & \to & B_1 & \to \cdots \to & B_{n-1} & \to & C & \to 0 \\ & & \| & & \downarrow & \cdots & \downarrow & & \| & \\ S': & 0 \to & A & \to & B'_1 & \to \cdots \to & B'_{n-1} & \to & C & \to 0. \end{array}$$

If $n = 1$, then $\sim$ and $\equiv$ agree by Corollary 1.3.12. We redefine $\operatorname{Ext}^n_{\mathcal{A}}(C, A)$ to be the set $E^n_{C,A}/\equiv$. We also redefine $\operatorname{Ext}^0_{\mathcal{A}}(C, A) := \mathcal{A}(C, A)$.

For $f \in \mathcal{A}(A, A')$ and $E \in E^1_{C,A}$, the short exact sequence obtained by pushing-out E by f is an element of $E^1_{C,A'}$, which we denote by fE. If $E \equiv E'$, then we have $fE \equiv fE'$. Hence, the map $\mathcal{A}(A, A') \times \operatorname{Ext}^1_{\mathcal{A}}(C, A) \to \operatorname{Ext}^1_{\mathcal{A}}(C, A')$ is defined by $(f, E) \mapsto fE$. Dually, the product $\operatorname{Ext}^1_{\mathcal{A}}(C, A) \times \operatorname{Hom}_{\mathcal{A}}(C', C) \to \operatorname{Ext}^1_{\mathcal{A}}(C', A)$ is also defined via pull-back. The associativity law $(fE)g = f(Eg)$ holds in $\operatorname{Ext}^1_{\mathcal{A}}(C', A')$.

Let $m, n \geq 1$, $S = (f_0, \dots, f_n) \in E_{C,A}$, and $S' = (g_0, \dots, g_m) \in E_{D,C}$. Then defining

$$S \circ S' := (f_0, \dots, f_{n-1}, f_n g_0, g_1, \dots, g_m),$$

we have $S \circ S' \in E_{D,A}$, and the product $\operatorname{Ext}^n_{\mathcal{A}}(C, A) \times \operatorname{Ext}^m_{\mathcal{A}}(D, C) \to \operatorname{Ext}^{n+m}_{\mathcal{A}}(D, A)$ is well-defined.

Obviously, the iterated product map from the (not necessarily small) set

$$\coprod_{L_1, \dots, L_{n-1}} \operatorname{Ext}^1_{\mathcal{A}}(L_1, A) \times \operatorname{Ext}^1_{\mathcal{A}}(L_2, L_1) \times \cdots \times \operatorname{Ext}^1_{\mathcal{A}}(L_{n-1}, L_{n-2}) \times \operatorname{Ext}^1_{\mathcal{A}}(C, L_{n-1})$$

to $\operatorname{Ext}^n_{\mathcal{A}}(C, A)$ is surjective. As a matter of fact, $\operatorname{Ext}^n_{\mathcal{A}}(C, A)$ is canonically isomorphic to the set above modulo the symmetric transitive closure relation of the relation

$$(E_1, \dots, E_i f, E_{i+1}, \dots, E_n) \sim (E_1, \dots, E_i, f E_{i+1}, \dots, E_n).$$

Thus, the product $\mathcal{A}(A, A') \times \operatorname{Ext}^n_{\mathcal{A}}(C, A) \to \operatorname{Ext}^n_{\mathcal{A}}(C, A')$ is well-defined by

$$f(E_1 \circ \cdots \circ E_n) := (fE_1) \circ \cdots \circ (fE_n).$$

Combining observations above, we have that for any non-negative integers m and n, the product

$$\operatorname{Ext}^n_{\mathcal{A}}(C, A) \times \operatorname{Ext}^m_{\mathcal{A}}(D, C) \to \operatorname{Ext}^{n+m}(D, A) \tag{1.5.3}$$

is well defined, and the associativity law $(S \circ S') \circ S'' = S \circ (S' \circ S'')$ is satisfied. We call this product the *Yoneda product.*

If S and S' are elements of $\mathrm{Ext}^n_{\mathcal{A}}(C, A)$ and $\mathrm{Ext}^n_{\mathcal{A}}(C', A')$, respectively, then $S \oplus S'$ is an element of $\mathrm{Ext}^n_{\mathcal{A}}(C \oplus C', A \oplus A')$.

For $S, S' \in \mathrm{Ext}^n_{\mathcal{A}}(C, A)$, we denote the element $\Delta_A(S \oplus S')\nabla_C$ by $S + S'$, where $\Delta_A : A \to A \oplus A$ is the diagonalization, and ∇_C is the codiagonalization. Obviously, $+$ is commutative and associative, and the distributive law $f(S + S')g = fSg + fS'g$ also holds. Thus, $\mathrm{Ext}^n_{\mathcal{A}}(C, A)$ is an $(\mathrm{End}_{\mathcal{A}} A, \mathrm{End}_{\mathcal{A}} C)$-bimodule with $+$ as its sum.

Theorem 1.5.4 *Let $\mathcal{A}$ be an exact category. If*

$$E : 0 \to A \xrightarrow{i} B \xrightarrow{p} C \to 0$$

is a short exact sequence and $D \in \mathcal{A}$, then the sequence of right $\mathrm{End}_{\mathcal{A}} D$-modules

$$0 \to \mathcal{A}(D, A) \xrightarrow{i} \mathcal{A}(D, B) \xrightarrow{p} \mathcal{A}(D, C) \xrightarrow{E} \mathrm{Ext}^1_{\mathcal{A}}(D, A) \xrightarrow{i} \cdots$$
$$\to \mathrm{Ext}^n_{\mathcal{A}}(D, A) \xrightarrow{i} \mathrm{Ext}^n_{\mathcal{A}}(D, B) \xrightarrow{p} \mathrm{Ext}^n_{\mathcal{A}}(D, C) \xrightarrow{E} \mathrm{Ext}^{n+1}_{\mathcal{A}}(D, A) \xrightarrow{i} \cdots$$

is exact, where all morphisms are Yoneda multiplication from the left.

(1.5.5) Assume that $\mathcal{A}$ is Karoubian. As a matter of fact, $\mathrm{Ext}^n_{\mathcal{A}}(C, A)$ agrees with that defined in the last subsection, and the Yoneda product is identified with the composition

$$\mathrm{Hom}_{D(\mathcal{A})}(C, T^n(A)) \otimes_{\mathbb{Z}} \mathrm{Hom}_{D(\mathcal{A})}(D, T^m(C))$$
$$\xrightarrow{T^m \otimes 1} \mathrm{Hom}_{D(\mathcal{A})}(T^m(C), T^{n+m}(A)) \otimes_{\mathbb{Z}} \mathrm{Hom}_{D(\mathcal{A})}(D, T^m(C))$$
$$\xrightarrow{\circ} \mathrm{Hom}_{D(\mathcal{A})}(D, T^{n+m}(A)).$$

This is proved utilizing [143, (2.3)].

(1.5.6) Let $F : \mathcal{A} \to \mathcal{A}'$ be an exact functor between exact categories. Then F maps an exact sequence in $\mathcal{A}$ to an exact sequence in $\mathcal{A}'$, and equivalent sequences are mapped to equivalent ones, so we have a canonical map

$$F : \mathrm{Ext}^n_{\mathcal{A}}(A, B) \to \mathrm{Ext}^n_{\mathcal{A}'}(FA, FB)$$

for $n \geq 0$ and $A, B \in \mathcal{A}$. This map obviously preserves the Yoneda product.

Lemma 1.5.7 *Let $\mathcal{A}$ be an exact category, and $\mathcal{B}$ a full subcategory of $\mathcal{A}$ closed under extensions. Then we have that the canonical map*

$$i : \mathrm{Ext}^n_{\mathcal{B}}(B, B') \to \mathrm{Ext}^n_{\mathcal{A}}(iB, iB')$$

is an isomorphism for $B, B' \in \mathcal{B}$ and $n = 0, 1$, where $i : \mathcal{B} \hookrightarrow \mathcal{A}$ is the canonical inclusion.

(1.5.8) Let R be a commutative ring and $\mathcal{A}$ an exact R-category. If $F : \mathcal{A} \to {}_R\mathbb{M}$ is an R-linear functor which satisfies the assumption of Theorem 1.4.9, then the canonical map

$$R^nF : \operatorname{Hom}_{D(\mathcal{A})}(\mathbb{F}, T^m(\mathbb{G})) \to \operatorname{Hom}_A(R^nF\mathbb{F}, R^{n+m}F\mathbb{G})$$

is E-linear, where $E := \operatorname{End}_{K(\mathcal{A})}(\mathbb{F})$. Hence, an R-linear pairing

$$\operatorname{Ext}^m(\mathbb{F}, \mathbb{G}) \otimes_E R^nF\mathbb{F} \to R^{n+m}F\mathbb{G}$$

is induced. We also call this pairing the *Yoneda product.*

(1.5.9) As we have seen, $\operatorname{Ext}^n_{\mathcal{A}}$ can be constructed for an exact category $\mathcal{A}$. However, this notation $\operatorname{Ext}^n_{\mathcal{A}}$ is confusing when we consider an exact structure such as in **4′** in Exercise 1.3.10, which is different from the usual one (cf. Exercise 1.3.10, **1**) for an abelian category $\mathcal{A}$.

So, we only consider the usual exact structure for an abelian category $\mathcal{A}$ unless otherwise specified. When we consider an exact structure $\mathcal{E}(F)$ determined by an exact functor $F : \mathcal{A} \to \mathcal{B}$ as in Exercise 1.3.10, **4′**, we use the notation $\operatorname{Ext}^n_{(\mathcal{A},F)}$ or Ext^n_F. If F is obvious from the context, then we also use the notation $\operatorname{Ext}^n_{\mathcal{A}/\mathcal{B}}$. We call a sequence in $\mathcal{E}(F)$ an F*-admissible* short exact sequence. We say that an additive functor Q from $\mathcal{A}$ to an abelian category $\mathcal{A}'$ is F*-exact* if $Q(E)$ is a short exact sequence in $\mathcal{A}'$ for $E \in \mathcal{E}(F)$. Similarly, an F-left exact functor and an F-right exact functor are defined.

1.6 The cobar resolution

Throughout this subsection, $\mathcal{A}$ and $\mathcal{B}$ denote categories, and $F : \mathcal{A} \to B$ denotes a functor with the right adjoint G. The unit (resp. counit) of adjunction is denoted by u (resp. ε).

(1.6.1) We define $\mathbb{R} := GF$. We can construct a cosimplicial object of the category of endofunctors of $\mathcal{A}$ using $u : \operatorname{Id}_{\mathcal{A}} \to \mathbb{R}$ and $\mu = G\varepsilon F : \mathbb{R}^2 = G(FG)F \to GF = \mathbb{R}$. As the relations (associativity and unit law)

$$\mu \circ \mathbb{R}\mu = \mu \circ \mu\mathbb{R} \quad (\mathbb{R}^3 \to \mathbb{R}), \qquad \mu \circ u\mathbb{R} = 1_{\mathbb{R}} = \mu \circ \mathbb{R}u \quad (\mathbb{R} \to \mathbb{R})$$

hold, we obtain a cosimplicial object $\operatorname{Cosimp}(\mathbb{R})$ defined by

(1.6.2) $\quad d^i_n := \mathbb{R}^i u \mathbb{R}^{n-i} : \ \mathbb{R}^n \to \mathbb{R}^{n+1} \quad (i = 0, 1, \ldots, n)$

(1.6.3) $\quad s^i_n := \mathbb{R}^i \mu \mathbb{R}^{n-i-2} : \ \mathbb{R}^n \to \mathbb{R}^{n-1} \quad (i = 0, 1, \ldots, n-2).$

If $\mathcal{A}$ is additive, then letting the alternating sum

$$\partial := d^0_n - d^1_n + \cdots + (-1)^n d^n_n : \mathbb{R}^n(A) \to \mathbb{R}^{n+1}(A)$$

be the boundary map, we obtain an augmented complex

$$(1.6.4)\ \operatorname{Cobar}_F(A) := 0 \to A \xrightarrow{u} C_F^0(A) \xrightarrow{\partial^0} C_F^1(A) \xrightarrow{\partial^1} C_F^2(A) \to \cdots,$$

where $C_F^i(A) := \mathbb{R}^{i+1}(A)$.

Lemma 1.6.5 *If $\mathcal{A}$ and $\mathcal{B}$ are additive and F is additive, then the complex $0 \to FA \xrightarrow{Fu} F(\operatorname{Cobar}_F(A))$ is split exact.*

Proof. We define $s^i : F\mathbb{R}^{i+1} \to F\mathbb{R}^i$ to be $\varepsilon F\mathbb{R}^i$. Then we have $s^0 Fu = 1$ and $s^{i+1}\partial + \partial s^i = 1$ by (1.2.2). □

(1.6.6) Let $\mathcal{A}$ and $\mathcal{B}$ be abelian, and $F : \mathcal{A} \to B$ be an exact functor.

Lemma 1.6.7 *For $I \in \mathcal{A}$, the following are equivalent.*

1 *$\mathcal{A}(?, I)$ is F-exact.*

2 *$\operatorname{Ext}_F^1(A, I) = 0$ for $A \in \mathcal{A}$.*

3 *$\operatorname{Ext}_F^n(A, I) = 0$ $(i > 0)$ for $A \in \mathcal{A}$.*

4 *I is an injective object of the exact category $(\mathcal{A}, \mathcal{E}(F))$.*

This is obvious from the argument in the last subsection. We say that I is *F-injective* if the conditions above are satisfied. The notion of F-projectivity is defined in a similar way.

An object $\mathcal{A}$ which is isomorphic to $G(B)$ for some $B \in \mathcal{B}$ is called an *F-cofree* object.

Lemma 1.6.8 *An F-cofree object is F-injective.*

Proof. Let E be an F-admissible short exact sequence. As we have

$$\mathcal{A}(E, G(B)) \cong \mathcal{A}(F(E), B)$$

and $F(E)$ is split exact, the right-hand side is exact. Hence, $\mathcal{A}(?, G(B))$ is F-exact. □

Let $A \in \mathcal{A}$. We say that

$$0 \to A \xrightarrow{\iota} I^0 \xrightarrow{d^1} I^1 \xrightarrow{d^2} I^2 \to \cdots$$

is an F-admissible resolution of A if the sequence is exact, and ι and all d^i's are F-admissible. If each I^i is F-injective, then it is called an F-injective resolution of A.

Lemma 1.6.9 *If F is faithful exact, then* $\mathrm{Cobar}_F(A)$ *is an F-injective resolution of A.*

Proof. As the ith term $C^i(A)$ of $\mathrm{Cobar}_F(A)$ is $G(F\mathbb{R}^i A)$ by definition, it is F-cofree. Hence, it is F-injective by Lemma 1.6.8.

By Lemma 1.6.5, we have $F(A) \xrightarrow{Fu} F(\mathrm{Cobar}_F(A))$ is split exact. As F is faithful exact, $A \to \mathrm{Cobar}_F(A)$ is exact and each morphism is F-admissible. $\square$

(1.6.10) From now on, we assume that F is faithful exact.

Let $\mathcal{A}'$ be an abelian category, and $Q : \mathcal{A} \to \mathcal{A}'$ an additive functor. We may consider Q as a functor on the exact category $(\mathcal{A}, \mathcal{E}(F))$, and its derived functor is given by $R^iQA = H^i(Q(\mathrm{Cobar}_F(A)))$ by Lemma 1.6.9. We denote the derived functor in this sense by $R^i_F Q(A)$ or $R^i_{\mathcal{A}/\mathcal{B}}Q(A)$, and call it the ith *F-right derived functor* of Q. As cobar resolutions are functorial on A, we may equally well take $R^i_F Q(A) := H^i(Q(\mathrm{Cobar}_F(A)))$ as its definition.

Lemma 1.6.11 *Let $B \in \mathcal{B}$, and $B \to \mathbb{I}$ be an injective resolution of B. If G is exact, then $GB \to G\mathbb{I}$ is an injective resolution of GB. In particular, we have* $\mathrm{inj.dim}_{\mathcal{A}} GB \leq \mathrm{inj.dim}_{\mathcal{B}} B$.

Proof. Obvious.

Lemma 1.6.12 *Let G be exact and $M \in \mathcal{A}$. Then for any $B \in \mathcal{B}$, we have*

$$\mathrm{Ext}^i_{\mathcal{A}}(M, GB) \cong \mathrm{Ext}^i_{\mathcal{B}}(FM, B).$$

In particular, we have

$$\mathrm{Ext}^i_{\mathcal{A}}(M, N) \cong \mathrm{Ext}^i_F(M, N)$$

for any $N \in \mathcal{A}$, if FM is $\mathcal{B}$-projective.

Proof. Take an injective resolution $\mathbb{I}$ of $B \in \mathbb{B}$. Then we have

$$\mathrm{Ext}^i_{\mathcal{A}}(M, GB) = H^i(\mathcal{A}(M, G\mathbb{I})) \cong H^i(\mathcal{B}(FM, \mathbb{I})) \cong \mathrm{Ext}^i_{\mathcal{B}}(FM, B).$$

Hence, if FM is $\mathcal{B}$-projective, then $\mathrm{Ext}^i_{\mathcal{A}}(M, GB) = 0$ for $i > 0$. This means that $\mathrm{Cobar}_F N$ is a $\mathrm{Hom}_{\mathcal{A}}(M, ?)$-acyclic resolution of N for any $N \in \mathcal{A}$, and the last assertion follows. $\square$

We say that F is *relatively acyclic* if any object of $\mathcal{A}$ is F-injective. If F is fully faithful, then $\mathcal{E}(F)$ consists only of split exact sequences, and clearly F is relatively acyclic.

Lemma 1.6.13 *Assume that G is exact. If I is an F-injective object of $\mathcal{A}$, then we have*

$$\text{inj.dim}_{\mathcal{A}} I \leq \text{inj.dim}_{\mathcal{B}} FI.$$

In particular, if F is relatively acyclic, then we have $\text{gl.dim}\,\mathcal{A} \leq \text{gl.dim}\,\mathcal{B}$.

Proof. The unit map $u : I \to GFI$ splits, as u is an admissible mono and I is F-injective. Hence, we have $\text{inj.dim}_{\mathcal{A}} I \leq \text{inj.dim}_{\mathcal{A}} GFI \leq \text{inj.dim}_{\mathcal{B}} FI$. □

Example 1.6.14 We show an example of the dual to above. Let R be a commutative ring, A an R-algebra, $\mathcal{B} := {}_A\mathbb{M}$, and $\mathcal{A} := {}_R\mathbb{M}$. We define $G : \mathcal{B} \to \mathcal{A}$ to be the restriction, and $F : \mathcal{A} \to \mathcal{B}$ to be the inflation $A \otimes ?$. Note that F is a left adjoint functor of G, and the bar resolution, which is defined as the dual of the cobar resolution, is the usual one. If G is relatively acyclic, then we say that A is R-semisimple [78]. To say that F is exact is the same as to say that A is R-flat. So if A is R-flat R-semisimple, then we have $\text{gl.dim}\,{}_A\mathbb{M} \leq \text{gl.dim}\,{}_R\mathbb{M}$ by the dual of Lemma 1.6.13. This was shown by Hochschild [82]. Note that $A = R/I$ is R-semisimple if I is an ideal of R. In fact, G is fully faithful in this case. Thus, an R-semisimple R-algebra is not necessary good even if R is good. We refer the reader to [38] for more information.

1.7 Grothendieck categories

(1.7.1) We say that an abelian category satisfies the (AB3) condition if any direct sum (indexed by a small set) exists. If (AB3) is satisfied, the category has inductive limits. We say that (AB4) is satisfied if (AB3) is satisfied and any direct sum is an exact functor. We say that (AB5) is satisfied if (AB3) is satisfied and any filtered inductive limit is exact. We say that an abelian category $\mathcal{A}$ satisfies (AB3*), (AB4*), and (AB5*), respectively, if $\mathcal{A}^{\text{op}}$ satisfies (AB3), (AB4), and (AB5), respectively.

Lemma 1.7.2 (Grothendieck) *Let $\mathcal{A}$ be an abelian category which satisfies* (AB3), *and $\mathcal{U} = (U_i)_{i \in I}$ a subset of* $\text{ob}(\mathcal{A})$, *where I is a small set. When we set $U := \bigoplus_{i \in I} U_i$, the following are equivalent.*

1 *If $j : B \hookrightarrow A$ is a monomorphism of $\mathcal{A}$ which is not an isomorphism, then there exists some $i \in I$ and $f : U_i \to A$ such that f does not factor through j.*

1′ *If $j : B \hookrightarrow A$ is a monomorphism of $\mathcal{A}$ which is not an isomorphism, then there exists some morphism $f : U \to A$ such that f does not factor through j.*

2 *Any object of $\mathcal{A}$ is a homomorphic image of a direct sum of copies of U.*

3 *If $f : A \to B$ is a non-zero morphism of $\mathcal{A}$, then there exists some $i \in I$ and $g : U_i \to A$ such that $fg \neq 0$.*

If the equivalent conditions above are satisfied, then we say that $\mathcal{U}$ is a *small family of G-generators* of $\mathcal{A}$ (we will use the word 'generator' for a slightly different meaning later, and we use this uncommon expression 'G-generator' instead). We say that $U \in \mathcal{A}$ is a *G-generator* of $\mathcal{A}$ if $\{U\}$ is a small family of G-generators of $\mathcal{A}$.

Lemma 1.7.3 *Let $\mathcal{A}$ be a category with a small family of G-generators, and $A \in \mathcal{A}$. Then the set of subobjects of A has the same cardinality as that of a small set.*

A monomorphism $i : A \to B$ in an abelian category $\mathcal{A}$ is *essential* if a morphism $f : B \to C$ is a monomorphism when fi is a monomorphism. A composite of essential monomorphisms is again essential. The notion of essential epimorphism is the dual notion. We say that a morphism $i : A \to I$ is an *injective hull* (or sometimes we say that I is an injective hull of A) if i is an essential monomorphism and I is an injective object of $\mathcal{A}$. The injective hull I of A is uniquely determined by A up to isomorphisms. A *projective cover* is the dual notion.

Definition 1.7.4 An abelian category $\mathcal{A}$ is called *Grothendieck* if $\mathcal{A}$ satisfies (AB5) and $\mathcal{A}$ has a small family of G-generators.

The category of left A-modules ${}_A\mathbb{M}$ is a Grothendieck category which satisfies the (AB4*) condition for a ring A.

Lemma 1.7.5 *Let $\mathcal{A}$ be a Grothendieck category. For an object $I \in \mathcal{A}$, I is an injective object if and only if any essential monomorphism $I \to J$ is an isomorphism.*

Theorem 1.7.6 (Grothendieck) *Let $\mathcal{A}$ be a Grothendieck category, and $A \in \mathcal{A}$. Then A has an injective hull. In particular, $\mathcal{A}$ has enough injectives. Moreover, there is a functor $I : \mathcal{A} \to \mathcal{I}$ and a natural transformation $f : \mathrm{Id}_{\mathcal{A}} \to iI$ such that f_M is a monomorphism for any $M \in \mathcal{A}$, where $\mathcal{I}$ denotes the full subcategory of $\mathcal{A}$ consisting of injective objects of $\mathcal{A}$, and $i : \mathcal{I} \hookrightarrow \mathcal{A}$ the inclusion.*

(1.7.7) Let $\mathcal{A}$, $\mathcal{I}$, i and I be as in the theorem. Define $J^0 := iI$, $d^{-1} := f$, $J^n := iI(\operatorname{Coker} d^{n-2})$, and d^n to be the composite natural map

$$d^n : J^n \xrightarrow{\operatorname{coker} d^{n-1}} \operatorname{Coker} d^{n-1} \xrightarrow{f \operatorname{Coker} d^{n-1}} iI(\operatorname{Coker} d^{n-1}) = J^{n+1}.$$

Then

$$\mathbb{J} : 0 \to \mathrm{Id} \xrightarrow{d^{-1}} J^0 \xrightarrow{d^0} J^1 \xrightarrow{d^1} J^1 \xrightarrow{d^2} \cdots$$

is a *functorial* injective resolution. Namely, $\mathbb{J}(M)$ is an injective resolution of M for any $M \in \operatorname{ob}(\mathcal{A})$, and $\mathbb{J}(\varphi) : \mathbb{J}(M) \to \mathbb{J}(N)$ covers φ for any $\varphi \in \mathcal{A}(M, N)$.

1.8 Grothendieck topology and sheaf theory

Let $\mathcal{A}$ be a category, and $f : I \to I'$ a functor between svelte categories.

(1.8.1) We define $f^P : \mathcal{P}' \to \mathcal{P}$ to be $f^P := \operatorname{Func}(f^{\mathrm{op}}, \mathcal{A})$, where $\mathcal{P} = \operatorname{Func}(I^{\mathrm{op}}, \mathcal{A})$ and $\mathcal{P}' = ((I')^{\mathrm{op}}, \mathcal{A})$.

If $\mathcal{A}$ has inductive limits, then f^P has a left adjoint. This is shown as follows. For $i' \in I'$, we define a category $I_{i'} = I^f_{i'}$ as follows: An object of $I_{i'}$ is a pair (i, φ) with $i \in I$ and $\varphi \in I'(i', fi)$, and we define

$$I_{i'}((i, \varphi), (j, \psi)) := \{\phi \in I(i, j) \mid f\phi \circ \varphi = \psi\}.$$

Then $I_{i'}$ is a svelte category, and for any $F \in \mathcal{P}$, when we define $F_{i'} : I_{i'} \to \mathcal{A}$ by $F_{i'}(i, \varphi) := Fi$, there exists some inductive limit $\varinjlim F_{i'}$. Letting $f_P(F)(i') := \varinjlim F_{i'}$, the functor $f_P : \mathcal{P} \to \mathcal{P}'$ is a left adjoint of f^P.

(1.8.2) If $\mathcal{A}$ is a Grothendieck category and I is a svelte category, then the category $\mathcal{P} = \operatorname{Func}(I^{\mathrm{op}}, \mathcal{A})$ of $\mathcal{A}$-valued presheaves over I is Grothendieck. In fact, $F \to F' \to F''$ is an exact sequence in $\mathcal{P}$ if and only if $F(A) \to F'(A) \to F''(A)$ is exact for any $A \in \mathcal{A}$. Inductive limits are also given by $(\varinjlim F_i)(A) = \varinjlim F_i A$, and they are exact if filtered. The construction of a small family of G-generators is given in [8, pp. 16–17].

If moreover $\mathcal{A}$ satisfies (AB4*), then direct products are also defined by $(\prod_i F_i)(A) = \prod_i F_i A$ and are exact. Hence, $\mathcal{P}$ also satisfies (AB4*) in this case.

(1.8.3) We say that a pair $T = (\operatorname{Cat} T, \operatorname{Cov} T)$ is a *site* if $\operatorname{Cat} T$ is a category, and $\operatorname{Cov} T$ is a set consisting of families $\{U_i \to U\}_{i \in I}$ of morphisms of $\operatorname{Cat} T$ (the codomain U in each family must be the same one), subject to the conditions

1 If φ is an isomorphism, then $\{\varphi\} \in \operatorname{Cov}(T)$.

2 If $\{U_i \to U\} \in \operatorname{Cov} T$ and $\{V_{ij} \to U_i\}_{j \in J_i} \in \operatorname{Cov} T$ for each $i \in I$, then $\{V_{ij} \to U\}_{i \in I, j \in J_i} \in \operatorname{Cov} T$.

3 If $\{U_i \to U\} \in \operatorname{Cov} T$ and $V \to U$ is any morphism of $\operatorname{Cat} T$, then $U_i \times_U V$ exists for any $i \in I$ and we have $\{U_i \times_U V \to V\} \in \operatorname{Cov} T$.

We say that $\operatorname{Cov} T$ gives a *Grothendieck topology* of $\operatorname{Cat} T$. By abuse of notation, we sometimes denote $\operatorname{Cat} T$ simply by T. An element $\{U_i \to U\}$ of $\operatorname{Cov} T$ is called a *covering* of U. The set of coverings of U is denoted by $\operatorname{cov} U$ for $U \in \operatorname{Cat} T$. Let $\mathcal{U} = \{U_i \xrightarrow{\varphi_i} U\}$ and $\mathcal{V} = \{V_j \xrightarrow{\psi_j} U\}$ be coverings of $U \in \operatorname{Cat} T$. We say that $\mathcal{V}$ is a *refinement* of $\mathcal{U}$ if there exist a map between the index sets $f : J \to I$ and morphisms $\eta_j : V_j \to U_{fj}$ for all $j \in J$ such that $\psi_j = \varphi_{fj} \circ \eta_j$. If we set $\mathcal{V} \geq \mathcal{U}$ when $\mathcal{V}$ is a refinement of $\mathcal{U}$, then $\operatorname{cov} U$ is a (not necessarily small) preordered set. As a category, $\operatorname{cov} U$ is filtered.

(1.8.4) Let T be a site, and $\mathcal{C}$ a category with direct products and finite limits. Let $F : T^{\mathrm{op}} \to \mathcal{C}$ be a presheaf on T with values in $\mathcal{C}$. We say that F is a *sheaf* on T with values in $\mathcal{C}$ if

$$F(U) \to \prod_{i \in I} F(U_i) \rightrightarrows \prod_{i,j \in J} F(U_i \times_U U_j)$$

is exact, in other words, if the left arrow is a difference kernel of the two arrows on the right for any $\{U_i \to U\} \in \operatorname{Cov} T$, where the arrows on the right are the morphisms induced by the first and second projections, respectively. We denote the set of sheaves on T with values in $\mathcal{C}$ by $\operatorname{sh}(T, \mathcal{C})$.

(1.8.5) We say that T has a *small topology* if there exists some small full subcategory T_0 of T such that any object $U \in T$ has a covering consisting only of morphisms with the source objects in T_0. If this is the case, $\operatorname{cov} U$ has a small final subcategory. Moreover, the composite

$$\operatorname{sh}(T, \mathcal{C}) \hookrightarrow \operatorname{Func}(T^{\mathrm{op}}, \mathcal{C}) \xrightarrow{\mathrm{res}} \operatorname{Func}(T_0^{\mathrm{op}}, \mathcal{C}) \tag{1.8.6}$$

is fully faithful. In particular, $\operatorname{sh}(T, \mathcal{C})$ is a category with small hom sets in this case. Assume moreover that T_0 is closed under fiber products. Then T_0 is a site with the same topology as that of T, and we have that the functor (1.8.6) induces an equivalence

$$\operatorname{sh}(T, \mathcal{C}) \xrightarrow{\mathrm{res}} \operatorname{sh}(T_0, \mathcal{C}).$$

(1.8.7) From now, until the end of this subsection, let $\mathcal{A}$ be a Grothendieck category which satisfies the (AB3*) condition, and T a site with a small topology.

(1.8.8) For a covering $\mathfrak{U} := \{U_i \to U\}$ of $U \in T$ and $\mathcal{F} \in \mathrm{Func}(T^{\mathrm{op}}, \mathcal{A})$, the Čech cohomology $\check{H}^i(\mathfrak{U}, \mathcal{F})$ is defined as in [71], see [8]. We define $\check{H}^i(U, \mathcal{F}) := \varinjlim \check{H}^i(\mathfrak{U}, \mathcal{F})$, where $\mathfrak{U}$ runs through $\mathrm{cov}(U)$. When we define $\mathcal{F}^+ \in \mathrm{Func}(T^{\mathrm{op}}, \mathcal{A})$ by $\mathcal{F}^+(U) := \check{H}^0(U, \mathcal{F})$, then we have that $\mathcal{F}^{++} \in \mathrm{sh}(T, \mathcal{A})$, and a functor $a : \mathrm{Func}(T^{\mathrm{op}}, \mathcal{A}) \to \mathrm{sh}(T, \mathcal{A})$ is defined by $a(\mathcal{F}) := \mathcal{F}^{++}$. We call $a(\mathcal{F})$ the *sheafification* of $\mathcal{F}$.

Lemma 1.8.9 *The functor a is left adjoint to the canonical embedding*

$$\mathrm{sh}(T, \mathcal{A}) \to \mathrm{Func}(T^{\mathrm{op}}, \mathcal{A}).$$

For the proof, see [8].

Corollary 1.8.10 *The category* $\mathrm{sh}(T, \mathcal{A})$ *is a Grothendieck category which satisfies* (AB3*). *The sheafification* $a : \mathrm{Func}(T^{\mathrm{op}}, \mathcal{A}) \to \mathrm{sh}(T, \mathcal{A})$ *is exact. The embedding* $\mathrm{sh}(T, \mathcal{A}) \to \mathrm{Func}(T^{\mathrm{op}}, \mathcal{A})$ *preserves injective objects and projective limits.*

Proof. Note that $\mathcal{S} := \mathrm{sh}(T, \mathcal{A})$ is a full subcategory of $\mathcal{P} := \mathrm{Func}(T^{\mathrm{op}}, \mathcal{A})$. We denote the embedding $\mathcal{S} \to \mathcal{P}$ by i. By Lemma 1.2.6, $\mathcal{S}$ has inductive limits. As projective limits preserve projective limits, any projective limit of sheaves as a presheaf is a sheaf, and hence is a projective limit as a sheaf. Hence, $\mathcal{S}$ has projective limits, and i preserves projective limits. In particular, $\mathcal{S}$ has kernels and cokernels, and clearly it is additive.

As a is a left adjoint, it preserves inductive limits. In particular, it preserves cokernels. As the functor $(?)^+$ is left exact, we have that ia is also left exact by construction. Since $\mathcal{S}$ has kernels, both ia and i preserve kernels, and i is faithful, it is easy to see that a preserves kernels.

Let φ be a morphism of $\mathcal{S}$. Then the canonical morphism $i \operatorname{Coker} \ker \varphi \to i \operatorname{Ker} \operatorname{coker} \varphi$ is identified with $ia \operatorname{Coker} \ker i\varphi \to ia \operatorname{Ker} \operatorname{coker} i\varphi$, since the counit $ai \to \mathrm{Id}$ is an isomorphism and a preserves both kernels and cokernels. This is an isomorphism as $\mathcal{P}$ is abelian. As i is faithful, this shows that $\mathcal{S}$ is also abelian. Now it is clear that a is exact. Again by Lemma 1.2.6, i preserves injective objects.

The category $\mathcal{S}$ satisfies (AB5). In fact, the inductive limit in $\mathcal{S}$ is nothing but the composite $a \varinjlim (i?)$, and hence it is left exact if filtered.

As it is easy to see that $(a\mathcal{F}_i)$ is a small family of G-generators of $\mathcal{S}$ if $(\mathcal{F}_i)$ is a small family of G-generators of $\mathcal{P}$, we have that $\mathcal{S}$ is Grothendieck. □

(1.8.11) Let T and T' be sites. We say that $f : T \to T'$ is a *continuous functor* if $f : \operatorname{Cat} T \to \operatorname{Cat} T'$ is a functor such that for any $\{U_i \to U\} \in \operatorname{Cov} T$, we have $\{fU_i \to fU\} \in \operatorname{Cov} T'$ and the canonical map

$$f(U_i \times_U U_j) \to fU_i \times_{fU} fU_j.$$

is an isomorphism for any i, j. If $f : T \to T'$ is a continuous functor and $\mathcal{F} \in \operatorname{sh}(T', \mathcal{A})$, then $f^P(\mathcal{F})$ is a sheaf. Hence, we obtain a functor $f^S = f^P : \operatorname{sh}(T', \mathcal{A}) \to \operatorname{sh}(T, \mathcal{A})$. Note also that $f_S := af_P : \operatorname{sh}(T, \mathcal{A}) \to \operatorname{sh}(T', \mathcal{A})$ is left adjoint to f^S.

Example 1.8.12 Let $\mathcal{B}$ be an abelian category. When we define that $\{B_i \to B\}_{i \in I}$ is a covering if I is a finite set (we allow the case I is empty) and $\bigoplus_i B_i \to B$ is an epimorphism (resp. isomorphism), then $\mathcal{B}$ is a site, which we denote by $\mathcal{B}_1$ (resp. $\mathcal{B}_0$). The category $\mathcal{S}_1 := \operatorname{sh}(\mathcal{B}_1, \mathcal{A})$ (resp. $\mathcal{S}_0 := \operatorname{sh}(\mathcal{B}_0, \mathcal{A})$) is nothing but $\operatorname{Sex}(\mathcal{B}^{\mathrm{op}}, \mathcal{A})$ (resp. the category of contravariant additive functors from $\mathcal{B}$ to $\mathcal{A}$). In particular, as we are assuming that $\mathcal{A}$ is Grothendieck and satisfies (AB3*), $\operatorname{Sex}(\mathcal{B}^{\mathrm{op}}, \mathcal{A})$ is also Grothendieck and satisfies (AB3*). Moreover, $\operatorname{Id}_{\mathcal{B}} : \mathcal{B}_0 \to \mathcal{B}_1$ is clearly continuous.

(1.8.13) Let $\tilde{T}$ be a site with fiber products, and T a subcategory of $\tilde{T}$ such that $\operatorname{ob}(T) = \operatorname{ob}(\tilde{T})$. We assume that any isomorphism of $\tilde{T}$ is a morphism of T, and any base change of a morphism of T is again a morphism of T. Moreover, we assume that $U_i \to U$ is a morphism of T for any $\{U_i \to U\} \in \operatorname{Cov}(\tilde{T})$.

For $X \in \tilde{T}$, the full subcategory of $\tilde{T}/X$ consisting of morphisms of T (with the fixed codomain X) is denoted by T_X. Note that T_X is a site with the topology of $\tilde{T}$. If $\varphi : X' \to X$ is a morphism of $\tilde{T}$, then a continuous functor $\varphi^\# : T/X \to T/X'$ is defined by $\varphi^\#(Y) = X' \times_X Y$. We denote $(\varphi^\#)^P$, $(\varphi^\#)_P$, $(\varphi^\#)^S$, and $(\varphi^\#)_S$ respectively by φ_p, φ^p, φ_*, and φ^{-1}, and call them the *direct image*, *presheaf inverse image*, *direct image*, and *inverse image* functor of φ, respectively.

Example 1.8.14 Let $\tilde{T}$ be the category of schemes, and let us define that $\{U_i \to U\}$ is a covering if each $g_i : U_i \to U$ is an open immersion (resp. étale morphism of finite presentation, flat morphism locally of finite presentation) and $\bigcup_i g_i(U_i) = U$ holds. Then $\tilde{T}$ is a site. The corresponding Grothendieck topology of $\tilde{T}$ is called the Zariski, étale, and fppf topology, respectively. When we define morphisms of T to be open immersions, étale morphisms, and morphisms locally of finite presentation respectively, then T_X has small topologies. An object of $\operatorname{sh}(T_X, \mathcal{C})$ is called a $\mathcal{C}$-valued Zariski sheaf, étale sheaf, and *fppf sheaf* (or *faisceau*), respectively.

1.9 Noetherian categories and locally noetherian categories

(1.9.1) Let $\mathcal{A}$ be an abelian category, and $M \in \mathrm{ob}(\mathcal{A})$. We say that M is a noetherian (resp. artinian) object if the set of subobjects of M satisfies the ascending (resp. descending) chain condition.

An abelian category is called *noetherian* (resp. *artinian*) if it is svelte and any object in it is noetherian (resp. artinian). An abelian category is called *locally noetherian* (resp. *locally artinian, locally finite*) if it satisfies (AB5), and has a small family of G-generators consisting of noetherian (resp. artinian, finite length) objects.

Note that the full subcategory of noetherian objects in an abelian category is very thick. Any noetherian category $\mathcal{C}$ is embedded in $\mathcal{D} := \mathrm{Sex}(\mathcal{C}^{\mathrm{op}}, \underline{\mathrm{Ab}})$ by the Gabriel–Quillen embedding. By Example 1.8.12, $\mathcal{D} = \mathrm{Sex}(\mathcal{C}^{\mathrm{op}}, \underline{\mathrm{Ab}})$ is locally noetherian and satisfies (AB3*). Moreover, $\mathcal{C}$ is equivalent to the full subcategory of $\mathcal{D}$ consisting of its noetherian objects via the Gabriel–Quillen embedding. In fact, any skeleton of $\mathcal{C}$ is a noetherian generator of $\mathcal{D}$.

(1.9.2) Conversely, for a given locally noetherian category $\mathcal{D}$, its full subcategory $\mathcal{D}_f$ of $\mathcal{D}$ consisting of noetherian objects of $\mathcal{D}$ is noetherian, and the Gabriel–Quillen functor $\mathcal{D} \to \mathrm{Sex}(\mathcal{D}_f^{\mathrm{op}}, \underline{\mathrm{Ab}})$ is an equivalence. In particular, any locally noetherian category satisfies (AB3*). Moreover, noetherian categories and locally noetherian categories are in one-to-one correspondence as above. We always use the symbol $\mathcal{D}_f$ to denote the full subcategory of noetherian objects of a locally noetherian category $\mathcal{D}$. For example, ${}_R\mathbb{M}$ is locally noetherian if R is a noetherian commutative ring. In this case, ${}_R\mathbb{M}_f$ is nothing but the category of R-finite modules. Hence, ${}_R\mathbb{M} \to \mathrm{Sex}({}_R\mathbb{M}_f^{\mathrm{op}}, \underline{\mathrm{Ab}})$ is an equivalence. More generally, if X is a noetherian scheme, then the category $\mathrm{Qco}(X)$ is locally noetherian, and $\mathrm{Qco}(X)_f$ is nothing but the category $\mathrm{Coh}(X)$ of coherent $\mathcal{O}_X$-modules, see [58].

Let $\mathcal{D}$ be a locally noetherian category.

Note that the presheaf direct product, projective limit, and *filtered* inductive limit of objects in $\mathrm{Sex}(\mathcal{C}^{\mathrm{op}}, \underline{\mathrm{Ab}})$ are left exact again, and hence are the direct product, projective limit, filtered inductive limit in $\mathrm{Sex}(\mathcal{C}^{\mathrm{op}}, \underline{\mathrm{Ab}})$, respectively. In particular, we have:

Lemma 1.9.3 *Let $\mathcal{D}$ be a locally noetherian category, $Y \in \mathcal{D}_f$, and let (X_i) be a filtered inductive system in $\mathcal{D}$. Then the canonical map*

$$\varinjlim \mathrm{Hom}_{\mathcal{D}}(Y, X_i) \to \mathrm{Hom}_{\mathcal{D}}(Y, \varinjlim X_i)$$

is an isomorphism. In particular, a filtered inductive limit (e.g., a direct sum) of injective objects in $\mathcal{D}$ is again injective.

By the lemma, we have the following

Lemma 1.9.4 (Gabriel) *Let $\mathcal{D}$ be as above. Then for $F \in \mathcal{D}$, viewed also as an object of* $\mathrm{Sex}(\mathcal{D}_f^{\mathrm{op}}, \underline{\mathrm{Ab}})$*, the following are equivalent.*

1 *F is a contravariant exact functor on $\mathcal{D}_f$.*

2 *F is an injective object of $\mathcal{D}$ (in other words, $\mathcal{D}(?, F)$ is exact on $\mathcal{D}$).*

3 *For any $M \in \mathcal{D}_f$, $\mathrm{Ext}^1_{\mathcal{D}}(M, F) = 0$.*

Lemma 1.9.5 *Let $\mathcal{D}$, Y, and (X_i) be as in* Lemma 1.9.3*. Then the canonical map*

$$\varinjlim \mathrm{Ext}^j_{\mathcal{D}}(Y, X_i) \to \mathrm{Ext}^j_{\mathcal{D}}(Y, \varinjlim X_i)$$

is an isomorphism for $j \geq 0$.

Proof. As $\mathcal{D}$ is Grothendieck, there is a functorial injective resolution $\mathbb{J}$ in $\mathcal{D}$, see (1.7.7). By Lemma 1.9.3,

$$\varinjlim \mathrm{Hom}^\bullet_{\mathcal{D}}(Y, \mathbb{J}(X_i)) \to \mathrm{Hom}^\bullet_{\mathcal{D}}(Y, \varinjlim \mathbb{J}(X_i))$$

is an isomorphism of complexes of abelian groups. Since $\varinjlim \mathbb{J}(X_i)$ is an injective resolution of $\varinjlim X_i$ by the (AB5) condition and the last assertion of Lemma 1.9.3, the assertion follows immediately, taking the cohomology. □

1.10 Semisimple objects in a Grothendieck category

(1.10.1) Let $\mathcal{A}$ be a Grothendieck category, and $\mathcal{B}$ a full subcategory of $\mathcal{A}$ closed under direct sums, subobjects, and quotient objects. Note that $\mathcal{B}$ is abelian, and the canonical embedding $i : \mathcal{B} \hookrightarrow \mathcal{A}$ is exact. Note also that $\mathcal{B}$ is Grothendieck, which is less trivial. In fact, if U is a generator of $\mathcal{A}$, $B \in \mathcal{B}$ and $f \in \mathcal{A}(U, B)$, then we have $\mathrm{Im}\, f \subset B$ and hence we have $\mathrm{Im}\, f \in \mathcal{B}$. So the set $(V_i)_{i \in I}$ of quotient objects of U which lies in $\mathcal{B}$ (by Lemma 1.7.3, we can take I to be small) is a small family of G-generators of $\mathcal{B}$. As the inductive limit in $\mathcal{A}$ of an inductive system consisting of objects in $\mathcal{B}$ lies in $\mathcal{B}$ by assumption, it is also an inductive limit in $\mathcal{B}$. Hence the (AB5) condition holds in $\mathcal{B}$. As is easily seen from the proof, $\mathcal{B}$ is locally noetherian if $\mathcal{A}$ is.

Lemma 1.10.2 *The embedding $i : \mathcal{B} \hookrightarrow \mathcal{A}$ has a right adjoint j. The unit $u : \mathrm{Id} \to ji$ is an isomorphism.*

Proof. For $A \in \mathcal{A}$, the set of subobjects of A is indexed by a small set (Lemma 1.7.3). So we may form the sum $j(A)$ of all subobjects of A which lie in B. As $\mathcal{B}$ is closed under inductive limits in $\mathcal{A}$, we have $j(A) \in \mathcal{B}$. Note that $j(A)$ is the largest subobject of A which lies in $\mathcal{B}$. For $B \in \mathcal{B}$ and $f \in \mathcal{A}(B, A)$, we have that f factors through $j(A)$, as $\operatorname{Im} f \in \mathcal{B}$. Hence, we have an isomorphism

$$\Phi_{B,A} : \mathcal{A}(i(B), A) \cong \mathcal{B}(B, j(A)).$$

In particular, for $A, A' \in \mathcal{A}$ and $g \in \mathcal{A}(A, A')$, the restriction of g to $j(A)$ factors through $j(A')$, and we have an induced morphism $j(g) \in \mathcal{B}(j(A), j(A'))$. It is easy to verify that j is a functor with this definition, and $\Phi_{B,A}$ is natural on B and A. As i is fully faithful, the last assertion follows from Lemma 1.2.6. □

As j has an exact left adjoint i, it preserves projective limits (in particular, kernels) and injective objects.

Lemma 1.10.3 *Let j be the right adjoint of i as above. If $\mathcal{A}$ is locally noetherian, then the canonical map*

$$f : \varinjlim j(A_\lambda) \to j(\varinjlim A_\lambda)$$

is an isomorphism for any filtered inductive system (A_λ) in $\mathcal{A}$.

Proof. Note that if $B \in \mathcal{B}_f$, then $i(B) \in \mathcal{A}_f$. Hence, using Lemma 1.9.3, we have isomorphisms of functors on $\mathcal{B}_f$:

$$\begin{aligned} \mathcal{B}(?, \varinjlim j(A_\lambda)) &\cong \varinjlim \mathcal{B}(?, j(A_\lambda)) \cong \varinjlim \mathcal{A}(i(?), A_\lambda) \\ &\cong \mathcal{A}(i(?), \varinjlim A_\lambda) \cong \mathcal{B}(?, j(\varinjlim A_\lambda)). \end{aligned}$$

As the canonical functor $y : \mathcal{B} \to \operatorname{Sex}(\mathcal{B}_0^{\mathrm{op}}, \underline{\mathrm{Ab}})$ is an equivalence and the composite isomorphism above is nothing but $y(f) : y(\varinjlim j(A_\lambda)) \to y(j(\varinjlim A_\lambda))$, f is an isomorphism. □

(1.10.4) Let $\mathcal{A}$ be a Grothendieck category. We say that $A \in \mathcal{A}$ is a *simple object* if there are exactly two subobjects of A. In other words, A is simple if and only if $A \ncong 0$, and any monomorphism into A is either zero or an isomorphism.

We say that $A \in \mathcal{A}$ is *semisimple* if A is isomorphic to a direct sum of simple objects. The following is well-known, and is proved in [123] in the case of modules over an algebra.

Lemma 1.10.5 *The full subcategory $\mathcal{A}_{\mathrm{ss}}$ of $\mathcal{A}$ consisting of semisimple objects of $\mathcal{A}$ is closed under inductive limits, subobjects, and quotient objects in $\mathcal{A}$.*

By the lemma, the sum of all semisimple subobjects of $A \in \mathcal{A}$ is the maximum semisimple subobject of A. This object is called the *socle* of A, and we denote it by $\operatorname{soc} A$. Thus, $\operatorname{soc} : \mathcal{A} \to \mathcal{A}_{\text{ss}}$ is a right adjoint functor of the canonical embedding $\mathcal{A}_{\text{ss}} \hookrightarrow \mathcal{A}$. Note that $\operatorname{soc} A$ is also the sum of all simple subobjects of A. As it is a right adjoint, soc preserves projective limits (e.g., kernels).

Assume moreover that $\mathcal{A}$ satisfies the (AB3*) condition. Then for $A \in \mathcal{A}$, we set

$$\operatorname{rad} A := \bigcap_{B \subset A,\ A/B:\text{semisimple}} B$$

and call $\operatorname{rad} A$ the *radical* of A. We denote $A/\operatorname{rad} A$ by $\operatorname{top} A$, and call it the *top* of A. Note that $\operatorname{top} A$ is not necessarily semisimple. However, if A is an artinian object, then $\operatorname{top} A$ is semisimple, and hence is the largest semisimple quotient of A.

Any non-zero artinian (resp. noetherian) object admits a simple subobject (resp. quotient object). Hence, we have

Lemma 1.10.6 *Let $\mathcal{A}$ be a Grothendieck category which satisfies the* (AB3*) *condition. If A is an artinian (resp. noetherian) object of $\mathcal{A}$ and* $\operatorname{soc} A = 0$ *(resp.* $\operatorname{top} A = 0$*), then we have $A = 0$.*

The following is also trivial.

Lemma 1.10.7 *Let $\mathcal{D}$ be a locally noetherian category, and (D_λ) a filtered inductive system in $\mathcal{D}$. Then the canonical map* $\varinjlim \operatorname{soc}(D_\lambda) \to \operatorname{soc}(\varinjlim D_\lambda)$ *is an isomorphism. If $\mathcal{A}$ is a locally artinian category and $0 \neq A \in \mathcal{A}$, then we have* $\operatorname{soc} A \neq 0$.

(1.10.8) We say that a ring A is a *division ring* if $A \neq 0$ and any non-zero element of A is a unit. The following is well-known as Wedderburn's theorem [123].

Theorem 1.10.9 *Let A be a ring. The following are equivalent.*

1 *The A-module ${}_AA$ is a semisimple object of ${}_A\mathbb{M}$.*

1* *The right A-module A_A is a semisimple object of $\mathbb{M}_A$.*

2 *A is a finite direct product $\prod_{i=1}^r \operatorname{Mat}_{n_i}(D_i)$ of matrix rings over division rings.*

If the conditions above are satisfied, then A is called a semisimple ring. A semisimple ring is both left and right artinian.

1.11 Full subcategories of an abelian category

Let $\mathcal{A}$ be an abelian category, and $\mathcal{X}$ a subset of $\mathrm{ob}(\mathcal{A})$. We define some full subcategories of $\mathcal{A}$.

1 We denote the full subcategory of $\mathcal{A}$ consisting of objects isomorphic to a direct summand of a finite direct sum of objects of $\mathcal{X}$ by $\mathrm{add}\,\mathcal{X}$. If $\mathcal{X} = \emptyset$, then $\mathrm{add}(\mathcal{X})$ consists of null objects of $\mathcal{A}$. Obviously, $\mathrm{add}(\mathcal{X})$ is a Karoubian additive category. If $\mathcal{X}$ is closed under extensions in $\mathcal{A}$, then so is $\mathrm{add}(\mathcal{X})$.

2 The full subcategory of $\mathcal{A}$ consisting of objects $A \in \mathcal{A}$ such that there exists some $r \geq 0$ and a filtration

$$0 = A_0 \subset A_1 \subset \cdots \subset A_r = A$$

such that A_i/A_{i-1} is isomorphic to some object in $\mathcal{X}$ for $i = 1, 2, \ldots, r$ is denoted by $\mathcal{F}(\mathcal{X})$. Note that $\mathcal{F}(\mathcal{X})$ is closed under extensions in $\mathcal{A}$. Note also that $\mathcal{F}(\mathcal{X})$ is not closed under direct summands in general even if $\mathcal{X}$ is so.

3 Let $\mathcal{X}$ be an additive full subcategory of $\mathcal{A}$. The full subcategory of $\mathcal{A}$ consisting of $A \in \mathcal{A}$ such that there exists some exact sequence

$$0 \to X_h \to \cdots \to X_1 \to X_0 \to A \to 0 \tag{1.11.1}$$

with $X_i \in \mathcal{X}$ is denoted by $\hat{\mathcal{X}}$. An exact sequence of the form (1.11.1) is called a finite $\mathcal{X}$-resolution of A. The smallest non-negative integer i such that $X_{i+1} = 0$ is called the length of the $\mathcal{X}$-resolution (1.11.1). For $A \in \hat{\mathcal{X}}$, we call the minimum length of $\mathcal{X}$-resolutions of A the *$\mathcal{X}$-resolution dimension* of A, and denote it by $\mathcal{X}\text{-resol.dim}\,A$. If $A \notin \hat{\mathcal{X}}$, then we define $\mathcal{X}\text{-resol.dim}\,A := \infty$.

3* Let $\mathcal{X}$ be an additive full subcategory of $\mathcal{A}$. Then we define $\check{\mathcal{X}} := (\widehat{\mathcal{X}^{\mathrm{op}}})^{\mathrm{op}}$. In other words, $\check{\mathcal{X}}$ consists of $A \in \mathcal{A}$ such that there exists some exact sequence

$$0 \to A \to X^0 \to X^1 \to \cdots \to X^h \to 0 \tag{1.11.2}$$

with $X^i \in \mathcal{X}$. An exact sequence as in (1.11.2) is called a finite *$\mathcal{X}$-coresolution* of A. The minimum non-negative integer i such that $X^{i+1} = 0$ is called the length of the $\mathcal{X}$-coresolution (1.11.2). For $A \in \check{\mathcal{X}}$, the minimum length of $\mathcal{X}$-coresolutions of A is called the *$\mathcal{X}$-coresolution dimension* of A, and we denote it by $\mathcal{X}\text{-cores.dim}\,A$. If $A \notin \check{\mathcal{X}}$, then we define $\mathcal{X}\text{-cores.dim}\,A := \infty$.

4 For $A \in \mathcal{A}$, we define

$$\mathcal{X}\text{-inj.dim}\, A := \sup(\{i \geq 0 \mid \operatorname{Ext}^i_{\mathcal{A}}(\mathcal{X}, A) \neq 0\} \cup \{0\}),$$

and we call $\mathcal{X}$-inj.dim A the *$\mathcal{X}$-injective dimension* of A. If $\mathcal{X}$-inj.dim $A = 0$, then we say that A is *$\mathcal{X}$-injective*. The full subcategory of $\mathcal{A}$ consisting of $\mathcal{X}$-injective objects in $\mathcal{A}$ is denoted by $\mathcal{X}^{\perp}$. Similarly, *$\mathcal{X}$-projective dimension* is defined, and we denote it by $\mathcal{X}$-proj.dim A. We also define an *$\mathcal{X}$-projective object* in a similar way, and the full subcategory of $\mathcal{X}$-projective objects of $\mathcal{A}$ is denoted by ${}^{\perp}\mathcal{X}$. Note that $\mathcal{X}^{\perp}$ is closed under extensions, direct summands, and monocokernels. Note also that ${}^{\perp}\mathcal{X}$ is closed under extensions, direct summands, and epikernels.

1.12 $\mathcal{X}$-approximations and the Auslander–Buchweitz theory

Let $\mathcal{A}$ be an abelian category. We say that a morphism $p : M \to N$ of $\mathcal{A}$ is *right minimal* if $\varphi \in \operatorname{End}_{\mathcal{A}}(M)$ and $p\varphi = p$ imply that φ is an isomorphism. *Left minimality* is the dual notion of right minimality. In other words, a morphism of $\mathcal{A}$ is called left minimal if it is right minimal as a morphism of the dual category $\mathcal{A}^{\text{op}}$. Let $\mathcal{X}$ be a full subcategory of $\mathcal{A}$. A morphism $f : X \to M$ of $\mathcal{A}$ is called a *right $\mathcal{X}$-approximation* of M if $X \in \mathcal{X}$, and for any $X' \in \mathcal{X}$ and any $g \in \mathcal{A}(X', M)$, there exists some $h \in \mathcal{A}(X', X)$ such that $fh = g$. It is equivalent to say that $\mathcal{A}(?, X) \in \operatorname{Func}(\mathcal{X}, \underline{\text{Ab}})$ is representable, and $\mathcal{A}(?, f) : \mathcal{A}(?, X) \to \mathcal{A}(?, M)$ is an epimorphism of the functor category $\operatorname{Func}(\mathcal{X}, \underline{\text{Ab}})$. *Left $\mathcal{X}$-approximation* is the dual notion. A right (resp. left) minimal right (resp. left) $\mathcal{X}$-approximation is simply called a right (resp. left) minimal $\mathcal{X}$-approximation. A right minimal $\mathcal{X}$-approximation of M is unique up to isomorphisms as an object of $\mathcal{A}/M$, if it exists.

Let $\mathcal{B}$ be a full subcategory of $\mathcal{A}$ which contains $\mathcal{X}$. We say that $\mathcal{X}$ is *contravariantly finite* (resp. *covariantly finite*) in $\mathcal{B}$ (or $\mathcal{B}$ has right (resp. left) $\mathcal{X}$-approximations) if any object in $\mathcal{B}$ has a right (resp. left) $\mathcal{X}$-approximation. If any object in $\mathcal{B}$ admits a right (resp. left) minimal $\mathcal{X}$-approximation, then we say that $\mathcal{B}$ has right (resp. left) minimal $\mathcal{X}$-approximations.

Lemma 1.12.1 *Let*

$$0 \to Y \xrightarrow{i} X \xrightarrow{p} M \to 0$$

be an exact sequence in $\mathcal{A}$ such that $X \in \mathcal{X}$, and assume that $\operatorname{Ext}^1_{\mathcal{A}}(\mathcal{X}, Y) = 0$. *Then p is a right $\mathcal{X}$-approximation of M.*

Proof. For any $X' \in \mathcal{X}$, we have that the sequence

$$\mathcal{A}(X', X) \to \mathcal{A}(X', M) \to \operatorname{Ext}^1_{\mathcal{A}}(X', Y) = 0$$

is exact. □

Lemma 1.12.2 (Wakamatsu's lemma) *Let*

$$0 \to Y \xrightarrow{i} X \xrightarrow{p} M$$

be an exact sequence in $\mathcal{A}$ such that p is a right minimal $\mathcal{X}$-approximation of M. If $\mathcal{X}$ is closed under extensions, then $\operatorname{Ext}^1_{\mathcal{A}}(\mathcal{X}, Y) = 0$.

For the proof, see [146, Lemma 2.1.1].

(1.12.3) Let A be a ring. The radical $\operatorname{rad}_A A$ of A as a left A-module and the radical $\operatorname{rad} A_A$ of A as a right A-module agree, and it is simply denoted by $\operatorname{rad} A$, and called the *Jacobson radical* of A. Note that $\operatorname{rad} A$ is a two-sided ideal of A. If $0 \neq M \in {}_A\mathbb{M}$ is A-finite module, then $M \neq (\operatorname{rad} A)M$ (Nakayama's lemma). For basics on Jacobson radicals, see [123].

Lemma 1.12.4 *Let A be a ring, and $I = \operatorname{rad} A$. The following are equivalent.*

1 *Any finitely generated A-module has a projective cover (in the category ${}_A\mathbb{M}$).*

1* *Any finitely generated right A-module has a projective cover (in the category ${}_{A^{\mathrm{op}}}\mathbb{M}$).*

2 *The ring A/I is semisimple, and for any idempotent $\bar{e}$ of A/I, there exists some idempotent e of A such that e modulo I equals $\bar{e}$.*

We say that a ring A is *semiperfect* if the equivalent conditions in the lemma are satisfied.

Lemma 1.12.5 *Let A be a semiperfect ring, $I = \operatorname{rad} A$, and $p : P \to M$ an A-linear map between A-modules. Assume that M is A-finite. Then p is a projective cover if and only if P is A-finite projective and the induced map $P/IP \to M/IM$ is an isomorphism.*

We say that a ring A is *local* if $A/\operatorname{rad} A$ is a division ring. A ring A is local if and only if $A \neq 0$ and the set of non-units of A is closed under addition. We say that a commutative noetherian local ring R is *Henselian* if any R-module finite algebra is semiperfect. If moreover the residue field of R is separably closed, then it is called *strictly Henselian.* Note that a complete local ring is Henselian [110, 111].

If a semiperfect ring $A \neq 0$ does not have any non-trivial idempotent, then $A/\operatorname{rad} A$ is a division ring by Theorem 1.10.9.

Let $\mathcal{C}$ be a Karoubian additive category, and C an object of $\mathcal{C}$. We say that C is *indecomposable* if $C \neq 0$, and for any split monomorphism $i : C' \to C$, it holds either $i = 0$ or i is an isomorphism. If $A := \operatorname{End}_{\mathcal{C}} C$ is local, then C is indecomposable, since A does not have any non-trivial idempotent. Conversely, if $A := \operatorname{End}_{\mathcal{C}} C$ is semiperfect and C is indecomposable, then A is local. The following Krull–Schmidt theorem holds.

Lemma 1.12.6 *Let $\mathcal{C}$ be a Karoubian additive category. Then the following are equivalent.*

1 *For any $M \in \mathcal{C}$, there exists some decomposition $M \cong M_1 \oplus \cdots \oplus M_r$ ($r \geq 0$) such that $\operatorname{End}_{\mathcal{C}} M_i$ is local for each i (in particular, each M_i is indecomposable).*

2 *For any object $M \in \mathcal{C}$, we have $\operatorname{End}_{\mathcal{C}} M$ is semiperfect.*

Moreover, if these conditions are satisfied, $M \in \mathcal{C}$, and there are two decompositions $M \cong M_1 \oplus \cdots \oplus M_r \cong N_1 \oplus \cdots \oplus N_s$ such that M_i and N_j are indecomposable, then we have $r = s$, and there exists some permutation $\sigma \in \mathfrak{S}_r$ such that $N_i \cong M_{\sigma i}$ for any i.

Here we only sketch the proof of **1**$\Rightarrow$**2**. Take a decomposition $M = M_1 \oplus \cdots \oplus M_r$ and set $E := \operatorname{End}_{\mathcal{C}} M$. For each i, set $\sigma_i : M_i \hookrightarrow M$ to be the inclusion, and $\pi_i : M \to M_i$ to be the projection. First prove that

$$J := \{\varphi \in E \mid \forall i, j \ \pi_j \varphi \sigma_i \text{ is not an isomorphism}\}$$

is a two-sided ideal of E. An argument similar to [123, 5.4., Lemma b] shows that $1 + J \subset E^\times$. It follows that $J \subset \operatorname{rad} E$. It is easy to see that E/J is semisimple, and we have $J = \operatorname{rad} E$. The lifting of idempotents is derived from Nakayama's lemma.

The other parts are (almost) proved in [124, 5.1]. □

Lemma 1.12.7 *Let $\mathcal{A}$ be an abelian category, and M an object of $\mathcal{A}$ which admits a composition series. Then $\operatorname{End}_{\mathcal{A}} M$ is semiperfect.*

This lemma follows from [124, (5.7.5)] and its proof, and Lemma 1.12.6. The next theorem is a variation of a theorem of Miyachi [112, Theorem 3.4].

Proposition 1.12.8 *Let $\mathcal{A}$ be an abelian category, and $p : M \to N$ a morphism of $\mathcal{A}$. Set $i : K \to M$ to be the kernel of f. Consider the following two conditions.*

1 *The morphism p is right minimal.*

2 *The objects K and M do not have any direct summands in common through i. In other words, if $M_0 \in \mathcal{A}$, $l \in \mathcal{A}(M_0, K)$, and $\pi \in \mathcal{A}(M, M_0)$, then $\pi \circ i \circ l$ is not an isomorphism.*

Then **1**$\Rightarrow$**2**. *If moreover,* $\mathrm{End}_{\mathcal{A}}(M)$ *is semiperfect, then there exists some decomposition $M = M_0 \oplus M_1$ such that $M_0 \subset \mathrm{Im}\, i$ and $M_1 \to N$ is right minimal. In particular,* **2**$\Rightarrow$**1** *in this case.*

Proof. Assume that $M \cong M_0 \oplus M_1$, $M_0 \neq 0$, and $M_0 \subset \mathrm{Im}\, i$. Then when we define φ to be the projector to M_1, then φ is not an isomorphism, but we have $p\varphi = p$. Hence **1**$\Rightarrow$**2**.

Next, assume that $E := \mathrm{End}_{\mathcal{A}}(M)$ is semiperfect. We denote the functor $\mathcal{A}(M, ?) : \mathcal{A} \to {}_{E^{\mathrm{op}}}\mathbb{M}$ by e_E. We have an exact sequence of right E-modules

$$0 \to e_E(K) \xrightarrow{e_E(i)} E \xrightarrow{e_E(p)} e_E(N).$$

As the right E-module $C := \mathrm{Im}\, e_E(p)$ is finitely generated, there is a projective cover $\pi : P \to C$ of C. As E is also E-projective, the map $e_E(p) : E \to C$ lifts to $\rho : E \to P$ so that $\pi\rho = e_E(p)$, and ρ is surjective by the definition of projective cover. Hence, ρ splits, and there exists some $\sigma : P \to E$ such that $\rho\sigma = 1_P$. The map $e := \sigma\rho \in \mathrm{End}_E E_E \cong E$ is a projector, and hence is an idempotent. When we set $M_0 := \mathrm{Im}(1 - e)$ and $M_1 := \mathrm{Im}\, e$, then $M = M_0 \oplus M_1$. As

$$pe = e_E(p)(e) = \pi\rho\sigma\rho = \pi\rho = p,$$

$M_0 \subset \mathrm{Ker}\, p = \mathrm{Im}\, i$.

We prove that $pj : M_1 \to N$ is right minimal, where $j : M_1 \hookrightarrow M$ is the canonical inclusion. Let ψ be an endomorphism of M_1 such that $pj\psi = pj$. Applying e_E, we have that $e_E(pj)e_E(\psi) = e_E(pj)$. As $e_E(pj)$ factors through the projective cover $eE \hookrightarrow E \to C$ of C, we have $e_E(\psi)$ is an isomorphism. As $e_E : \mathrm{add}\, M \to \mathrm{add}\, E_E$ is an equivalence [12, Proposition II.2.1], ψ is an isomorphism. Hence, $pj : M_1 \to N$ is right minimal. □

Corollary 1.12.9 *Let $\mathcal{X}$ be a full subcategory of $\mathcal{A}$ closed under direct summands, and assume that the endomorphism ring of any object of $\mathcal{X}$ is semiperfect. If $M \in \mathcal{A}$ has a right (resp. left) $\mathcal{X}$-approximation, then M has a unique right (resp. left) minimal $\mathcal{X}$-approximation.*

Let $\mathcal{A}$ be an abelian category, and $\mathcal{X}$ be a full subcategory of $\mathcal{A}$. We say that a full subcategory ω of $\mathcal{X}$ is a *cogenerator* of $\mathcal{X}$ if for any $X \in \mathcal{X}$, there exists some short exact sequence of $\mathcal{A}$

$$0 \to X \to W \to X' \to 0$$

such that $W \in \omega$ and $X' \in \mathcal{X}$. We say that ω is an *injective cogenerator* of $\mathcal{X}$ if it is a cogenerator of $\mathcal{X}$ and $\omega \subset \mathcal{X}^{\perp}$.

The following theorem is due to Auslander and Buchweitz [10]. Some part is taken from [11].

Theorem 1.12.10 *Let $\mathcal{A}$ be an abelian category, and $\mathcal{X}$, $\mathcal{Y}$, and ω be full subcategories of $\mathcal{X}$ such that*

AB1 *$\mathcal{X}$ is closed under extensions, epikernels and direct summands in $\mathcal{A}$.*

AB2 *$\mathcal{Y}$ is closed under monocokernels, extensions and direct summands in $\mathcal{A}$, and we have $\mathcal{Y} \subset \hat{\mathcal{X}}$.*

AB3 *$\omega = \mathcal{X} \cap \mathcal{Y}$, and ω is an injective cogenerator of $\mathcal{X}$.*

Then the following hold:

1 $\hat{\omega} = \mathcal{Y}$.

2 *If $\omega' \subset \mathcal{X}$ is an injective cogenerator of $\mathcal{X}$, then* $\operatorname{add} \omega' = \omega$.

3 *For $M \in \hat{\mathcal{X}}$, the following hold.*

i *($\mathcal{X}$-approximation) There exists some exact sequence of $\mathcal{A}$*

$$0 \to Y \to X \xrightarrow{p} M \to 0$$

such that $X \in \mathcal{X}$ and $Y \in \mathcal{Y}$.

ii *($\mathcal{Y}$-hull) There exists some exact sequence of $\mathcal{A}$*

$$0 \to M \xrightarrow{\iota} Y \to X \to 0$$

such that $X \in \mathcal{X}$ and $Y \in \mathcal{Y}$.

4 *For $M \in \hat{\mathcal{X}}$, the following are equivalent.*

i $M \in \mathcal{X}$ **ii** $\operatorname{Ext}^i_{\mathcal{A}}(M, \mathcal{Y}) = 0$ $(i > 0)$
ii′ $\operatorname{Ext}^1_{\mathcal{A}}(M, \mathcal{Y}) = 0$ **iii** $\operatorname{Ext}^i_{\mathcal{A}}(M, \omega) = 0$ $(i > 0)$.

Hence, the morphism p in the short exact sequence in **3**, **i** *is a right $\mathcal{X}$-approximation of M.*

5 *For $M \in \hat{\mathcal{X}}$, the following are equivalent.*

i $M \in \mathcal{Y}$ **ii** $\operatorname{Ext}^i_{\mathcal{A}}(\mathcal{X}, M) = 0$ $(i > 0)$ **ii′** $\operatorname{Ext}^1_{\mathcal{A}}(\mathcal{X}, M) = 0$
iii $\mathcal{X}$-inj.dim $M < \infty$, and $M \in \omega^{\perp}$.

Hence, the morphism ι in the short exact sequence in **3**, **ii** *is a left $\mathcal{Y}$-approximation of M.*

6 *For $M \in \hat{\mathcal{X}}$, we have*

$$\mathcal{X}\text{-resol.dim}(M) = \mathcal{Y}\text{-proj.dim}(M) = \omega\text{-proj.dim}(M).$$

7 *For $Y \in \mathcal{Y}$, we have ω-resol.dim$(Y) = \mathcal{X}$-resol.dim(Y).*

8 *If $0 \to M_1 \to M_2 \to M_3 \to 0$ is an exact sequence of $\mathcal{A}$ and two of M_1, M_2, and M_3 belong to $\hat{\mathcal{X}}$, then the third also belongs to $\hat{\mathcal{X}}$.*

We call an exact sequence as in **3**, **i** an $\mathcal{X}$*-approximation.* It is called *minimal* if p in **3**, **i** is right minimal. We call an exact sequence as in **3**, **ii** a $\mathcal{Y}$*-hull.* It is called minimal if ι in **3**, **ii** is left minimal.

We say that a triple $(\mathcal{X}, \mathcal{Y}, \omega)$ of full subcategories of an abelian category $\mathcal{A}$ is a *weak Auslander–Buchweitz context* in $\mathcal{A}$ if the conditions **AB1–3** in the theorem are satisfied. If moreover $\hat{\mathcal{X}} = \mathcal{A}$, then it is called an *Auslander–Buchweitz context.* If $(\mathcal{X}, \mathcal{Y}, \omega)$ is an Auslander–Buchweitz context in $\mathcal{A}$, then any of $\mathcal{X}$, $\mathcal{Y}$, and ω determines the others.

We list some useful lemmas related to (weak) Auslander–Buchweitz contexts.

Lemma 1.12.11 *Let $\mathcal{A}$ be an abelian category, and*

$$\begin{aligned} \alpha :&\quad 0 \to M_0 \to M_1 \to M_2 \to 0, \\ \beta_0 :&\quad 0 \to Y_0 \to X_0 \to M_0 \to 0, \\ \beta_2 :&\quad 0 \to Y_2 \to X_2 \xrightarrow{\psi} M_2 \to 0 \end{aligned}$$

exact sequences in $\mathcal{A}$. If $\operatorname{Ext}^2_{\mathcal{A}}(X_2, Y_0) = 0$, then we have a commutative diagram

$$\begin{array}{ccccccccc} & & 0 & & 0 & & 0 & & \\ & & \downarrow & & \downarrow & & \downarrow & & \\ 0 & \to & Y_0 & \to & Y_1 & \to & Y_2 & \to & 0 \\ & & \downarrow & & \downarrow & & \downarrow & & \\ 0 & \to & X_0 & \to & X_1 & \to & X_2 & \to & 0 \\ & & \downarrow & & \downarrow & & \downarrow \psi & & \\ 0 & \to & M_0 & \to & M_1 & \to & M_2 & \to & 0 \\ & & \downarrow & & \downarrow & & \downarrow & & \\ & & 0 & & 0 & & 0 & & \end{array}$$

with exact rows and columns such that the first and third columns agree with β_0 and β_2, respectively, and the third row agrees with α. If moreover we have $\operatorname{Ext}^1_{\mathcal{A}}(X_2, Y_0) = 0$, then such a diagram is unique up to isomorphisms of diagrams.

Proof. Set

$$\alpha' : 0 \to M_0 \xrightarrow{\varphi} M_1' \to X_2 \to 0$$

to be the exact sequence $\alpha\psi$, the pull-back of α by ψ. From α', we have a long exact sequence

$$\mathrm{Ext}^1_{\mathcal{A}}(X_2, Y_0) \to \mathrm{Ext}^1_{\mathcal{A}}(M_1', Y_0) \xrightarrow{\varphi^{\#}} \mathrm{Ext}^1_{\mathcal{A}}(M_0, Y_0) \to \mathrm{Ext}^2_{\mathcal{A}}(X_2, Y_0).$$

If we have an exact sequence

$$\beta_1' : 0 \to Y_0 \to X_1 \to M_1' \to 0$$

such that $\varphi^{\#}(\beta_1') = \beta_0$, then letting Y_1 be the kernel of the composite morphism $X_1 \to M_1' \to M_1$, the construction of a diagram in question is easy.

Conversely, if a diagram as in the lemma exists, then taking M_1' to be the cokernel of the composite morphism $Y_0 \to Y_1 \to X_1$, we obtain β_1' such that $\varphi^{\#}(\beta_1') = \beta_0$, and giving β_1' is the same thing as to construct such a diagram.

Such a β_1' does exist if $\mathrm{Ext}^2_{\mathcal{A}}(X_2, Y_0) = 0$, and it is unique up to equivalence if

$$\mathrm{Ext}^1_{\mathcal{A}}(X_2, Y_0) = 0.$$

Hence, the assertion of the lemma follows. □

Proposition 1.12.12 *Let $\mathcal{A}$ be an abelian category, and $\mathcal{X}$, $\mathcal{Y}$, $\mathcal{Z}_0$, and $\mathcal{Z}$ be full subcategories of $\mathcal{A}$. Assume that $\mathcal{X}$ and $\mathcal{Y}$ are closed under extensions, and* $\mathrm{Ext}^2_{\mathcal{A}}(\mathcal{X}, \mathcal{Y}) = 0$. *Moreover, assume one of the following:*

a $\mathcal{Z} = \mathcal{F}(\mathcal{Z}_0)$

b $\mathcal{Z} = \mathrm{add}\,\mathcal{F}(\mathcal{Z}_0)$ *and* $\mathcal{Z} \subset \mathcal{X}$.

If for any $Z \in \mathcal{Z}_0$ there exists some exact sequence

$$0 \to Y \to X \to Z \to 0$$

such that $Y \in \mathcal{Y}$ and $X \in \mathcal{X}$, then there exists some exact sequence of the same type for any $Z \in \mathcal{Z}$. If for any $Z \in \mathcal{Z}_0$ there exists some exact sequence

$$0 \to Z \to Y \to X \to 0$$

such that $Y \in \mathcal{Y}$ and $X \in \mathcal{X}$, then there exists some exact sequence of the same type for any $Z \in \mathcal{Z}$.

Proof. Assume **a**. The first assertion follows easily from Lemma 1.12.11, and the last assertion is the dual assertion of the first. The case where **b** holds is proved easily from **a**, and is left to the reader. □

Corollary 1.12.13 *Let $\mathcal{A}$ be an abelian category, and $\mathcal{X}_0$ a full subcategory of $\mathcal{A}$. We set $\mathcal{X} := \operatorname{add} \mathcal{F}(\mathcal{X}_0)$. Let ω be a full subcategory of $\mathcal{X} \cap \mathcal{X}^{\perp}$ closed under direct sums. If for any $X \in \mathcal{X}_0$ there exists some exact sequence*

$$0 \to X \to W \to X' \to 0$$

such that $W \in \omega$ and $X' \in \mathcal{X}$, then ω is an injective cogenerator of $\mathcal{X}$.

Lemma 1.12.14 *Let $\mathcal{X}$ and $\mathcal{Y}$ be full subcategories of $\mathcal{A}$ which satisfy* **AB1, AB2**. *We assume that $\mathcal{Y} \subset \mathcal{X}^{\perp}$. Let ω be an injective cogenerator of $\mathcal{X}$ such that $\omega \subset \mathcal{X} \cap \mathcal{Y}$ and $\operatorname{add} \omega = \omega$. Then $(\mathcal{X}, \mathcal{Y}, \omega)$ is a weak Auslander–Buchweitz context in $\mathcal{A}$.*

Proof. By assumption, $\mathcal{X} \cap \mathcal{Y}$ is an injective cogenerator of $\mathcal{X}$. Hence, we have that $(\mathcal{X}, \mathcal{Y}, \mathcal{X} \cap \mathcal{Y})$ is a weak Auslander–Buchweitz context. By Theorem 1.12.10, **2**, we have $\omega = \mathcal{X} \cap \mathcal{Y}$. □

Remark 1.12.15 Some important terminology in category theory is not unified in its usage. The definition of Grothendieck category in these notes is the same as that in [33]. In Freyd's book [56], a Grothendieck category means an abelian category which satisfies the (AB5) condition. For the definition of an exact category, we follow Quillen [126]. There is some old literature in which the same expression 'exact category' is used for different meanings. For the definition of generator and cogenerator, we follow Auslander–Buchweitz [10]. A family of generators (resp. generator) in the sense of Grothendieck [63] is called a family of G-generators (resp. G-generator) in these notes for clarity. A family of G-generators and a generator (in our sense) are one and the same thing for an abelian category. Moreover, the words thick and épaisse, and sheaf and faisceau are used for different meanings in these notes. Épaisse is used only for triangulated categories, and a faisceau is a sheaf in the fppf topology. The definition of Grothendieck topology and a site is a little more restrictive than that in [9]. There, a Grothendieck topology in our sense is called a pretopology, and a site in our sense is a site whose topology comes from a pretopology. The definition of sheaf in (1.8.4) is the same as that in [9, (II.6.1)], but we put an unnecessary restrictive hypothesis on the value category $\mathcal{C}$ (i.e., the existence of limits) for simplicity. Similarly for the definition of continuous functor (1.8.11), see [9, (III.1.6)]. The definition of contravariant and covariant finiteness of $\mathcal{X}$ in (1.12) may be different from that in [13], if $\mathcal{X} \neq \operatorname{add} \mathcal{X}$.

Notes and References. This section is merely a survey of keywords for later use, and there is no new result. For basics on category theory, relative homological algebra, and cobar resolutions, we refer the reader to [144, 105,

106, 56, 63, 58]. For exact categories, see [126, 141, 120] for more. For triangulated categories and derived categories, see [69, 143, 130, 121]. For Grothendieck topology and sheaf theory, see [8, 111, 9]. For Auslander–Buchweitz theory, see [10, 11, 112, 146, 13].

2 From commutative ring theory

This section is devoted to giving a summary of commutative ring theory used later.

2.1 Flat modules and pure maps

An R-module M is said to be *R-flat* if $? \otimes_R M$ is an exact functor. It is said to be *R-faithfully flat* if $? \otimes_R M$ is faithful exact. An R-algebra A is called R-flat or R-faithfully flat if the same holds for A as an R-module. For any multiplicatively closed subset S of R, the localization R_S is R-flat. The following is well-known.

Lemma 2.1.1 *Let M be an R-module. Then the following are equivalent.*

1 *M is R-flat.*

2 *For any R-module N and any $i > 0$, $\mathrm{Tor}_i^R(N, M) = 0$.*

2′ *For any finitely generated ideal I of R, $\mathrm{Tor}_1^R(R/I, M) = 0$.*

3 *M is an inductive limit of an inductive system of R-finite free modules parameterized by a directed set.*

3′ *M is a filtered inductive limit of R-flat modules.*

4 *For any commutative R-algebra S, $S \otimes_R M$ is S-flat.*

4′ *For any $\mathfrak{m} \in \operatorname{Max} R$, $M_\mathfrak{m}$ is $R_\mathfrak{m}$-flat.*

The proof of **1⇒3** is due to Lazard [100]. By Lazard's proof, the following holds.

Lemma 2.1.2 *Let R be a noetherian ring, and F a countably generated R-flat module. Then F is an inductive limit of an inductive system of R-finite free modules parameterized by the ordered set $\mathbb{N}$.*

The following lemma is also well-known.

Lemma 2.1.3 *Let M be an R-module. Then the following are equivalent.*

1 *M is R-faithfully flat.*

2 *M is R-flat, and $M \neq \mathfrak{m}M$ for any $\mathfrak{m} \in \operatorname{Max} R$.*

An R-linear map $f : N \to M$ of R-modules is said to be *pure* if $1_V \otimes f : V \otimes N \to V \otimes M$ is injective for any R-module V. If f is a split monomorphism, or f is injective and $\operatorname{Coker} f$ is R-flat, then f is pure. Note that a pure R-linear map is injective. Let N be an R-submodule of M. We say that N is a *pure submodule* of M, if the canonical injection $N \hookrightarrow M$ is R-pure.

Lemma 2.1.4 *Let R be noetherian, P and F be R-flat modules, and $f : P \to F$ be an R-linear map. Consider the following conditions.*

1 *f is injective and $\operatorname{Coker} f$ is R-flat.*

2 *f is pure.*

3 *For $\mathfrak{p} \in \operatorname{Spec} R$, $\kappa(\mathfrak{p}) \otimes f : \kappa(\mathfrak{p}) \otimes P \to \kappa(\mathfrak{p}) \otimes F$ is injective.*

4 *For any $\mathfrak{m} \in \operatorname{Max} R$, $\kappa(\mathfrak{m}) \otimes f : \kappa(\mathfrak{m}) \otimes P \to \kappa(\mathfrak{m}) \otimes F$ is injective.*

Then **1–3** *are equivalent. If moreover P is R-projective, then* **1–4** *are equivalent.*

Proof. The direction **1⇒2⇒3⇒4** is easy.

First we prove **1**, assuming **3**, or that P is R-finite projective and **4** holds. We assume the contrary for contradiction. Then there exists some $\mathfrak{m} \in \operatorname{Max} R$ such that $f_{\mathfrak{m}}$ is not injective or $\operatorname{Coker} f_{\mathfrak{m}}$ is not R-flat. Hence, we may assume that $(R, \mathfrak{m})$ is local. There exists some ideal I of R such that $R/I \otimes f$ is not injective or $R/I \otimes \operatorname{Coker} f$ is not R/I-flat, because $I = 0$ is one of them. As R is noetherian, we can take a maximal such I. Replacing R by R/I, we may and shall assume $I = 0$ is maximal among such. We set $C := \operatorname{Coker} f$ and $K := \operatorname{Ker} f$. Note that $R/I \otimes f$ is injective for any non-zero ideal I of R.

Let $I \neq 0$ be an ideal of R. From the short exact sequence

$$0 \to P/K \to F \to C \to 0, \qquad (2.1.5)$$

we get a long exact sequence

$$0 \to \operatorname{Tor}_1^R(R/I, C) \to R/I \otimes P/K \to F/IF \to C/IC \to 0.$$

As $R/I \otimes f : P/IP \to F/IF$ is injective, we have $R/I \otimes P \cong R/I \otimes P/K$ and $\operatorname{Tor}_1^R(R/I, C) = 0$. Hence, C is R-flat. By the short exact sequence (2.1.5), P/K is also R-flat. Hence, $K \hookrightarrow P$ is pure and K is R-flat. In particular, $R/I \otimes K = 0$ for any non-zero ideal I of R.

Note that $K = 0$ leads to a contradiction as we already know that C is R-flat. We prove $K = 0$. If R is not an integral domain, then we have $R/\mathfrak{p} \otimes K = 0$ for $\mathfrak{p} \in \operatorname{Spec} R$, and R admits a finite filtration

$$0 \subset R_1 \subset \cdots \subset R_r = R$$

of R-modules such that for any i, R_i/R_{i-1} is isomorphic to $R/\mathfrak{p}$ for some $\mathfrak{p} \in \operatorname{Spec} R$. This shows that $K = R \otimes K = 0$ for this case.

So we may assume that R is an integral domain. First, consider the case that **3** holds. Then the canonical map $K \to \kappa(0) \otimes K = 0$ is injective, and hence $K = 0$. Next, consider the case that P is R-finite projective and **4** holds. In this case, we have $K/\mathfrak{m}K = 0$, and K is R-finite. By Nakayama's lemma, we have $K = 0$.

Finally, we prove that if P is R-projective and **4** holds, then **1** holds, and this completes the proof of the lemma. We may assume that R is local, and in this case, P is R-free by Kaplansky's theorem [92]. We fix a basis B of P, and we denote the set of finite subsets of B by Λ. Note that Λ is a directed set with respect to the incidence relation. For $\lambda \in \Lambda$, we define P_λ to be the free summand of P generated by λ. We denote the composite map $P_\lambda \hookrightarrow P \to F$ by f_λ. Then we have $K = \varinjlim \operatorname{Ker} f_\lambda = 0$, and $C = \varinjlim \operatorname{Coker} f_\lambda$ is R-flat. □

Corollary 2.1.6 *Let R be noetherian, and P an R-flat module. If $P \otimes \kappa(\mathfrak{p}) = 0$ for any $\mathfrak{p} \in \operatorname{Spec} R$, then we have $P = 0$.*

Proof. Applying Lemma 2.1.4 to the map $P \to 0$, we have that this map is injective. □

Lemma 2.1.7 *Let A be an R-algebra, M and N be A-modules, and H be an R-module. Then the map*

$$\rho : \operatorname{Hom}_A(M, N) \otimes H \to \operatorname{Hom}_A(M, N \otimes H) \qquad f \otimes h \mapsto (m \mapsto fm \otimes h)$$

is an R-linear map (an A-linear map if A is commutative) which is natural with respect to M, N and H. It is an isomorphism if one of the following holds.

a *H is R-flat and M is A-finitely presented.*

b *M is A-projective and H is finitely presented.*

c *H is R-finite projective.*

d *M is A-finite projective.*

Proof. We only prove **a**. As both $\mathrm{Hom}_A(?, N) \otimes H$ and $\mathrm{Hom}_A(?, N \otimes H)$ are left exact, we may assume that $M = A^n$ by the five lemma, which case is trivial. □

Corollary 2.1.8 *Let R be noetherian, and assume that I is an injective R-module, and F a flat R-module. Then $I \otimes F$ is R-injective.*

Proof. Note that the category ${}_R\mathbb{M}$ of R-modules is locally noetherian, and ${}_R\mathbb{M}_f$ is nothing but the full subcategory of finitely generated R-modules. We have an isomorphism

$$\mathrm{Hom}(?, I \otimes F) \cong \mathrm{Hom}(?, I) \otimes F$$

of functors on ${}_R\mathbb{M}_f$ by the lemma, and hence $\mathrm{Hom}(?, I \otimes F)$ is exact. By Lemma 1.9.4, $I \otimes F$ is R-injective. □

(2.1.9) Let $\mathcal{F}$ be the full subcategory of R-flat modules in ${}_R\mathbb{M}$. For an R-module M, the $\mathcal{F}$-resolution dimension (1.11) $\mathcal{F}\text{-resol.dim}\, M$ of M is called the *R-flat dimension* (or R-weak dimension) of M, and is denoted by $\mathrm{flat.dim}_R\, M$.

Lemma 2.1.10 *If P is an R-finitely presented R-flat module, then P is R-projective.*

Proof. Let $f : V \to W$ be a surjective R-linear map. We are to prove that

$$\mathrm{Hom}(P, f) : \mathrm{Hom}(P, V) \to \mathrm{Hom}(P, W)$$

is surjective. By Lemma 2.1.7, **a**, this is checked after localization at maximal ideals of R, and we may assume that $(R, \mathfrak{m})$ is local.

As a local ring is semiperfect, P admits a projective cover $p : F \to P$. Note that F is finite free, and $K := \mathrm{Ker}\, p$ is finitely generated by assumption (see [110, Theorem 2.6]).

Since P is flat,

$$0 \to K/\mathfrak{m}K \to F/\mathfrak{m}F \xrightarrow{p \otimes R/\mathfrak{m}} P/\mathfrak{m}P \to 0$$

is exact. As $p \otimes R/\mathfrak{m}$ is an isomorphism by Lemma 1.12.5, we have $K/\mathfrak{m}K = 0$. Hence, $K = 0$ by Nakayama. We have $P \cong F$ is projective. □

Lemma 2.1.11 *Let R be a commutative ring, P an R-projective module, and M an R-finite pure submodule of P. Then P/M is R-projective, and hence $M \hookrightarrow P$ splits.*

Proof. As P is a direct summand of an R-free module, we may assume that P is an R-free module with a basis B. As M is R-finite, there exists some finite subset B_0 of B such that M is contained in the R-span P_0 of B_0. If we denote the R-span of $B \setminus B_0$ by P_1, then we have $P/M \cong P_0/M \oplus P_1$. Hence, replacing P by P_0, we may assume that P is R-finite free. Then P/M is R-finitely presented and R-flat, and hence is R-projective. □

Similarly, the following holds.

Exercise 2.1.12 Let R be noetherian, and $f : M \to P$ be an R-linear map. Assume that P is R-finite projective and M is R-finite. If $f \otimes \kappa(\mathfrak{m})$ is injective for any $\mathfrak{m} \in \operatorname{Max} R$, then f is a split monomorphism.

The proof is left to the reader.

2.2 Mittag-Leffler modules

We review the theory of Mittag-Leffler modules after [128]. Throughout this subsection, Λ denotes a directed set.

(2.2.1) We say that a projective system of R-modules $\mathcal{P} = (P_\lambda, f_{\lambda\mu})_{\lambda\in\Lambda,\,\mu\geq\lambda}$ indexed by Λ satisfies the *Mittag-Leffler condition* if for any $\lambda \in \Lambda$, there exists some $\mu \geq \lambda$ such that for any $\gamma \geq \mu$, we have $\operatorname{Im} f_{\lambda\gamma} = \operatorname{Im} f_{\lambda\mu}$.

Lemma 2.2.2 (Grothendieck) *Let $0 \to \mathcal{P}' \to \mathcal{P} \to \mathcal{P}'' \to 0$ be an exact sequence of projective systems of R-modules indexed by Λ. Then we have*

1 *If $\mathcal{P}'$ and $\mathcal{P}''$ satisfy the Mittag-Leffler condition, then so does $\mathcal{P}$.*

2 *If $\mathcal{P}$ satisfies Mittag-Leffler condition, then so does $\mathcal{P}''$.*

3 *Assume that Λ has a final countable subset. If P' satisfies the Mittag-Leffler condition, then the sequence*

$$0 \to \varprojlim \mathcal{P}' \to \varprojlim \mathcal{P} \to \varprojlim \mathcal{P}'' \to 0$$

is exact.

For the proof, see [64, Proposition O.13.2]. From now on, until the end of this subsection, any projective or inductive system is assumed to be indexed by a directed set.

Lemma 2.2.3 *Let $\mathcal{P}$ be a projective system of R-modules which satisfies the Mittag-Leffler condition. If $F : {}_R\mathbb{M} \to \underline{\mathrm{Ab}}$ is a right exact functor, then $F(\mathcal{P})$ satisfies the Mittag-Leffler condition. In particular, for any R-module M, the projective system $M \otimes \mathcal{P}$ satisfies the Mittag-Leffler condition.*

Proof. Obvious. □

Lemma 2.2.4 *Let (P_λ) be an inductive system of finite free R-modules such that the projective system (P_λ^*) satisfies the Mittag-Leffler condition. If M is an R-module, then the projective system $(\mathrm{Hom}_R(P_\lambda, M))$ satisfies the Mittag-Leffler condition.*

Proof. Obvious by Lemma 2.2.3 and Lemma 2.1.7, **d**. □

Definition 2.2.5 We say that an R-module M is *R-Mittag-Leffler* if there exists some inductive system (F_λ) of finite free R-modules such that $M \cong \varinjlim F_\lambda$ and the projective system (F_λ^*) satisfies the Mittag-Leffler condition.

By Lemma 2.1.1, an R-Mittag-Leffler module is R-flat. We list some properties of Mittag-Leffler modules. For the proof, see [128].

From Lemma 2.2.4, Lemma 2.1.2 and Lemma 2.2.2, we have the following.

Proposition 2.2.6 *Let R be noetherian. If M is a Mittag-Leffler R-module of countable type, then M is R-projective.*

The following criterion for the Mittag-Leffler property of an R-module is due to Raynaud–Gruson [129].

Proposition 2.2.7 *For an R-flat module M, the following are equivalent.*

1 *M is R-Mittag-Leffler.*

2 *For any inductive system (F_λ) of finite free R-modules such that $\varinjlim F_\lambda \cong M$, the projective system (F_λ^*) satisfies the Mittag-Leffler condition.*

3 *For any finite free R-module Q and any $x \in Q \otimes M$, there is a smallest R-submodule Q' of Q such that $x \in Q' \otimes M$.*

Corollary 2.2.8 *A pure submodule of a Mittag-Leffler R-module is Mittag-Leffler.*

Lemma 2.2.9 *Let (M_λ) be a family of R-modules. Then $\bigoplus_\lambda M_\lambda$ is Mittag-Leffler if and only if M_λ is Mittag-Leffler for any λ.*

Corollary 2.2.10 *An R-projective module is R-Mittag-Leffler.*

Lemma 2.2.11 *Let (M_λ) be an inductive system of R-Mittag-Leffler modules consisting of R-pure maps. Then $\varinjlim M_\lambda$ is R-Mittag-Leffler.*

Lemma 2.2.12 *Let (M_λ) be a family of R-modules. Then for any finitely presented R-module N, the canonical map*

$$N \otimes \prod_\lambda M_\lambda \to \prod_\lambda N \otimes M_\lambda$$

is an isomorphism.

Proof. As both sides are right exact with respect to N, we may assume that $N = R$, which case is trivial. □

Corollary 2.2.13 *If moreover R is noetherian in the lemma, then there exists some isomorphism*

$$\mathrm{Tor}_i^R(N, \prod_\lambda M_\lambda) \cong \prod_\lambda \mathrm{Tor}_i^R(N, M_\lambda)$$

for $i \geq 0$.

Corollary 2.2.14 *Let R be noetherian, and (M_λ) a family of R-modules. Then $\prod_\lambda M_\lambda$ is R-flat (resp. Mittag-Leffler) if and only if the same is true of M_λ for any λ.*

An argument similar to above shows the following.

Proposition 2.2.15 *Let R be noetherian, (P_λ) a projective system of R-flat modules (resp. R-Mittag-Leffler modules) indexed by a countable directed set which satisfies the Mittag-Leffler condition. Then $\varprojlim P_\lambda$ is R-flat (resp. R-Mittag-Leffler).*

A projective module over a noetherian commutative ring is characterized as follows:

Theorem 2.2.16 *Let R be noetherian, and P an R-module. Then the following are equivalent.*

1 *P is a direct sum of countable Mittag-Leffler R-modules.*

2 *P is R-projective.*

Proof. **1⇒2** follows from Theorem 2.2.6. **2⇒1** is well-known as Kaplansky's theorem [92]. □

Exercise 2.2.17 Let R be a noetherian, and F a countably generated flat R-module. Prove that we have $\mathrm{proj.dim}_R F \leq 1$.

Let

$$\mathbb{F} : \cdots \to F_n \to F_{n-1} \to \cdots$$

be a chain complex of R-modules. We say that $\mathbb{F}$ is an R-free (resp. R-projective, R-flat) complex if each F_n is R-free (resp. R-projective, R-flat). A free complex $\mathbb{F}$ is said to be finite free (resp. finite projective) if each F_n is R-finite free (resp. R-finite projective) and $\mathbb{F}$ is bounded. Sometimes an R-finite projective complex is referred as a *perfect complex.*

2.3 Faithfully flat morphisms and descent theory

(2.3.1) Let $f : A \to B$ be a homomorphism of commutative rings, and assume that f is faithfully flat. Then by definition, $F := f_\# = B \otimes_A ? : {}_A\mathbb{M} \to {}_B\mathbb{M}$ is faithful exact, and it has a right adjoint $G := f^\# = \mathrm{Hom}_B(B, ?) : {}_B\mathbb{M} \to {}_A\mathbb{M}$. Then for an A-module V, the cobar resolution (1.6) $\mathrm{Cobar}_F(V)$ of V with respect to the adjoint pair (F, G) is as follows:

$$0 \to V \xrightarrow{\varepsilon} B \otimes_A V \xrightarrow{d^0} B^{\otimes 2} \otimes_A V \xrightarrow{d^1} B^{\otimes 3} \otimes_A V \to \cdots,$$

where

$$d^i(b_0 \otimes \cdots \otimes b_i \otimes v) = \sum_{j=0}^{i+1} (-1)^j b_0 \otimes \cdots \otimes b_{j-1} \otimes 1 \otimes b_j \otimes \cdots \otimes b_i \otimes v$$

and $\varepsilon(v) = 1 \otimes v$.

More generally, if $\varphi : Y \to X$ is a faithfully flat morphism of schemes, then $\varphi^* : {}_X\mathbb{M} \to {}_Y\mathbb{M}$ is faithful exact, and φ_* is its right adjoint. For an $\mathcal{O}_X$-module $\mathcal{M}$, applying $\Gamma(X, ?)$ on $\mathrm{Cobar}_{\varphi^*}(\mathcal{M})$, we have an exact sequence

$$0 \to \Gamma(X, \mathcal{M}) \to \Gamma(Y, \varphi^* M) \to \Gamma(Y \times_X Y, \psi^* M), \tag{2.3.2}$$

where $\psi : Y \times_X Y \to X$ is the canonical map.

(2.3.3) For a commutative ring R and an R-scheme X, the representable functor

$$\mathrm{Hom}_X(?, \mathbb{A}^1_X) : Y \mapsto \Gamma(Y, \mathcal{O}_Y)$$

is a faisceau of R-algebras (1.8.14). We denote it by $\mathcal{O}_X^{\mathrm{fl}}$ or simply by $\mathcal{O}_X$ (by abuse of notation). We say that an $\mathcal{O}_X^{\mathrm{fl}}$-module faisceau $\mathcal{M}$ is *quasi-coherent* if for any X-morphism of affine schemes over X of finite type $\mathrm{Spec}\, B \to \mathrm{Spec}\, A$, the canonical map $B \otimes_A \mathcal{M}(\mathrm{Spec}\, A) \to \mathcal{M}(\mathrm{Spec}\, B)$ is an isomorphism.

(2.3.4) The restriction of a quasi-coherent $\mathcal{O}_X^{\mathrm{fl}}$-module $\mathcal{M}$ to the Zariski site of X is a quasi-coherent $\mathcal{O}_X$-module in the usual sense. Conversely, if M is a quasi-coherent $\mathcal{O}_X$-module, then defining an $\mathcal{O}_X^{\mathrm{fl}}$-module $W(M)$ by $W(M)(f) := \Gamma(Y, f^*M)$ for any $f : Y \to X$, $W(M)$ is a faisceau by (2.3.2), and is clearly quasi-coherent. With this correspondence above, we identify a (usual) quasi-coherent $\mathcal{O}_X$-module and a quasi-coherent $\mathcal{O}_X^{\mathrm{fl}}$-module. If $X = \mathrm{Spec}\, A$ is affine and $M = \tilde{N}$, then $W(M)$ is also denoted by N_a. For a quasi-coherent $\mathcal{O}_X$-module M and any X-scheme $f : Y \to X$, we sometimes denote $\Gamma(Y, f^*M)$ simply by $\Gamma(Y, M)$.

(2.3.5) Let $p : Y \to X$ be a faithfully flat morphism of schemes. Let $p_i : Y \times_X Y \to Y$ be the ith projection, $p_{ji} : Y \times_X Y \times_X Y \to Y \times_X Y$ be the morphism given by $(y_1, y_2, y_3) \mapsto (y_i, y_j)$, and $q_i : Y \times_X Y \times_X Y \to Y$ be the ith projection.

We call a pair $(\mathcal{M}, \phi)$ a *descent datum* of a quasi-coherent sheaf with respect to p, if $\mathcal{M}$ is a quasi-coherent $\mathcal{O}_Y$-module, and $\phi : p_1^*\mathcal{M} \to p_2^*\mathcal{M}$ is an isomorphism such that the identity $p_{31}^*\phi = p_{32}^*\phi \circ p_{21}^*\phi$ holds.

Let $\mathcal{N}$ be a quasi-coherent $\mathcal{O}_X$-module. We set $\mathcal{M} := p^*\mathcal{N}$. Then we denote the composite of the canonical maps

$$p_1^*\mathcal{M} = (p \circ p_1)^*\mathcal{N} = (p \circ p_2)^*\mathcal{N} = p_2^*\mathcal{M}$$

by $\phi = \phi_{\mathcal{N}} : p_1^*\mathcal{M} \to p_2^*\mathcal{M}$. Then it is easy to verify that we have the identity $p_{31}^*\phi = p_{32}^*\phi \circ p_{21}^*\phi$ of maps from $q_1^*\mathcal{M}$ to $q_3^*\mathcal{M}$. Hence, from $\mathcal{N}$, we obtain a descent datum $(\mathcal{M}, \phi) = (p^*\mathcal{N}, \phi_{\mathcal{N}})$ of a quasi-coherent sheaf with respect to p. When we define h_1 and h_2 by

$$h_1 : p_*\mathcal{M} \to p_*(p_1)_*p_1^*\mathcal{M} \cong p_*(p_2)_*p_1^*\mathcal{M} \xrightarrow{(p \circ p_2)_*\phi} p_*(p_2)_*p_2^*\mathcal{M}$$

and

$$h_2 : p_*\mathcal{M} \xrightarrow{\text{via the unit}} p_*(p_2)_*p_2^*\mathcal{M},$$

then $\mathcal{N}$ agrees with $\mathrm{Ker}(h_1 - h_2)$ by (2.3.2). Thus, $\mathcal{N}$ is recovered from the data $(p^*\mathcal{N}, \phi_{\mathcal{N}})$.

(2.3.6) For descent data $(\mathcal{M}, \phi)$ and $(\mathcal{M}', \phi')$ of quasi-coherent sheaves with respect to p, we say that an $\mathcal{O}_Y$-module map $f : \mathcal{M} \to \mathcal{M}'$ is a map of descent data if $\phi' \circ p_1^*f = p_2^*f \circ \phi$ is satisfied. If $\varphi : \mathcal{N} \to \mathcal{N}'$ is a map of quasi-coherent $\mathcal{O}_X$-modules, then $p^*\varphi : p^*\mathcal{N} \to p^*\mathcal{N}'$ is a map of descent data from $(p^*\mathcal{N}, \phi_{\mathcal{N}})$ to $(p^*\mathcal{N}', \phi_{\mathcal{N}'})$.

(2.3.7) Assume that p is quasi-compact. Let $(\mathcal{M}, \phi)$ be a given descent datum of a quasi-coherent sheaf with respect to p. Then we define h_1 and h_2 as above, and we define $\mathcal{N} := \mathrm{Ker}(h_1 - h_2)$. Then by [89, Proposition 1.51], $\mathcal{N}$ is a quasi-coherent $\mathcal{O}_X$-module.

Proposition 2.3.8 *The map $p^*\mathcal{N} \to \mathcal{M}$ which corresponds to $\mathcal{N} \to p_*\mathcal{M}$ is an isomorphism of descent data from $(p^*\mathcal{N}, \phi_{\mathcal{N}})$ to $(\mathcal{M}, \phi)$. Thus,* $\mathrm{Qco}(X)$ *and the category of descent data and maps of descent data with respect to p are equivalent.*

For the proof, see [116, Proposition 7.1.4]. The proposition above is a sketch of an important part of descent theory. The following is another side, which is also important, of descent theory.

Theorem 2.3.9 *Let $f : Y \to X$ and $g : X' \to X$ be morphisms of locally noetherian schemes. Assume that g is faithfully flat and quasi-compact. If we denote by $f' : Y' \to X'$ the base change of f by g, and if f' is quasi-compact (resp. of finite type, proper, an open immersion, affine, finite, quasi-finite, flat, smooth, étale), then so is f.*

For the proof, see [65, IV.2.6–2.7]. For the definitions of the properties of morphisms listed here, see (2.7).

Exercise 2.3.10 Let $A \to B$ be a faithfully flat homomorphism of commutative rings, and M an A-module. If $M \otimes_A B$ is finitely generated (resp. countably generated, flat, Mittag-Leffler, finitely presented, coherent) as a B-module, then so is M as an A-module. If A is noetherian and $M \otimes_A B$ is B-projective, then so is M as an A-module.

2.4 The I-depth

Let R be a commutative noetherian ring. The reference for this subsection is [110, Chapter 6].

(2.4.1) Let $M \neq 0$ be a finite R-module. The dimension $\dim M$ of M means the Krull dimension of the ring $R/\operatorname{ann} M$, where $\operatorname{ann} M := \{r \in R \mid rM = 0\}$ is the annihilator of M. Note that $\dim M$ is the same as the dimension of the closed subset

$$\operatorname{supp} M := \{P \in \operatorname{Spec} R \mid M_P \neq 0\} = \operatorname{Spec}(R/\operatorname{ann} M)$$

of $\operatorname{Spec} R$. Let I be an ideal of R. If $IM \neq M$, then we define

$$\operatorname{depth}_R(I, M) = \operatorname{depth}(I, M) := \min\{i \mid \operatorname{Ext}^i_R(R/I, M) \neq 0\},$$

and call $\operatorname{depth}_R(I, M)$ the *I-depth* of M. We define $\operatorname{depth}(I, M) := \infty$ if $IM = M$, as a convention. For a finitely generated R-module M, we have $IM \neq M$ if and only if $\operatorname{depth}(I, M) < \infty$ if and only if $\operatorname{depth}(I, M) \leq \dim M$.

(2.4.2) We say that a sequence of elements $a_1, \ldots, a_n$ of R is a poor M-sequence if the multiplication $a_i : M_{i-1} \to M_{i-1}$ is injective for $i = 1, \ldots, n$, where $M = M_0$, and $M_i = M/(a_1, \ldots, a_i)M$ for $i = 1, \ldots, n-1$. If moreover $M \neq (a_1, \ldots, a_n)M$, then we call it an *$M$-sequence*.

Theorem 2.4.3 ([110, Theorem 16.6]) *Let R be a noetherian ring, M a finitely generated R-module, and I an ideal of R. Assume that $M \neq IM$. For $n > 0$, the following are equivalent.*

1 $\operatorname{depth}(I, M) \geq n$.

2 *For any finitely generated R-module N such that* $\operatorname{supp} N \subset \operatorname{supp} R/I$*, we have* $\operatorname{Ext}^i_R(N, M) = 0$ $(i < n)$.

2′ *For some finitely generated R-module N such that* $\operatorname{supp} N = \operatorname{supp} R/I$*, we have* $\operatorname{Ext}^i_R(N, M) = 0$ $(i < n)$.

3 *There is an M-sequence $a_1, \ldots, a_n$ of length n consisting of elements of I.*

(2.4.4) An M-sequence $a_1, \ldots, a_n$ consisting of elements of I is called a *maximal M-sequence* if it cannot be extended to a longer M-sequence $a_1, \ldots, a_n, a$ with $a \in I$. By the theorem above, the length n of maximal M-sequences is independent of the choice of a sequence, and it agrees with $\operatorname{depth}(I, M)$.

(2.4.5) For an R-finite module M, we set $\operatorname{codim}_R M := \operatorname{ht} \operatorname{ann} M$, where ht denotes the height of an ideal. More generally, if X is a noetherian scheme and $\mathcal{M}$ a coherent sheaf of X, then we define $\operatorname{codim}_X \mathcal{M}$ to be $\inf_{x \in \operatorname{supp} \mathcal{M}} \operatorname{codim}_{\mathcal{O}_{X,x}} \mathcal{M}_x$. If $f : Y \to X$ is a finite morphism, then we define $\operatorname{codim}_X Y := \operatorname{codim}_X f_* \mathcal{O}_Y$.

(2.4.6) For an R-finite module M, we define $\operatorname{grade} M := \operatorname{depth}(\operatorname{ann} M, R)$. By the theorem above, we have

$$\operatorname{grade} M = \min\{i \mid \operatorname{Ext}^i_R(M, R) \neq 0\}.$$

In general, if $M \neq 0$, then we have

$$\text{(2.4.7)} \qquad \operatorname{proj.dim}_R M \geq \operatorname{codim} M \geq \operatorname{grade} M.$$

If $a_1, \ldots, a_r$ is an R-sequence consisting of elements of $\operatorname{ann} M$, then we have $\operatorname{codim} M \geq \operatorname{ht}(a_1, \ldots, a_r) = r$, and the second inequality follows. The first one is a consequence of the *New Intersection Theorem* proved by P. Roberts, stated below.

Theorem 2.4.8 ([133]) *Let $(R, \mathfrak{m})$ be a noetherian local ring, and $\mathbb{F}$ a finite free R-complex. If $\mathbb{F}$ is not exact, and the homology groups of $\mathbb{F}$ are of finite lengths, then the length of $\mathbb{F}$ is greater than or equal to the Krull dimension* $\dim R$ *of R.*

Let $h := \operatorname{codim} M$, and take a minimal prime P of $\operatorname{ann} M$ such that $\operatorname{ht} P = h$. Then M_P is a non-zero artinian A_P-module, and the length of the minimal free resolution (see (2.8)) of M_P is at least $h = \operatorname{ht} P = \dim A_P$ by the theorem. Hence, we have

$$\operatorname{proj.dim}_R M \geq \operatorname{proj.dim}_{R_P} M_P \geq h = \operatorname{codim} M,$$

as desired, and the proof of the inequalities (2.4.7) is complete.

(2.4.9) We say that an R-finite module M is a *perfect R-module* if $M \neq 0$ and $\text{proj.dim}_R M = \text{grade}\, M$. If M is perfect, then the inequalities in (2.4.7) are equalities of finite numbers, and we call $\text{codim}\, M$ the *codimension* of M. Let S be a noetherian commutative R-flat algebra such that $S \otimes M \neq 0$. Then we have

$$\text{grade}_R M \leq \text{grade}_S S \otimes M \leq \text{proj.dim}_S S \otimes M \leq \text{proj.dim}_R M.$$

Hence, if M is a perfect R-module and $S \otimes M \neq 0$, then we have $S \otimes_R M$ is perfect as an S-module.

(2.4.10) Let I be a proper ideal of R. Then sometimes we denote $\text{grade}\, R/I$ by $\text{grade}\, I$ (commonly used, although it is confusing). If R/I is a perfect module, then we say that I is a *perfect ideal.*

We say that an ideal $I \neq R$ of R is a *Gorenstein ideal* if I is a perfect ideal of R, and $\text{Ext}^h_R(R/I, R)$ is a rank-one R/I-projective module, where $h = \text{ht}\, I$ is the codimension.

An ideal generated by an R-sequence is called a *complete intersection ideal.* A complete intersection ideal is a Gorenstein ideal, see subsection 2.9.

(2.4.11) Let $f : S \to R$ be a surjective map of noetherian commutative rings, M a finite R-module, and J an ideal of S. Then $a_1, \ldots, a_n$ is a maximal M-sequence of J if and only if $fa_1, \ldots, fa_n$ is a maximal M-sequence of $JR = f(J)$. Hence

$$\text{depth}_S(J, M) = \text{depth}_R(JR, M).$$

2.5 Cohen–Macaulay, Gorenstein, and regular rings

(2.5.1) Let $(R, \mathfrak{m})$ be a noetherian local ring. For an R-finite module M, we denote $\text{depth}(\mathfrak{m}, M)$ by $\text{depth}\, M$, and call it the *depth* of M. If $M \neq 0$, then we have

$$\text{depth}\, M \leq \dim M \leq \dim R.$$

If $\text{depth}\, M = \dim M$, then we say that M is *Cohen–Macaulay.* If $\text{depth}\, M = \dim R$, then we say that M is *maximal Cohen–Macaulay* (*MCM*, for short). As a convention, we define that 0 is both Cohen–Macaulay and MCM. We say that R is a Cohen–Macaulay local ring if the R-module R is a Cohen–Macaulay module.

(2.5.2) The next theorem is well-known as the Auslander–Buchsbaum formula.

Theorem 2.5.3 ([110, Theorem 19.1]) *Let R be noetherian local, and $M \neq 0$ be an R-finite module. If* $\text{proj.dim}_R M < \infty$, *then we have*

$$\text{proj.dim}_R M + \text{depth}\, M = \text{depth}\, R.$$

By the theorem above, if M is an MCM R-module of finite projective dimension, then we have R is Cohen–Macaulay local, and M is R-free.

Let R be a Cohen–Macaulay local ring. Then we have $\operatorname{Ass} R = \operatorname{Min} R$. Moreover, for a non-zero R-module M, we have

$$(2.5.4) \qquad \dim M + \operatorname{grade} M = \dim M + \operatorname{codim} M = \dim R.$$

In particular, R is equidimensional (i.e., $\dim R = \dim R/P$ for any $P \in \operatorname{Min} R$), see [110, Theorem 17.4].

If R is a Cohen–Macaulay local ring and M is an R-finite module of finite projective dimension, then M is Cohen–Macaulay if and only if M is perfect. This is an immediate consequence of Theorem 2.5.3 and (2.5.4).

(2.5.5) The following corollary to Theorem 2.4.8 had long been known as Bass's conjecture until Theorem 2.4.8 was proved by P. Roberts.

Theorem 2.5.6 *Let $(R, \mathfrak{m})$ be a noetherian local ring, and assume that there exists some non-zero finite R-module of finite injective dimension. Then R is Cohen–Macaulay.*

For the implication from Theorem 2.4.8 to Theorem 2.5.6, we refer the reader to [83]. Conversely, a Cohen–Macaulay local ring has a non-zero finite module of finite injective dimension [110, p. 151].

(2.5.7) A noetherian local ring $(R, \mathfrak{m})$ is called a *Gorenstein local ring* if the R-module R is of finite injective dimension.

Theorem 2.5.8 ([110, Theorem 18.1]) *Let $(R, \mathfrak{m})$ be a d-dimensional noetherian local ring. Then the following are equivalent.*

1 *R is Gorenstein.*

1′ $\operatorname{inj.dim}_R R = d$.

2 *For $i \neq d$, we have* $\operatorname{Ext}^i_R(R/\mathfrak{m}, R) = 0$, *and* $\operatorname{Ext}^d_R(R/\mathfrak{m}, R) \cong R/\mathfrak{m}$.

3 *There exists some $i > d$ such that* $\operatorname{Ext}^i_R(R/\mathfrak{m}, R) = 0$.

4 $\operatorname{Ext}^i_R(R/\mathfrak{m}, R) = 0$ *for $i < d$, and* $\operatorname{Ext}^d_R(R/\mathfrak{m}, R) \cong R/\mathfrak{m}$.

4′ *R is Cohen–Macaulay, and* $\operatorname{Ext}^d_R(R/\mathfrak{m}, R) \cong R/\mathfrak{m}$.

(2.5.9) For a noetherian local ring $(R, \mathfrak{m})$, we denote the minimal number of generators $\dim_{R/\mathfrak{m}} \mathfrak{m}/\mathfrak{m}^2$ of $\mathfrak{m}$ by $\operatorname{emb.dim} R$, and call it the *embedding dimension* of R. We have $\operatorname{emb.dim} R \geq \dim R$ in general. If this inequality is an equality, then R is called a *regular* local ring. The following theorem is well-known as Serre's theorem.

Theorem 2.5.10 *For a d-dimensional noetherian local ring* $(R, \mathfrak{m})$, *the following are equivalent.*

1 *R is a regular local ring.*

2 $\mathrm{gl.dim}\, R < \infty$.

3 $\mathrm{gl.dim}\, R = d$.

4 $\mathrm{proj.dim}_R\, R/\mathfrak{m} < \infty$.

5 $\mathrm{proj.dim}_R\, R/\mathfrak{m} = d$.

As the assertion **5** shows that $\mathrm{Ext}_R^{d+1}(R/\mathfrak{m}, R) = 0$, so a regular local ring is Gorenstein. Moreover, any regular local ring is a UFD (proved by Auslander–Buchsbaum).

(2.5.11) If R is a Cohen–Macaulay (resp. Gorenstein, regular) local ring and P is a prime ideal of R, then the localization R_P is again Cohen–Macaulay (resp. Gorenstein, regular). If R is a noetherian local ring and M is an MCM R-module, then M_P is again MCM for any $P \in \mathrm{Spec}\, R$.

(2.5.12) Let R be a noetherian ring which is not necessarily local. We say that R is Cohen–Macaulay (resp. Gorenstein, regular, *normal*) if R_P is Cohen–Macaulay (resp. Gorenstein, regular, an integrally closed domain) for any $P \in \mathrm{Spec}\, R$. A locally noetherian scheme X is called Cohen–Macaulay (resp. Gorenstein, regular, normal) if the same is true of $\mathcal{O}_{X,x}$ for any $x \in X$. It is equivalent to say that so is $\mathcal{O}_{X,x}$ for any closed point x of X.

(2.5.13) Let R be noetherian, and $i \geq 0$. We say that an R-finite module M satisfies *Serre's* (S_i) *condition* if $\mathrm{depth}\, M_P < i$ implies that M_P is MCM as an R_P-module for any $P \in \mathrm{Spec}\, R$. A finite R-module is MCM if and only if (S_i) condition is satisfied for any $i \geq 0$. We say a ring R satisfies (S_i) condition if the R-module R does.

We say that R satisfies *Serre's* (R_i) *condition* if $\dim R_P < i$ implies that R_P is regular local for any $P \in \mathrm{Spec}\, R$.

For a noetherian ring R, R is reduced if and only if R satisfies (R_0) and (S_1). R is normal if and only if (R_1) and (S_2) are satisfied (Krull–Serre–Nagata's theorem).

2.6 Local cohomology

The references for this subsection are [26] and [70].

(2.6.1) Let X be a topological space, Y a closed subset of X, $U := X - Y$, and $\mathcal{F}$ an abelian sheaf on X. Then we denote the kernel of the natural restriction map $\Gamma(X, \mathcal{F}) \to \Gamma(U, \mathcal{F})$ by $\Gamma_Y(X, \mathcal{F})$. Note that $\Gamma_Y(X, ?)$ is a left exact functor. We denote its ith right derived functor by $H^i_Y(X, ?)$, and call it the ith local cohomology functor with support in Y.

What we are interested in here is the following case. R is a noetherian ring, $X = \operatorname{Spec} R$, I is an ideal of R, $Y = V(I) = \operatorname{Spec} R/I$, M is an R-module, and $\mathcal{F} = \tilde{M}$. In this case, we denote $\Gamma_Y(X, \mathcal{F})$ by $\Gamma_I(M)$. As is easily verified, we have

$$\Gamma_I(M) = \{m \in M \mid \exists n \;\; I^n m = 0\} = \varinjlim \operatorname{Hom}_R(R/I^n, M),$$

and hence $\Gamma_I(M)$ is an R-module in a natural way. As $\tilde{J}$ is flabby for an injective R-module J [71, Proposition 3.4], and we have $H^i_Y(X, \mathcal{F}) = 0$ $(i > 0)$ for any flabby sheaf $\mathcal{F}$ [70, Proposition 1.10], the derived functor of $\Gamma_I(M)$ in the category of R-modules agrees with the local cohomology, which we denote by $H^i_I(M)$. As ${}_R\mathbb{M}$ satisfies the (AB5) condition, we have

$$H^i_I(M) \cong \varinjlim \operatorname{Ext}^i_R(R/I^n, M).$$

If $\dim R = n$, then we have $H^i_I(M) = 0$ $(i > n)$ [70, Proposition 1.12].

Lemma 2.6.2 *Let R be a noetherian ring, I a proper ideal of R, M an R-finite module, and $n \geq 0$. Then the following are equivalent.*

1 $H^i_I(M) = 0$ $(i < n)$.

2 $\operatorname{depth}(I, M) \geq n$.

For the proof, see [70, Theorem 3.8].

2.7 Ring-theoretic properties of morphisms

The references for this subsection are [65], [17], [18], [19], [15] and [20].

(2.7.1) Let k be a field. We say that a k-algebra R is geometrically regular (resp. normal) over k if $K \otimes_k R$ is regular (resp. normal) for any finite algebraic extension K of k. If k is a perfect field, then R is geometrically regular (resp. normal) if and only if R is regular (resp. normal).

Let $\varphi : X \to Y$ be a morphism of locally noetherian schemes. We say that φ is *regular* (resp. *normal, flat*) at $x \in X$ if $\mathcal{O}_{X,x}$ is flat over $\mathcal{O}_{Y,\varphi(x)}$ and $\mathcal{O}_{X,x}/\mathfrak{m}_{Y,\varphi(x)}\mathcal{O}_{X,x}$ is geometrically regular (resp. geometrically normal, flat) over $\kappa(\varphi(x))$. We say that φ is regular (resp. normal, flat) if φ has that property at every point of X. A regular morphism locally of finite type is called a *smooth morphism.* Although this definition (and some others which are equivalent) is commonly used, some authors use the same word 'smooth' for slightly different meanings.

(2.7.2) Let $\varphi : X \to Y$ be a morphism locally of finite type of locally noetherian schemes. For $x \in X$, we define $\dim_\varphi(x)$ to be $\inf_{x \in U} \dim(U \cap \varphi^{-1}(\varphi(x)))$, where U runs through quasi-compact open neighborhoods of x. The function $x \mapsto \dim_\varphi(x)$ is upper-semicontinuous [65, (IV.13.1.3)]. We say that φ is *quasi-finite* at x if $\dim_\varphi(x) = 0$. If φ is quasi-finite at any point of X, then we say that φ is quasi-finite. A quasi-finite smooth morphism is said to be *étale*.

Let $\varphi : X \to Y$ be a morphism of finite type of noetherian schemes. We say that φ is *equidimensional* if $\varphi(Z)$ is dense in some irreducible component of Y for each irreducible component Z of X, and the function $x \mapsto \dim_\varphi(x)$ is locally constant. If this is the case, the value $e = \dim_\varphi(x)$ is constant on each connected component X_0 of X, and we say that the *relative dimension* of φ over X_0 is e.

(2.7.3) If $\varphi : X \to Y$ is a flat morphism of schemes locally of finite presentation, then φ is an open map [65, IV.1.8.4].

If $\varphi : X \to Y$ is a morphism of finite type between noetherian schemes all of whose local rings are equidimensional, and φ is an open map, then φ is equidimensional [65, IV.§14].

If Y is a regular scheme, X is a noetherian Cohen–Macaulay scheme, and $\varphi : X \to Y$ is equidimensional (of finite type), then φ is flat. This follows from the corresponding statement for local rings [110, Theorem 23.1] and the dimension formula [110, Theorem 15.6], see [65, (IV.13.3.6)].

(2.7.4) Let $R \to S$ be a ring homomorphism of commutative rings. We denote the kernel of the multiplication map $m : S \otimes_R S \to S$ $(a \otimes b \mapsto ab)$ by I. The S-module I/I^2 is denoted by $\Omega_{S/R}$. If $R \to R'$ is a ring homomorphism of commutative rings, then we have

$$\Omega_{S'/R'} \cong R' \otimes_R \Omega_{S/R} \cong S' \otimes_S \Omega_{S/R},$$

where $S' = R' \otimes_R S$. Moreover, for any multiplicatively closed subset Γ of S, we have $\Omega_{S_\Gamma/R} \cong S_\Gamma \otimes_S \Omega_{S/R}$ [110, p. 198]. Hence, as the patching process goes well, a quasi-coherent $\mathcal{O}_X$-module $\Omega_{X/Y}$ is defined for a morphism of schemes $X \to Y$. We call $\Omega_{X/Y}$ the *sheaf of Kähler differentials.* If $X \to Y$ is a morphism locally of finite type between locally noetherian schemes, then $\Omega_{X/Y}$ is coherent.

Proposition 2.7.5 *Let $\varphi : X \to Y$ be a morphism locally of finite type between locally noetherian schemes. Then the following are equivalent.*

1 *φ is smooth.*

2 *φ is flat, $\Omega_{X/Y,x}$ is $\mathcal{O}_{X,x}$-free, and* $\operatorname{rank} \Omega_{X/Y,x} = \dim_\varphi(x)$ *for $x \in X$.*

For the proof, see [66, (II.5.5)]. In particular, if φ is smooth, then φ is equidimensional and its relative dimension is $\operatorname{rank} \Omega_{X/Y}$ over each component. Let $\varphi : X \to Y$ be a smooth morphism of locally noetherian schemes. We denote the invertible sheaf $\bigwedge^{\mathrm{top}} \Omega_{X/Y}$ by $\omega_{X/Y}$.

Theorem 2.7.6 *Let $X \xrightarrow{f} Y \xrightarrow{g} Z$ be a sequence of morphisms locally of finite type of locally noetherian schemes. Then we have*

1 *There exists an exact sequence of the form*

$$f^*\Omega_{Y/Z} \to \Omega_{X/Z} \to \Omega_{X/Y} \to 0.$$

2 *If, moreover, f is smooth, then there is an exact sequence*

$$0 \to f^*\Omega_{Y/Z} \to \Omega_{X/Z} \to \Omega_{X/Y} \to 0.$$

3 *If f is a closed immersion, and $\mathcal{I}$ is the defining ideal sheaf of X, then there exists an exact sequence of the form*

$$\mathcal{I}/\mathcal{I}^2 \to f^*\Omega_{Y/Z} \to \Omega_{X/Z} \to 0.$$

4 *If, moreover, X is smooth over Z in* **3**, *then there exists some exact sequence of the form*

$$0 \to \mathcal{I}/\mathcal{I}^2 \to f^*\Omega_{Y/Z} \to \Omega_{X/Z} \to 0.$$

If gf in **4** is an isomorphism, then we have $f^*\Omega_{Y/Z} \cong \mathcal{I}/\mathcal{I}^2$.

(2.7.7) So far we have defined Cohen–Macaulay and Gorenstein properties of rings. We extend these notions to morphisms of finite flat dimensions.

For two commutative local rings $(R, \mathfrak{m})$ and $(S, \mathfrak{n})$, we say that a map $\varphi : R \to S$ is a *local homomorphism* if φ is a ring homomorphism such that $\varphi(\mathfrak{m}) \subset \mathfrak{n}$. Let $\varphi : (R, \mathfrak{m}) \to (S, \mathfrak{n})$ be a local homomorphism of local rings.

Theorem 2.7.8 ([20, (1.1)]) *Let $\varphi : (R, \mathfrak{m}) \to (S, \mathfrak{n})$ be a local homomorphism of noetherian local rings. If S is complete, then there exist some complete noetherian local ring T and local homomorphisms $\tau : R \to T$ and $\sigma : T \to S$ such that τ is flat, $T/\mathfrak{m}T$ is regular, σ is surjective, and $\sigma \circ \tau = \varphi$.*

We call such a decomposition $\sigma \circ \tau$ of φ a *Cohen factorization* of φ. $\operatorname{Ker} \sigma$ is a perfect ideal (resp. Gorenstein ideal, complete intersection ideal, ideal of finite projective dimension) for some Cohen factorization if and only if this is true for any Cohen factorization (because of the comparison of Cohen factorization, see [20, (1.2)]). If a Cohen factorization of the composite map

$$R \xrightarrow{\varphi} S \xrightarrow{\text{completion}} \hat{S}$$

satisfies the condition, we say that φ is a *Cohen–Macaulay* (resp. *Gorenstein*, *complete intersection* (*c.i.*, for short), *of finite flat dimension*) local homomorphism. As we have

$$\text{flat.dim}_R\, S = \text{flat.dim}_R\, \hat{S} < \infty \iff \text{proj.dim}_T\, \hat{S} < \infty,$$

$\text{flat.dim}_R\, S < \infty$ if and only if φ is of finite flat dimension.

If S is R-flat, then φ is Cohen–Macaulay (resp. locally Gorenstein) if and only if the fiber ring $S/\mathfrak{m}S$ is Cohen–Macaulay (resp. Gorenstein).

(2.7.9) We say that a morphism $f : X \to Y$ of locally noetherian schemes is Cohen–Macaulay (resp. Gorenstein, complete intersection, of finite flat dimension) at $x \in X$ if the local homomorphism $\mathcal{O}_{Y,f(x)} \to \mathcal{O}_{X,x}$ is Cohen–Macaulay (resp. Gorenstein, complete intersection, of finite flat dimension). We say that f is Cohen–Macaulay (or locally Cohen–Macaulay) (resp. Gorenstein (or locally Gorenstein), local complete intersection (l.c.i., for short), locally of finite flat dimension) if f satisfies the corresponding property at any point of X.

A ring homomorphism $R \to S$ of noetherian rings is said to be locally Cohen–Macaulay (resp. locally Gorenstein, local complete intersection, regular, locally of finite flat dimension) if the corresponding property is satisfied by the associated morphism $\operatorname{Spec} S \to \operatorname{Spec} R$. Note that a local homomorphism of local rings $\varphi : R \to S$ is locally of finite flat dimension if and only if it is of finite flat dimension.

We say that a noetherian ring R is a *G-ring* if the completion $R_P \to \widehat{R_P}$ is regular for any $P \in \operatorname{Spec} R$. It is known that R is a G-ring if and only if the completion $R_\mathfrak{m} \to \widehat{R_\mathfrak{m}}$ is regular for any $\mathfrak{m} \in \operatorname{Max} R$ [110, Theorem 32.4]. Note that a complete local ring is a G-ring. The ring $\mathbb{Z}$ is also a G-ring. A ring essentially of finite type over a G-ring is also a G-ring.

Let $\varphi : R \to S$ be a local homomorphism of noetherian commutative local rings. If R is a G-ring, then φ is locally Cohen–Macaulay if and only if it is Cohen–Macaulay as a local homomorphism. This is not true without the G-property in general. Similarly for the Gorenstein and c.i. properties.

A locally noetherian scheme X is Cohen–Macaulay (resp. Gorenstein) if and only if X is Cohen–Macaulay (resp. Gorenstein) over $\operatorname{Spec} \mathbb{Z}$.

We say that a noetherian local ring $(A, \mathfrak{m})$ is a *complete intersection* if A is complete intersection over $\mathbb{Z}_{\mathfrak{m}\cap\mathbb{Z}}$. We say that a locally noetherian scheme X is a local complete intersection (l.c.i.) if X is l.c.i. over $\operatorname{Spec} \mathbb{Z}$. X is l.c.i. if and only if $\mathcal{O}_{X,x}$ is a complete intersection for any closed point x of X. If a local homomorphism $\varphi : (R, \mathfrak{m}) \to (S, \mathfrak{n})$ is flat, then φ is c.i. (at $\mathfrak{n}$) if and only if $S/\mathfrak{m}S$ is a complete intersection. Note that a regular scheme is l.c.i., and an l.c.i. scheme is Gorenstein.

Theorem 2.7.10 *Let*

$$X \xrightarrow{\varphi} Y \xrightarrow{\psi} Z$$

be a sequence of morphisms of locally noetherian schemes. Then the following hold.

1 *If ψ and φ are Cohen–Macaulay (resp. Gorenstein, l.c.i., regular), then so is $\psi\varphi$.*

1′ *If Y and φ are Cohen–Macaulay (resp. Gorenstein, l.c.i., regular), then so is X.*

2 *If φ and ψ are locally of finite flat dimension and $\psi\varphi$ is Cohen–Macaulay (resp. Gorenstein, l.c.i.), then so is φ, and so is ψ at each point of $\varphi(X)$.*

2′ *If φ is locally of finite flat dimension and X is Cohen–Macaulay (resp. Gorenstein, l.c.i.), then so is φ, and so is Y at each point of $\varphi(X)$.*

2″ *If φ is faithfully flat and $\psi\varphi$ is Cohen–Macaulay (resp. Gorenstein, l.c.i.), then φ and ψ are Cohen–Macaulay (resp. Gorenstein, l.c.i.).*

3 *If ψ and $\psi\varphi$ are flat (resp. Cohen–Macaulay, Gorenstein, l.c.i.) and the morphism between the fibers $X_z \to Y_z$ is flat for any $z \in Z$, then φ is flat (Cohen–Macaulay, Gorenstein, l.c.i.).*

3′ *If φ is faithfully flat, and X is Cohen–Macaulay (resp. Gorenstein, l.c.i., regular), then so is Y.*

4 *Assume that $\mathcal{O}_{Y,y}$ is a G-ring for any point y of Y. If φ is Cohen–Macaulay (resp. Gorenstein, l.c.i.) at each closed point of X, then so is φ at any point of X.*

Theorem 2.7.11 *Let $\varphi : X \to Y$ and $\psi : Y' \to Y$ be morphisms between locally noetherian schemes. We assume that φ is locally of finite type and ψ is flat. We define $\varphi' : X' := Y' \times_Y X \to Y'$ to be the first projection. Let $x' \in X'$, and set x to be the image of x' in X. Then the following are equivalent.*

1 *φ is Cohen–Macaulay (resp. Gorenstein, l.c.i., regular) at x.*

2 *φ' is Cohen–Macaulay (resp. Gorenstein, l.c.i., regular) at x'.*

We have seen that there is a hierarchy of properties of noetherian commutative rings:

$$\text{regular} \Rightarrow \text{l.c.i.} \Rightarrow \text{Gorenstein} \Rightarrow \text{Cohen–Macaulay.}$$

It is not overstating to say that this hierarchy has been playing the central role in the modern theory of commutative rings.

2.8 Betti numbers, Bass numbers and complete intersections

(2.8.1) Let $(R, \mathfrak{m})$ be a noetherian local ring, and M an R-finite module. We denote $\dim_{R/\mathfrak{m}} \operatorname{Tor}_i^R(R/\mathfrak{m}, M)$ by $\beta_i^R(M)$, and call it the ith *Betti number* of M. We denote $\dim_{R/\mathfrak{m}} \operatorname{Ext}_R^i(R/\mathfrak{m}, M)$ by $\mu_R^i(M)$, and call it the ith Bass number of M. Note that $\beta_i^R(M) < \infty$ and $\mu_R^i(M) < \infty$ for any $i \in \mathbb{Z}$. We call the power series $P_M^R(t) := \sum_{i \geq 0} \beta_i^R(M) t^i \in \mathbb{Z}[[t]]$ the *Poincaré series* (or *Betti series*) of M. The power series $Q_R^M(t) := \sum_{i \geq 0} \mu_R^i(M) t^i \in \mathbb{Z}[[t]]$ is called the *Bass series* of M.

We say that an R-free complex

(2.8.2) $$\mathbb{F} : \cdots \to F_i \xrightarrow{\partial_i} F_{i-1} \to \cdots$$

is *minimal* if $\partial_i(F_i) \subset \mathfrak{m} F_{i-1}$ for each i.

Lemma 2.8.3 *Let $(R, \mathfrak{m})$ be a noetherian local ring, M a finite R-module, and $\mathbb{F}$ (2.8.2) an R-free resolution of M. Then the following are equivalent.*

1 *$\mathbb{F}$ is minimal.*

2 *$F_0 \to M$ and $F_i \to \operatorname{Im} \partial_i$ $(i \geq 1)$ are projective covers.*

3 *For $i \geq 0$, we have $\operatorname{rank} F_i = \beta_i^R(M)$.*

There exists some minimal free resolution of M, uniquely up to isomorphisms of complexes.

Proof. Follows easily from Lemma 1.12.5 and Lemma 1.12.4. □

If $\mathbb{F}$ is a minimal free resolution of $R/\mathfrak{m}$, then we have

$$\mu_R^i(R/\mathfrak{m}) = \dim_{R/\mathfrak{m}} H^i(\operatorname{Hom}_R(\mathbb{F}, R/\mathfrak{m}))$$
$$= \dim_{R/\mathfrak{m}} (R/\mathfrak{m} \otimes_R F_i)^* = \operatorname{rank} F_i = \beta_i^R(R/\mathfrak{m}).$$

Proposition 2.8.4 *Let $(R, \mathfrak{m})$ be a d-dimensional noetherian local ring, and $e = \operatorname{emb.dim} R$. Then we have*

1 $e = \beta_1^R(R/\mathfrak{m})$.

2 $\beta_2^R(R/\mathfrak{m}) \geq \binom{e}{2} + e - d$.

3 *The following are equivalent.*

a $\beta_2^R(R/\mathfrak{m}) = \binom{e}{2} + e - d$

b $P_{R/\mathfrak{m}}^R(t) = (1+t)^e / (1-t^2)^{e-d}$

b′ *There exist some $e' \geq 0$ and $c' \geq 0$ such that $P^R_{R/\mathfrak{m}}(t) = (1+t)^{e'}/(1-t^2)^{c'}$.*

c *R is a complete intersection.*

For the proof, see [14, Theorem 7.3.3]. What is important is that the complete intersection property of R is determined only by the Betti series of $R/\mathfrak{m}$.

Lemma 2.8.5 *If M is a non-zero finite R-module, then the following hold:*

1 $\operatorname{depth}_R M = \inf\{i \mid \mu^i_R(M) \neq 0\}$;

2 $\operatorname{inj.dim}_R M = \sup\{i \mid \mu^i_R(M) \neq 0\}$.

(2.8.6) Let M be a Cohen–Macaulay R-module. Then we call $\mu^d_R(M)$ the *Cohen–Macaulay type* (or sometimes simply the type) of M, and denote it by $\operatorname{type} M$, where $d = \dim M = \operatorname{depth} M$. We say that an R-module M is *Gorenstein* if M is a Cohen–Macaulay R-module with $\operatorname{type} M = 1$. Do not confuse a module of Gorenstein dimension 0, which is sometimes erroneously referred to as a Gorenstein module in the literature, with a Gorenstein module.

2.9 Resolutions of perfect modules

Let R be noetherian.

Lemma 2.9.1 *Let $M \neq 0$ be a perfect R-module of codimension h, and N a finite R-module. Then $\operatorname{Ext}^h_R(M,R)$ is perfect of codimension h, and*

$$\operatorname{Ext}^i_R(M,N) \cong \operatorname{Tor}^R_{h-i}(\operatorname{Ext}^h_R(M,R),N)$$

for $i \geq 0$.

Proof. Let $\mathbb{F}$ be a finite projective resolution of M of length h. Then as $\operatorname{Ext}^i_R(M,R) = 0$ $(i \neq h)$, the complex $\mathbb{F}^*[h]$ is a finite projective resolution of $\operatorname{Ext}^h_R(M,R)$ of length h. Hence,

$$\operatorname{Ext}^i_R(\operatorname{Ext}^h_R(M,R),R) \cong H^i((\mathbb{F}^*[h])^*) \cong H^{i-h}(\mathbb{F}) = H_{h-i}(\mathbb{F}),$$

and it is 0 for $i \neq h$, and is M for $i = h$. As $M \neq 0$, we have

$$\operatorname{grade} \operatorname{Ext}^h_R(M,R) = h \geq \operatorname{proj.dim}_R \operatorname{Ext}^h_R(M,R) \geq \operatorname{grade} \operatorname{Ext}^h_R(M,R),$$

and $\operatorname{Ext}^h_R(M,R)$ is also perfect of codimension h. Moreover,

$$\operatorname{Ext}^i_R(M,N) \cong H^i(\operatorname{Hom}_R(\mathbb{F},N)) \cong H^i(\mathbb{F}^* \otimes N) \cong \operatorname{Tor}^R_{h-i}(\operatorname{Ext}^h_R(M,R),N).$$

□

Corollary 2.9.2 *Let R be a noetherian ring, I a Gorenstein ideal of R of codimension h, and N an R-module. Then we have*

$$\operatorname{Ext}^h_R(R/I, R) \otimes_{R/I} \operatorname{Tor}^R_i(R/I, N) \cong \operatorname{Ext}^{h-i}_R(R/I, N)$$

for any $i \in \mathbb{Z}$.

The next theorem is called the *depth sensitivity* of a resolution of a perfect module.

Proposition 2.9.3 *Let R be a noetherian ring, M a perfect R-module of codimension h, and N a finite R-module. Assume that $M \otimes N \neq 0$. Then we have*

$$\inf\{i \mid \operatorname{Ext}^i_R(M, N) \neq 0\} + \sup\{j \mid \operatorname{Tor}^R_j(M, N) \neq 0\} = h.$$

(2.9.4) There is a nice resolution of a complete intersection ideal, called a *Koszul complex*. Let $a_1, \ldots, a_h \in R$. We set $F := R^h$, and let $e_1, \ldots, e_h$ be an R-free basis of F. We define $\operatorname{Kos}(a_1, \ldots, a_h)$ to be the complex

$$0 \to \textstyle\bigwedge^h F \xrightarrow{\partial_h} \bigwedge^{h-1} F \to \cdots \to \bigwedge^1 F \xrightarrow{\partial_1} R \to 0,$$

where the boundary map is given by

$$\partial_l(e_{i_1} \wedge \cdots \wedge e_{i_l}) = \sum_{j=1}^{l} (-1)^{j+1} a_{i_j} (e_{i_1} \wedge \cdots \overset{j}{\check{}} \cdots \wedge e_{i_l}).$$

For an R-module M, we denote $\operatorname{Kos}(a_1, \ldots, a_h) \otimes^\bullet_R M$ by $\operatorname{Kos}(a_1, \ldots, a_h; M)$.

The next lemma, which is standard [110, Theorem 16.8], is called the depth sensitivity of a Koszul complex.

Lemma 2.9.5 *Let R be a noetherian ring, $a_1, \ldots, a_h \in R$, and M be a finite R-module. Set $I := (a_1, \ldots, a_h) \subset R$. If $M \neq IM$, then we have*

$$\operatorname{depth}_R(I, M) = h - \inf\{i \mid H_i(\operatorname{Kos}(a_1, \ldots, a_h; M)) \neq 0\}.$$

In particular, $\operatorname{Kos}(a_1, \ldots, a_h; M)$ is a resolution of M/IM if and only if $\operatorname{depth}(I, M) = h$ if and only if $a_1, \ldots, a_h$ is an M-sequence.

Assume that $a_1, \ldots, a_h$ is an R-sequence so that $I := (a_1, \ldots, a_h)$ is a complete intersection ideal. Then $\mathbb{F} := \operatorname{Kos}(a_1, \ldots, a_h; R)$ is a resolution of R/I. As we have $\mathbb{F}^*[h] \cong \mathbb{F}$, we see that I is a Gorenstein ideal.

The next lemma is called the rigidity of a Koszul complex.

Lemma 2.9.6 *Let R be noetherian, I a complete intersection ideal of R of codimension h, and M an R-module. If $i \geq 0$ and $\operatorname{Tor}^R_i(R/I, M) = 0$, then we have $\operatorname{Tor}^R_j(R/I, M) = 0$ for any $j \geq i$. In particular, we have $\operatorname{depth}(I, M) > h - i$.*

Note that even the following holds.

Theorem 2.9.7 (Lichtenbaum [101]) *Let R be a regular ring, M and N finite R-modules, and $i \geq 0$. If $\operatorname{Tor}_i^R(M,N) = 0$, then $\operatorname{Tor}_j^R(M,N) = 0$ for $j \geq i$.*

2.10 Dualizing complexes and canonical modules

For dualizing complexes, see [69] and [80].

(2.10.1) Let X be a noetherian scheme. We say that a complex of quasi-coherent $\mathcal{O}_X$-modules $I^\bullet$ is a *dualizing complex* of X if $I^\bullet$ is bounded, each term of $I^\bullet$ is an injective $\mathcal{O}_X$-module, each cohomology group of $I^\bullet$ is coherent, and the canonical map

$$\mathcal{O}_X \to \underline{\operatorname{Hom}}^\bullet_{\mathcal{O}_X}(I^\bullet, I^\bullet)$$

is a quasi-isomorphism. Usually, a dualizing complex is regarded as an object of the derived category $D^+({}_X\mathbb{M})$, and hence any object isomorphic to a dualizing complex in $D^+({}_X\mathbb{M})$ is also called a dualizing complex. Note that a quasi-coherent $\mathcal{O}_X$-module I is an injective $\mathcal{O}_X$-module if and only if its stalk I_x at x is an injective $\mathcal{O}_{X,x}$-module for any $x \in X$ [69, Proposition II.7.17].

If $I^\bullet$ is a dualizing complex of X, for any complex $F^\bullet$ of $\mathcal{O}_X$-modules with coherent cohomology groups, the canonical map

$$F^\bullet \to \operatorname{Hom}^\bullet_R(\operatorname{Hom}^\bullet_R(F^\bullet, I^\bullet), I^\bullet)$$

is a quasi-isomorphism [69, Proposition V.2.1].

The dualizing complex is unique, in the following sense.

Theorem 2.10.2 ([69, Theorem V.3.1]) *Let X be a connected noetherian scheme, $I^\bullet$ a dualizing complex of X, and $I'^\bullet$ a complex of $\mathcal{O}_X$-modules bounded above with coherent cohomology groups. Then $I'^\bullet$ is dualizing if and only if there exists some invertible sheaf L and some integer n such that $I'^\bullet$ is isomorphic to $I^\bullet \otimes_{\mathcal{O}_X} L[n]$ in $D({}_X\mathbb{M})$. In this case, L and n are determined by*

$$L[n] \cong R\,\underline{\operatorname{Hom}}^\bullet_{\mathcal{O}_X}(I^\bullet, I'^\bullet).$$

(2.10.3) A complex $I^\bullet$ of quasi-coherent $\mathcal{O}_X$-modules is called a *fundamental dualizing complex* if $I^\bullet$ is bounded with coherent cohomology groups, and

$$\bigoplus_{i \in \mathbb{Z}} I^p \cong \bigoplus_{x \in X} J(x)$$

is satisfied, where $J(x)$ denotes the constant sheaf $(i_x)_*(E_{\mathcal{O}_{X,x}}(\kappa(x)))^{\sim}$, where $E_{\mathcal{O}_{X,x}}(\kappa(x))$ denotes the injective hull of the $\mathcal{O}_{X,x}$-module $\kappa(x)$, and $i_x : \operatorname{Spec} \mathcal{O}_{X,x} \to X$ is the canonical map. A fundamental dualizing complex is a dualizing complex, and if there is a dualizing complex of X, then there is a fundamental dualizing complex of X [69, V.2.3, V.7.3].

(2.10.4) If there is a dualizing complex of X, then X is finite dimensional. A bounded-below complex $I^\bullet$ of quasi-coherent $\mathcal{O}_X$-modules with coherent cohomology groups over a locally noetherian scheme X is called *pointwise dualizing* if $I^\bullet_x$ is a dualizing complex of $\operatorname{Spec} \mathcal{O}_{X,x}$ for any $x \in X$. A dualizing complex of a noetherian scheme is pointwise dualizing. Conversely, a pointwise dualizing complex of a finite dimensional noetherian scheme is dualizing.

(2.10.5) Let $(R, \mathfrak{m})$ be a noetherian local ring, and $I^\bullet$ a dualizing complex of R, that is to say, the complex of $\mathcal{O}_X$-modules associated to $I^\bullet$ is a dualizing complex of X, where $X = \operatorname{Spec} R$. Then $\operatorname{Ext}^i_R(R/\mathfrak{m}, I^\bullet)$ is non-zero for one and only one i, and it is isomorphic to $R/\mathfrak{m}$. If the i such that $\operatorname{Ext}^i_R(R/\mathfrak{m}, I^\bullet) \neq 0$ is zero, then we say that $I^\bullet$ is a *normalized dualizing complex*. Note that a normalized fundamental dualizing complex is unique up to isomorphisms of R-complexes.

For a fundamental dualizing complex $I^\bullet$ of $(R, \mathfrak{m})$, any localization $I^\bullet_\mathfrak{p}$ at $\mathfrak{p} \in \operatorname{Spec} R$ is again a fundamental dualizing complex. If moreover $I^\bullet$ is normalized, then $I^\bullet_\mathfrak{p}[-\dim R/\mathfrak{p}]$ is normalized.

If R is a finite dimensional Gorenstein ring, then a minimal injective resolution $I^\bullet$ of the R-module R is a fundamental dualizing complex of R. If R is local and $d = \dim R$, then $I^\bullet[d]$ is normalized.

The following theorem due to T. Kawasaki is an affirmative answer to *Sharp's conjecture* for local rings.

Theorem 2.10.6 ([94, Corollary 6.2]) *Let R be a noetherian local ring. Then R has a dualizing complex if and only if R is a homomorphic image of a Gorenstein local ring.*

The following theorem is known as the local duality theorem.

Theorem 2.10.7 ([69, Theorem 6.2]) *Let $(R, \mathfrak{m})$ be a noetherian local ring, $I^\bullet$ a normalized dualizing complex of R, and M a finite R-module. Then there is an isomorphism*

$$H^i_\mathfrak{m}(M) \cong \operatorname{Hom}_R(\operatorname{Ext}^{-i}_R(M, I^\bullet), E_R(R/\mathfrak{m}))$$

which is natural with respect to M.

Proof. As $\mathrm{Hom}^\bullet_R(M, I^\bullet)$ is bounded with R-finite homology groups, it has a free resolution $F_\bullet$ with each term finite free. As $I^\bullet$ is dualizing, there exist quasi-isomorphisms

$$M \to \mathrm{Hom}^\bullet_R(\mathrm{Hom}^\bullet_R(M, I^\bullet), I^\bullet) \cong \mathrm{Hom}^\bullet_R(F_\bullet, I^\bullet).$$

By Corollary 2.1.8, $\mathrm{Hom}^\bullet_R(F_\bullet, I^\bullet)$ is an injective resolution of M. Hence, we have

$$H^i_{\mathfrak{m}}(M) \cong H^i(\Gamma_{\mathfrak{m}}(\mathrm{Hom}^\bullet_R(F_\bullet, I^\bullet))) \cong H^i(\mathrm{Hom}^\bullet_R(F_\bullet, \Gamma_{\mathfrak{m}}(I^\bullet))).$$

Note that we have $\Gamma_{\mathfrak{m}}(I^\bullet)$ is quasi-isomorphic to $E_R(R/\mathfrak{m})$, which is easily seen when we consider the case $I^\bullet$ is fundamental. As $E_R(R/\mathfrak{m})$ is an injective module, we have quasi-isomorphisms

$$\mathrm{Hom}^\bullet_R(F_\bullet, \Gamma_{\mathfrak{m}}(I^\bullet)) \\ \cong \mathrm{Hom}^\bullet_R(F_\bullet, E_R(R/\mathfrak{m})) \cong \mathrm{Hom}_R(\mathrm{Hom}^\bullet_R(M, I^\bullet), E_R(R/\mathfrak{m})).$$

Hence, we have the isomorphism in question, as desired. □

2.11 The duality of proper morphisms and rational singularities

(2.11.1) Let $\mathcal{A}$ be an abelian category, and $\mathcal{A}'$ its thick subcategory. We denote the full subcategory of $D^?(\mathcal{A})$ consisting of objects X such that $H^i(X) \in \mathcal{A}'$ for any i by $D^?_{\mathcal{A}'}(\mathcal{A})$, where ? is either b, $+$, $-$ or $\emptyset$. Obviously, $D^?_{\mathcal{A}'}(\mathcal{A})$ is a triangulated subcategory of $D^?(\mathcal{A})$. For a locally noetherian scheme X, we denote $D^?_{\mathrm{Coh}\,X}({}_X\mathbb{M})$ (resp. $D^?_{\mathrm{Qco}\,X}({}_X\mathbb{M})$) by $D^?_c(X)$ (resp. $D^?_{qc}(X)$). Note that the forgetful functor $D^?(\mathrm{Qco}\,X) \to D^?_{qc}(X)$ is an equivalence for a quasi-compact scheme X for $? = +, \emptyset$ [25, 6.7].

(2.11.2) Let X be a noetherian scheme. The following was proved by M. Nagata [118]. See also [103].

Theorem 2.11.3 *Let $f : Y \to X$ be a morphism of finite type between noetherian schemes. Then f is* compactifiable *in the sense that there exist some scheme $\tilde{Y}$, a proper morphism $p : \tilde{Y} \to X$, and an open immersion $i : Y \to \tilde{Y}$ such that $pi = f$.*

A factorization $pi = f$ as in the theorem is called a compactification of f. The following is known as the global duality theorem of proper morphisms, see [121] and [102]. See also [69] and its appendix by Deligne [41].

Theorem 2.11.4 *Let $p : Y \to X$ be a proper morphism between noetherian schemes. Then the derived functor $\underline{R}p_* : D_{qc}(Y) \to D_{qc}(X)$ (the unbounded derived functor, see [137]) has a right adjoint $p^{!,\mathrm{ub}} : D_{qc}(X) \to D_{qc}(Y)$. We have $p^{!,\mathrm{ub}}(D^+_{qc}(X)) \subset D^+_{qc}(Y)$, and the restriction of $p^{!,\mathrm{ub}}$ to $D^+_{qc}(X)$, which we denote by $p^!$, is a right adjoint functor of $\underline{R}^+p_*$.*

(2.11.5) Let $F(X)$ denote the category of X-schemes of finite type. Any morphism $f : Y \to Y'$ of $F(X)$ has a compactification $pi = f$. We define $f^! := i^* \circ p^! : D^+_{qc}(Y') \to D^+_{qc}(Y)$, where $p^!$ is the right adjoint of $\underline{R}^+ p_*$.

Proposition 2.11.6 *Under the notation as above, the following hold.*

1 *The definition of $f^!$ is independent of the compactification $pi = f$ of f, up to isomorphisms of functors.*

2 *For two morphisms f and g in $F(X)$, we have $(g \circ f)^! \cong f^! \circ g^!$, provided that $g \circ f$ is defined.*

3 (Residue isomorphism) *If $h : Y \to Y'$ is a smooth X-morphism of relative dimension d, then $h^!$ is isomorphic to $h^\# := h^* ? \otimes^L_{\mathcal{O}_Y} \omega_{Y'/Y}[d]$.*

4 *If $g : Y \to Y'$ is a finite X-morphism, then $g^!$ is isomorphic to $g^\natural := \bar{g}^* R\,\underline{\mathrm{Hom}}^\bullet_{\mathcal{O}_{Y'}}(g_*\mathcal{O}_Y, ?)$, where $\bar{g} : (Y, \mathcal{O}_Y) \to (Y', g_*\mathcal{O}_Y)$ is the canonical morphism of ringed spaces.*

5 *Let $f : Y \to Y'$ be a morphism of $F(X)$, and $g : Z' \to Y'$ a flat morphism of noetherian schemes. If $gf' = fg'$ is a fiber square, then we have a canonical isomorphism $(g')^* \circ f^! \cong (f')^! \circ g^*$.*

6 *If X has a dualizing complex I_X, then $I_Y := f^! I_X$ is a dualizing complex of Y for any morphism $f : Y \to X$ of finite type.*

Corollary 2.11.7 (Duality for proper morphisms) *Let $p : Y \to X$ be a proper morphism between noetherian schemes, $\mathbb{F} \in D_{qc}(Y)$ and $\mathbb{G} \in D^+_{qc}(X)$. Then there is an isomorphism*

$$\theta_p : \underline{R}p_* \underline{R}\,\underline{\mathrm{Hom}}^\bullet_{\mathcal{O}_Y}(\mathbb{F}, p^!\mathbb{G}) \cong \underline{R}\,\underline{\mathrm{Hom}}^\bullet_{\mathcal{O}_X}(\underline{R}p_*\mathbb{F}, \mathbb{G}),$$

which is functorial on $\mathbb{F}$ and $\mathbb{G}$.

Proof. (Sketch) Consider the composite of the canonical maps

$$\theta_p : \underline{R}p_* \underline{R}\,\underline{\mathrm{Hom}}^\bullet_{\mathcal{O}_Y}(\mathbb{F}, p^!\mathbb{G}) \to \underline{R}\,\underline{\mathrm{Hom}}^\bullet_{\mathcal{O}_X}(\underline{R}p_*\mathbb{F}, \underline{R}p_*p^!\mathbb{G}) \xrightarrow{\underline{R}\,\underline{\mathrm{Hom}}^\bullet_{\mathcal{O}_X}(\underline{R}p_*\mathbb{F}, e)} \underline{R}\,\underline{\mathrm{Hom}}^\bullet_{\mathcal{O}_X}(\underline{R}p_*\mathbb{F}, \mathbb{G}),$$

where the first arrow is the natural map given in [102, (3.5.4)], and $e : \underline{R}p_*p^! \to \mathrm{Id}$ is the counit of adjunction. It suffices to show that

$$\underline{R}\Gamma(U, \theta_p) : \underline{R}\,\mathrm{Hom}^\bullet_{\mathcal{O}_{U_Y}}(\mathbb{F}|_{U_Y}, (p^!\mathbb{G})|_{U_Y}) \to \underline{R}\,\mathrm{Hom}^\bullet_{\mathcal{O}_U}((\underline{R}p_*\mathbb{F})|_U, \mathbb{G}|_U)$$

is an isomorphism for any open set U of X, where $U_Y = p^{-1}(U)$. As we may identify $(p^!\mathbb{G})|_{U_Y}$ with $(p|_{U_Y})^!(\mathbb{G}|_U)$ by **5** of Proposition 2.11.6 and $(\underline{R}p_*\mathbb{F})|_U$ by $\underline{R}(p|_{U_Y})_*\mathbb{F}|_{U_Y}$, we may assume that $U = X$, after replacing X by U and p by $p|_{U_Y}$. Then the assertion is clear, because $p^!$ is the restriction of $p^{!,\mathrm{ub}}$, and $p^{!,\mathrm{ub}}$ is the right adjoint of $\underline{R}p_*$. □

(2.11.8) Usually, if the base scheme X and its dualizing complex I_X are obvious from the context, then we call $I_Y := f^!(I_X)$ 'the' dualizing complex of Y for any X-scheme $f : Y \to X$ of finite type. If $X = \operatorname{Spec} R$ is affine and R is a d-dimensional Gorenstein local ring, then we always consider that $I_X = R[d]$.

(2.11.9) Let X be a noetherian scheme with a fundamental dualizing complex I_X, and Y a connected X-scheme of finite type. We set the minimal $i \in \mathbb{Z}$ such that $H^i(I_Y) \neq 0$ to be r. We denote $H^r(I_Y)$ by ω_Y, and call it the *canonical sheaf* of Y. The coherent $\mathcal{O}_Y$-module ω_Y is determined only by Y (not depending on X or I_X) up to the tensor product with an invertible sheaf. If Y is disconnected, then we define ω_Y componentwise.

Let $(R, \mathfrak{m})$ be a complete noetherian local ring. In this case, the fundamental dualizing complex I_R of R is uniquely determined up to degree shifting. Hence, $\bar{\omega}_R$ is uniquely defined to be the non-zero cohomology group of I_R. We call $\bar{\omega}_R$ the canonical module of R. For a not necessarily complete noetherian local ring $(R, \mathfrak{m})$, if there exists a finite R-module K such that $\hat{K} \cong \bar{\omega}_{\hat{R}}$, then such a K is unique up to isomorphisms, where $\hat{R}$ is the $\mathfrak{m}$-adic completion of R, and $\hat{K} = \hat{R} \otimes K$. We usually denote this K by K_R, and call it the *canonical module* of R. If R has a dualizing complex, then we have $\bar{\omega}_R \cong K_R$. However, R may not have a dualizing complex even if there is a canonical module of R.

Lemma 2.11.10 *Let Y be a connected X-scheme of finite type. Then the following are equivalent.*

1 *For some $d \in \mathbb{Z}$, we have $\omega_Y \cong I_Y[-d]$ in $D(Y)$.*

2 *ω_Y has a finite injective dimension as an object of ${}_Y\mathbb{M}$.*

3 *Y is Cohen–Macaulay.*

If the conditions are satisfied, then we have $\operatorname{supp} \omega_Y = Y$, and in particular we have $\omega_{Y,y} \cong \bar{\omega}_{\mathcal{O}_{Y,y}}$ for any $y \in Y$.

Proof. **1⇒2** is trivial.

2⇒3 We set $\operatorname{supp} \omega_Y = Z$. We define r to be the minimum i such that $H^i(I_Y) \neq 0$. Let J be a fundamental dualizing complex which represents I_Y. Note that Z is the union of all irreducible components Y_i of Y such that $J(\eta_i)$ is a direct summand of J^r, where η_i is the generic point of Y_i. For $y \in Z$, $\omega_{Y,y}$ is a non-zero, finitely generated $\mathcal{O}_{Y,y}$-module of finite injective dimension by [69, Proposition II.7.20]. Hence, by Theorem 2.5.6, Y is Cohen–Macaulay at any point of Z. We denote by Z' the union of all irreducible components of Y not contained in Z. If $Z' \neq \emptyset$, then as Y is connected, there is a point $y \in Z \cap Z'$. As J_y is a fundamental dualizing complex of the

Cohen–Macaulay local ring $\mathcal{O}_{Y,y}$, we have that the positions s at which $E_{\mathcal{O}_{Y,\eta_i}}(\kappa(\eta_i))$ appears as a direct summand of J_y^s are equal for all generic points of irreducible components of Y which contain y. This contradicts $y \in Z \cap Z'$, and we have that $Z' = \emptyset$. Hence, we have $\operatorname{supp} \omega_Y = Z = Y$, and Y is Cohen–Macaulay.

We show **3⇒1**. An argument similar to above shows that $\operatorname{supp} \omega_Y = Y$. For each $y \in Y$, there is at most one i such that $H^i_{\mathfrak{m}_y}(\mathcal{O}_{Y,y}) \neq 0$. By the local duality, there is at most one i such that $H^i(I_{Y,y}) \neq 0$. As we have $\omega_{Y,y} \neq 0$, there exists some $d \in \mathbb{Z}$, which is independent of y, such that $H^{-d}(I_{Y,y}) = \omega_{Y,y} \neq 0$. Hence, we have $H^i(I_Y) = 0$ $(i \neq -d)$, and we are done. □

Corollary 2.11.11 *Let $f : Y \to Y'$ be a finite X-morphism between X-schemes of finite type. If Y is connected and Y' is Cohen–Macaulay, then we have*

$$\omega_Y \cong \bar{f}^* \underline{\operatorname{Ext}}^h_{\mathcal{O}_{Y'}}(f_*\mathcal{O}_Y, \omega_{Y'}),$$

where $h = \operatorname{codim}_{Y'} Y$.

Proof. We may assume that Y' is also connected. The morphism of ringed spaces $\bar{f} : (Y, \mathcal{O}_Y) \to (Y', f_*\mathcal{O}_Y)$ is flat. There exists an integer d such that $I_{Y'}[-d]$ is an injective resolution of $\omega_{Y'}$. Hence, we have

$$H^i(I_Y) \cong H^i(\bar{f}^*(\underline{\operatorname{Hom}}_{\mathcal{O}_{Y'}}(f_*\mathcal{O}_Y, I_{Y'}))) \cong \bar{f}^* \underline{\operatorname{Ext}}^{d+i}_{\mathcal{O}_{Y'}}(f_*\mathcal{O}_Y, \omega_{Y'}).$$

This cohomology group is equal to zero if and only if

$$\operatorname{Ext}^{d+i}_{\mathcal{O}_{Y',y'}}((f_*\mathcal{O}_Y)_{y'}, \omega_{Y',y'}) = 0$$

for any $y' \in Y'$. By the local duality theorem, this module is non-zero for $i = -d + \operatorname{codim}_{\mathcal{O}_{Y',y'}}(f_*\mathcal{O}_Y)_{y'}$, and is zero if i is smaller than this value, see [26, Corollary 3.5.11]. Hence, the minimum i such that $H^i(I_Y) \neq 0$ is $h - d$, and we are done. □

Lemma 2.11.12 *Let $f : Y \to Y'$ be a smooth X-morphism between noetherian X-schemes of finite type. Then we have*

$$\omega_Y \cong \omega_{Y'} \otimes \omega_{Y/Y'}.$$

Proof. Obvious. □

Proposition 2.11.13 *Let k be a field, X an x-dimensional Cohen–Macaulay normal k-variety, Y a k-scheme of finite type. Let $f : X \to Y$ be a proper k-morphism, and assume that $\mathcal{O}_Y \to f_*\mathcal{O}_X$ is an isomorphism, $R^i f_*\mathcal{O}_X = 0$ $(i > 0)$, and there exists an $r \geq 0$ such that $R^i f_*\omega_X = 0$ $(i \neq r)$. Then Y is an $(x - r)$-dimensional Cohen–Macaulay normal variety, f is surjective, and $R^r f_*\omega_X \cong \omega_Y$.*

Proof. Note that Y is connected, as X is connected and the support of $f_*\mathcal{O}_X$ is contained in exactly one connected component of Y.

Set $y := \dim Y$. As Y is of finite type over $\operatorname{Spec} k$, it is easy to see that $\omega_Y = H^{-y}(I_Y)$ (we consider that the current base scheme is $\operatorname{Spec} k$). By Corollary 2.11.7 and assumption, we have

$$H^i(I_Y) = \underline{\operatorname{Ext}}^i_{\mathcal{O}_Y}(f_* O_X, I_Y) \cong R^{i+x} f_* \omega_X.$$

By assumption and Lemma 2.11.10, we have Y is Cohen–Macaulay, $-y+x = r$, and $R^r f_* \omega_X \cong \omega_Y$.

It remains to show that Y is normal and f is surjective. As the question is local on Y, we may assume that $Y = \operatorname{Spec} A$, with A of finite type over k. As

$$A = H^0(X, \mathcal{O}_X) = \bigcap_{x \in X} \mathcal{O}_{X,x}$$

is an intersection of normal domains, A is a normal domain. As f is dominating and is a closed map, it is surjective. □

(2.11.14) A desingularization $f : X \to Y$ which satisfies the conditions $R^i f_* \mathcal{O}_X = 0$ $(i > 0)$, $f_* \mathcal{O}_X = \mathcal{O}_Y$, and $R^i f_* \omega_X = 0$ $(i > 0)$ is called a *rational resolution* of Y. By Proposition 2.11.13, if Y has a rational resolution, then it is a Cohen–Macaulay normal variety.

Assume that the characteristic of k is zero, and Y is integral. Then Y has a rational resolution if and only if any desingularization of Y is rational. If the equivalent conditions are satisfied, then we say that Y has (at most) *rational singularities*.

2.12 Summary of open loci results

Let R be a noetherian ring.

Definition 2.12.1 We say that a finite R-module M is of *Gorenstein dimension* 0 if M is reflexive (i.e., the canonical map $M \to M^{**}$ is an isomorphism), and $\operatorname{Ext}^i_R(M, R) = 0 = \operatorname{Ext}^i_R(M^*, R)$ for any $i \geq 1$. We set $\mathcal{G}$ to be the full subcategory of ${}_R\mathbb{M}_f$ consisting of modules of Gorenstein dimension 0. For $N \in {}_R\mathbb{M}_f$, we call $\mathcal{G}$-resol.dim N the *Gorenstein dimension* of N.

Lemma 2.12.2 *Let $R \to S$ be a homomorphism of commutative noetherian rings, N a finite R-module, M a finite S-module.*

1 *For each of the following conditions, the subset of* $\operatorname{Spec} R$ *consisting of* $\mathfrak{p} \in \operatorname{Spec} R$ *such that the condition is satisfied is Zariski open: $M_\mathfrak{p} = 0$, $M_\mathfrak{p}$ is $S_\mathfrak{p}$-free, $M_\mathfrak{p}$ is of Gorenstein dimension* 0 *as an $S_\mathfrak{p}$-module.*

2 *If S is of finite type over R, then the subset of* Spec S *consisting of* $P \in$ Spec S *such that* M_P *is* $R_{R \cap P}$*-flat is open.*

3 *If S is essentially of finite type over a complete local ring or S is essentially of finite type over $\mathbb{Z}$, then for each of the conditions Cohen–Macaulay, Gorenstein, l.c.i., and regular, the subset of* Spec S *consisting of* $P \in$ Spec S *such that the condition is satisfied for* S_P *is open.*

4 *If S has a dualizing complex, then for each of the conditions equidimensional, Cohen–Macaulay, and Gorenstein, the subset of* Spec S *consisting of* $P \in$ Spec S *such that the condition is satisfied for* S_P *is open.*

5 *If S is a homomorphic image of a Cohen–Macaulay ring, then the Cohen–Macaulay locus and the MCM locus of M are open. In particular, the Cohen–Macaulay locus of S is open.*

For the proof, see [110], [61], and [65].

Corollary 2.12.3 *Let R be a noetherian ring, and M a finite R-module. For each i with $1 \leq i \leq \infty$ and each of the following conditions, the subset of* Spec R *consisting of* $P \in$ Spec R *such that the condition is satisfied is open:* $\text{proj.dim}_{R_P} M_P < i$, $\mathcal{G}\text{-resol.dim}\, M_P < i$ *and* M_P *is zero or perfect of codimension i. If R is Cohen–Macaulay, then the subset of P such that* $\dim R_P - \text{depth}_{R_P} M_P < i$ *is also open.*

Corollary 2.12.4 *Let $\varphi : X \to Y$ be a morphism locally of finite type between locally noetherian schemes. Then for each $\mathbb{P}$ of the following properties, the subset $U(\mathbb{P}, \varphi)$ of X consisting of $x \in X$ such that $\mathbb{P}$ is satisfied at x is open: of finite flat dimension, Cohen–Macaulay, Gorenstein, c.i., flat, and smooth. If $f : Y' \to Y$ is a flat morphism of locally noetherian schemes, then we have $U(\mathbb{P}, \varphi') = Y' \times_Y U(\mathbb{P}, \varphi')$, where $\varphi' : X' := Y' \times_Y X \to Y'$ is the base change of φ by f.*

Proof. We prove the first assertion. The flatness assertion follows from Lemma 2.12.2.

As for the rest of the properties, we may assume that both $X = \operatorname{Spec} A$ and $Y = \operatorname{Spec} B$ are affine and connected, as the question is local on both X and Y. Furthermore, as φ is a composite of a closed immersion and an affine n-space, we may and shall assume that φ is a closed immersion by Theorem 2.7.10 (except for smoothness).

As φ is of finite flat dimension if and only if $\text{proj.dim}_A B < \infty$, the locus of finite flat dimension is open by Corollary 2.12.3. So we may assume that φ is a closed immersion of finite flat dimension, to prove the assertion for the Cohen–Macaulay property. Now the assertion follows from the open loci result for perfectness in Corollary 2.12.3. So we may assume that $B = A/I$, with I a perfect ideal of codimension $h < \infty$, say, to prove the Gorenstein and l.c.i. assertions. As the Gorenstein locus of φ is nothing but the rank-one free locus of the finite B-module $\operatorname{Ext}^h_A(B, A)$, it is open in Y. The l.c.i. locus of φ agrees with the free locus of I/I^2, see Lemma 2.13.2 below, hence it is open.

To prove the assertion for a smooth locus, we may assume by the discussion above that φ is a flat l.c.i. morphism of finite type between affine connected noetherian schemes. It is easy to see that such a morphism is equidimensional. Let e be the relative dimension of φ. Then by Proposition 2.7.5 φ is smooth at x if and only if $\Omega_{X,x}$ is free of rank e. Such a locus is clearly open.

The last assertion is obvious by Theorem 2.7.11. □

For more about open loci results, see [16, 61, 127].

2.13 Normal flatness

(2.13.1) Let R be a commutative ring, I an ideal of R, and M an R-module. We set $G := \operatorname{Gr}_I R$ to be the graded ring associated to the ideal I. That is to say, G is the R/I-module $\bigoplus_{i\geq 0} I^i/I^{i+1}$ equipped with the $\mathbb{N}_0$-graded R/I-algebra structure $I^i/I^{i+1} \otimes_{R/I} I^j/I^{j+1} \to I^{i+j}/I^{i+j+1}$ induced by the product of R. Note that $\operatorname{Gr}_I M := \bigoplus_{i\geq 0} I^i M/I^{i+1}M$ is a graded G-module in a natural way. If $\operatorname{Gr}_I M$ is R/I-flat (in other words, if $I^iM/I^{i+1}M$ is R/I-flat for any $i \in \mathbb{N}_0$), then we say that M is *normally flat* along I.

Let X be a scheme, $\mathcal{I}$ a quasi-coherent ideal sheaf of $\mathcal{O}_X$ which defines a closed subscheme Y of X, and $\mathcal{M} \in \operatorname{Qco}(X)$. Then we define $\operatorname{Gr}_{\mathcal{I}} \mathcal{M} := \bigoplus_{i\geq 0} \mathcal{I}^i\mathcal{M}/\mathcal{I}^{i+1}\mathcal{M}$. Note that $\operatorname{Gr}_{\mathcal{I}} \mathcal{O}_X$ is a sheaf of $\mathcal{O}_Y$-algebras, and $\operatorname{Gr}_{\mathcal{I}} \mathcal{M}$ is a $\operatorname{Gr}_{\mathcal{I}} \mathcal{O}_X$-module. If $(\operatorname{Gr}_{\mathcal{I}} \mathcal{M})_x$ is $\mathcal{O}_{X,x}/\mathcal{I}_x$-flat for all $x \in X$, then we say that $\mathcal{M}$ is normally flat along $\mathcal{I}$ (or along Y). We say that X is normally flat along Y if $\mathcal{O}_X$ is.

Lemma 2.13.2 *Let R be a noetherian local ring, and I a proper ideal of R. Then the following are equivalent.*

1 *I is a complete intersection ideal.*

2 $\operatorname{proj.dim}_R I < \infty$, *and I/I^2 is R/I-free.*

2′ $\operatorname{proj.dim}_R I < \infty$, *and R is normally flat along I.*

3 *I/I^2 is R/I-free, and the canonical map $\operatorname{Sym}_{R/I} I/I^2 \to \operatorname{Gr}_I R$ is an isomorphism.*

For the proof, see [110].

Theorem 2.13.3 ([110, Theorem 15.7]) *Let R be a noetherian local ring, and I a proper ideal of R. If we set $G = \operatorname{Gr}_I R$, then we have $\dim G = \dim R$.*

Although we will not use it later, the following is also important.

Theorem 2.13.4 *Let R be a noetherian local ring, and I a proper ideal of R. If G is Cohen–Macaulay (resp. Gorenstein, regular, normal), then so is R.*

For the proof, see [34, Theorem 3.9, Theorem 3.13]. This theorem is a corollary to Corollary II.2.4.3, see [16].

Lemma 2.13.5 *Let $(R, \mathfrak{m})$ be a noetherian local ring, and M a finite R-module. Let P be a prime ideal of R, and assume M is normally flat along P. Then the following hold:*

1 *If $M \neq 0$, then $\dim M = \dim M_P + \dim R/P$.*

2 *If R is normally flat along P, then M is R-free if and only if M_P is R_P-free.*

If, moreover, $\operatorname{Ext}^i_R(R/P, M)$ is R/P-free for all $i \geq 0$, then the following hold:

3 *We have $\operatorname{depth} M = \operatorname{depth} M_P + \operatorname{depth} R/P$. In particular, M is Cohen–Macaulay if and only if both M_P and R/P are Cohen–Macaulay. If this is the case, then we have*

$$\operatorname{type} M = \operatorname{type} M_P \cdot \operatorname{type} R/P,$$

where type denotes the Cohen–Macaulay type, see (2.8.6). In particular, M is Gorenstein if and only if both M_P and R/P are Gorenstein.

4 *Assume that R/P is Gorenstein. Then $\mu_R^{i+\dim R/P}(M) = \mu^i_{R_P}(M_P)$ for $i \geq 0$.*

Proof. **1** As $M/PM \neq 0$ and M/PM is R/P-free, we have $\dim R/P = \dim M/PM$, and hence

$$\dim M \geq \dim M_P + \dim R/P.$$

On the other hand,

$$\dim M = \dim \operatorname{Gr}_P M \leq \dim \kappa(\mathfrak{m}) \otimes \operatorname{Gr}_P M + \dim M/PM.$$

As $\operatorname{Gr}_P M$ is R/P-flat, we have

$$\dim \kappa(\mathfrak{m}) \otimes \operatorname{Gr}_P M = \dim \kappa(P) \otimes \operatorname{Gr}_P M = \dim \operatorname{Gr}_{PR_P} M_P = \dim M_P,$$

since the Hilbert function of $\kappa(\mathfrak{m}) \otimes \operatorname{Gr}_P M$ agrees with that of $\kappa(P) \otimes \operatorname{Gr}_P M$. Hence, $\dim M \leq \dim M_P + \dim R/P$.

We show **2**. The 'only if' part is trivial. We prove the 'if' part. By assumption, the canonical map

$$\gamma_n : (P^n/P^{n+1}) \otimes_{R/P} M/PM \to P^n M/P^{n+1}M$$

is a surjective map of finite free R/P-free modules for any n. The localized map $(\gamma_n)_P$ is an isomorphism by [110, Theorem 22.3], as we assume that M_P is R_P-free. As γ_n is a surjective map between finite free modules of the same rank, it is an isomorphism. Using the local criterion [110, Theorem 22.3] again, M is R-flat, and it is R-free.

We show **3**. We set $\operatorname{depth} M_P = q_0$ and $\operatorname{depth} R/P = p_0$. There is a spectral sequence

$$(2.13.6) \quad E_2^{p,q} = \operatorname{Ext}^p_{R/P}(R/\mathfrak{m}, \operatorname{Ext}^q_R(R/P, M)) \Rightarrow \operatorname{Ext}^{p+q}_R(R/\mathfrak{m}, M).$$

As $\operatorname{Ext}^q_R(R/P, M)$ is R/P-free, $\operatorname{Ext}^q_R(R/P, M) = 0$ if and only if

$$\operatorname{Ext}^q_R(R/P, M)_P = \operatorname{Ext}^q_{R_P}(\kappa(P), M_P) = 0.$$

Hence $E_2^{p,q} = 0$ for $q < q_0$. On the other hand, as $\operatorname{Ext}^q_R(R/P, M)$ is R/P-free, we have $E_2^{p,q} = 0$ for $p < p_0$. Hence $\operatorname{Ext}^i_R(R/\mathfrak{m}, M) = 0$ for $i < p_0 + q_0$. On the other hand

$$\begin{aligned}(2.13.7) \quad \operatorname{Ext}^{p_0+q_0}_R(R/\mathfrak{m}, M) &\cong E_\infty^{p_0,q_0} \cong E_2^{p_0,q_0} \\ &\cong \operatorname{Ext}^{p_0}_{R/P}(R/\mathfrak{m}, \operatorname{Ext}^{q_0}_R(R/P, M)) \neq 0.\end{aligned}$$

Hence

$$\operatorname{depth} M = p_0 + q_0 = \operatorname{depth} M_P + \operatorname{depth} R/P.$$

By **1**, M is Cohen–Macaulay if and only if both M_P and R/P are Cohen–Macaulay.

Assume that M is Cohen–Macaulay (and hence both M_P and R/P are Cohen–Macaulay), and set $\operatorname{type} M_P = r$ and $\operatorname{type} R/P = r'$. Then as we have $\operatorname{Ext}^{q_0}_R(R/P, M) \cong (R/P)^{\oplus r}$,

$$\operatorname{Ext}^{p_0+q_0}_R(R/\mathfrak{m}, M) \cong \operatorname{Ext}^{p_0}_{R/P}(R/\mathfrak{m}, (R/P))^{\oplus r} \cong (R/\mathfrak{m})^{\oplus rr'}$$

by (2.13.7). This shows $\operatorname{type} M = rr'$.

4 As we have $E_2^{p,q} = 0$ for $p \neq p_0 = \dim R/P$ in (2.13.6), the spectral sequence (2.13.6) collapses, and the assertion is clear. □

Remark 2.13.8 In [81] are listed some examples which show that even if R/P is regular, R_P is Cohen–Macaulay and R is normally flat along P, R may not be Cohen–Macaulay. The freeness of $\operatorname{Ext}^i_R(R/P, R)$ is really necessary.

Lemma 2.13.9 *Let $(R, \mathfrak{m})$ be a noetherian local ring, and $P \in \operatorname{Spec} R$. Assume that R is normally flat along P, and R/P is regular. Then we have:*

1 emb.dim $R - \dim R =$ emb.dim $R_P - \dim R_P$. *In particular, R is regular if and only if R_P is regular.*

2 *Assume moreover that $\operatorname{Ext}_R^i(R/P, R/P)$ is R/P-free for $i \geq 2$. Then R is a complete intersection if and only if R_P is a complete intersection.*

Proof. We take $b_1, \ldots, b_d \in \mathfrak{m}$ so that the image of $b_1, \ldots, b_d$ in R/P forms a regular system of parameters of R/P, where $d = \dim R/P$. We set $J := (b_1, \ldots, b_d)$. As the image of $b_1, \ldots, b_d$ in $\mathfrak{m}_{R/P}/\mathfrak{m}_{R/P}^2 = \mathfrak{m}/(\mathfrak{m}^2 + P)$ is linearly independent, its image in $\mathfrak{m}/\mathfrak{m}^2$ is also linearly independent, and we have emb.dim $R/J =$ emb.dim $R - d$. As $b_1, \ldots, b_d$ is an R/P-sequence, the Koszul complex $\operatorname{Kos}(b_1, \ldots, b_d; R/P)$ is acyclic. As P^i/P^{i+1} is R/P-free by assumption, $\operatorname{Kos}(b_1, \ldots, b_d; P^i/P^{i+1})$ is also acyclic. If $j > 0$ and α is a j-cycle of $\mathbb{F} := \operatorname{Kos}(b_1, \ldots, b_d; R)$, then $\alpha \in \bigcap_{i>0}(B_j(\mathbb{F}) + P^i\mathbb{F}_j) = B_j(\mathbb{F})$, and hence $\mathbb{F}$ is also acyclic, where $B_j(\mathbb{F})$ is the set of j-boundaries of $\mathbb{F}$. We have that $b_1, \ldots, b_d$ is an R-sequence, and $\mathbb{F}$ is a free resolution of R/J. In particular, $\dim R/J = \dim R - d$.

Next, take a free resolution $\mathbb{G}$ of R/P. As $b_1, \ldots, b_d$ is R/P-regular, we have $\operatorname{Tor}_i^R(R/P, R/J) = 0$ $(i > 0)$. Hence, $\mathbb{G} \otimes_R R/J$ is an R/J-free resolution of $R/\mathfrak{m} = R/(J + P)$. As we have an isomorphism

$$(\mathbb{G} \otimes_R \mathbb{G}) \otimes_R R/J \cong (\mathbb{G} \otimes_R R/J) \otimes_{R/J} (\mathbb{G} \otimes_R R/J),$$

we obtain a spectral sequence

$$E_{p,q}^2 = \operatorname{Tor}_p^R(\operatorname{Tor}_q^R(R/P, R/P), R/J) \Rightarrow \operatorname{Tor}_{p+q}^{R/J}(R/\mathfrak{m}, R/\mathfrak{m})$$

associated with the double complex $(\mathbb{G} \otimes_R \mathbb{G}) \otimes \mathbb{F}$.

As $\operatorname{Tor}_0^R(R/P, R/P) = R/P$ and $\operatorname{Tor}_1^R(R/P, R/P) \cong P/P^2$ are R/P-free by assumption, we have isomorphisms

$$\begin{aligned} \operatorname{Tor}_i^{R/J}(R/\mathfrak{m}, R/\mathfrak{m}) &\cong \operatorname{Tor}_i^R(R/P, R/P) \otimes_R R/J \\ &\cong \operatorname{Tor}_i^R(R/P, R/P) \otimes_{R/P} R/\mathfrak{m} \end{aligned}$$

for $i = 0, 1, 2$. As $\operatorname{Tor}_i^R(R/P, R/P)$ is R/P-free for $i = 0, 1$,

$$\begin{aligned} \text{emb.dim}\, R/J = \beta_1^{R/J}(R/\mathfrak{m}) &= \dim_{R/\mathfrak{m}} \operatorname{Tor}_1^R(R/P, R/P) \otimes_{R/P} R/\mathfrak{m} \\ &= \dim_{\kappa(P)} \operatorname{Tor}_1^{R_P}(\kappa(P), \kappa(P)) = \text{emb.dim}\, R_P. \end{aligned}$$

Combining this with Lemma 2.13.5, we have

$$\text{emb.dim}\, R - \dim R = \text{emb.dim}\, R/J + d - \dim R = \text{emb.dim}\, R_P - \dim R_P,$$

and we have proved **1**.

We show **2**. As

$$\mathrm{Hom}^\bullet_R(\mathbb{G}, \mathbb{G}) \otimes_R R/J \cong \mathrm{Hom}^\bullet_{R/J}(\mathbb{G} \otimes_R R/J, \mathbb{G} \otimes_R R/J)$$

and proj.dim$_R\, R/J < \infty$, there is a spectral sequence

$$E_2^{p,q} = \mathrm{Tor}^R_{-p}(\mathrm{Ext}^q_R(R/P, R/P), R/J) \Rightarrow \mathrm{Ext}^{p+q}_{R/J}(R/\mathfrak{m}, R/\mathfrak{m}),$$

see Lemma III.2.1.2.

As $\mathrm{Ext}^0_R(R/P, R/P) \cong R/P$ and

$$\mathrm{Ext}^1_R(R/P, R/P) \cong \mathrm{Hom}_{R/P}(P/P^2, R/P),$$

we have that $\mathrm{Ext}^q_R(R/P, R/P)$ is R/P-free for $q \geq 0$ by assumption. Hence $E_2^{p,q} = 0$ $(p \neq 0)$, and we have an isomorphism

$$\mathrm{Ext}^i_R(R/P, R/P) \otimes_{R/P} R/\mathfrak{m} \cong \mathrm{Ext}^i_{R/J}(R/\mathfrak{m}, R/\mathfrak{m}).$$

This shows

$$\beta_i^{R/J}(R/\mathfrak{m}) = \mu^i_{R/J}(R/\mathfrak{m}) = \mu^i_{R_P}(\kappa(P)) = \beta_i^{R_P}(\kappa(P))$$

for $i \geq 0$. By Proposition 2.8.4, R/J is a complete intersection if and only if R_P is. As J is a complete intersection ideal, R/J is a complete intersection if and only if R is. Hence, R is a complete intersection if and only if R_P is. □

Notes and References. There is no new result in this section, except for some of the lemmas followed by proofs. Although some important topics such as Cohen–Macaulay rings and perfect modules are reviewed from the first definitions, this section is merely a glossary on commutative ring theory. For basic notation, terminology and results on commutative ring theory, see [110] and [26]. Undefined terminology on algebraic geometry should be found in [71]. We treat Cohen–Macaulay approximations and related topics in subsection 4.10.

3 Hopf algebras over an arbitrary base

This section is devoted to reviewing Hopf algebras over an arbitrary commutative ring R. All results in this section are basic, and some non-trivial results can be found in [90, 145].

3.1 Coalgebras and bialgebras

(3.1.1) We say that A is an R-algebra, if A is a ring, and a ring homomorphism $u : R \to A$ such that $u(R) \subset Z(A)$ is given, where $Z(A)$ denotes the center of A. This is the same as to say that A is an R-module, R-linear maps $u : R \to A$ and $m : A \otimes A \to A$ are given, and the diagrams

$$\begin{array}{ccc} A\otimes A\otimes A & \xrightarrow{m\otimes 1_A} & A\otimes A \\ \downarrow{\scriptstyle 1_A\otimes m} & & \downarrow{\scriptstyle m} \\ A\otimes A & \xrightarrow{m} & A \end{array} \qquad \begin{array}{ccccc} & & A\otimes A & & \\ & \nearrow{\scriptstyle 1_A\otimes u} & \downarrow{\scriptstyle m} & \nwarrow{\scriptstyle u\otimes 1_A} & \\ A\otimes R & \xrightarrow{\cong} & A & \xleftarrow{\cong} & R\otimes A \end{array}$$

are commutative. In fact, if A is an R-algebra, then A is an R-module in a natural way, and if we define $m : A \otimes A \to A$ by $m(a \otimes a') := aa'$, then it is easy to see that the diagrams above are commutative. Conversely, if u and m are given so that the diagrams are commutative, then A is a ring with the product $aa' := m(a \otimes a')$, and A is an R-algebra by $u : R \to A$. We call $m = m_A : A \otimes A \to A$ the *product map* of A, and $u = u_A : R \to A$ the *unit map* of A.

The notion of R-*coalgebra* is the dual to that of R-algebra. Namely,

Definition 3.1.2 We say that a triple (C, Δ, ε) is an *R-coalgebra* if C is an R-module, and $\Delta : C \to C \otimes C$ and $\varepsilon : C \to R$ are R-linear maps such that the diagrams

$$\begin{array}{ccc} C\otimes C\otimes C & \xleftarrow{\Delta\otimes 1_C} & C\otimes C \\ \uparrow{\scriptstyle 1_C\otimes \Delta} & & \uparrow{\scriptstyle \Delta} \\ C\otimes C & \xleftarrow{\Delta} & C \end{array} \qquad \begin{array}{ccccc} & & C\otimes C & & \\ & \swarrow{\scriptstyle 1_C\otimes \varepsilon} & \uparrow{\scriptstyle \Delta} & \searrow{\scriptstyle \varepsilon\otimes 1_C} & \\ C\otimes R & \xleftarrow{\cong} & C & \xrightarrow{\cong} & R\otimes C \end{array}$$

are commutative. We sometimes say that C is an R-coalgebra if there is no confusion. The commutativity of the first (resp. second) diagram is called the *coassociativity law* (resp. *counit law*). We call $\Delta = \Delta_C$ the *coproduct* of C, and $\varepsilon = \varepsilon_C$ the *counit map* of C.

If, moreover, C is an R-algebra and Δ_C and ε_C are R-algebra maps, then we say that C is an *R-bialgebra*.

An R-coalgebra C is called *cocommutative* if $\tau \circ \Delta_C = \Delta_C$, where $\tau : C \otimes C \to C \otimes C$ is the R-linear map given by $\tau(c \otimes c') = c' \otimes c$ for $c, c' \in C$. Note that cocommutativity is the dual notion of the commutativity of an algebra.

Let A and A' be R-algebras. Then to say that $\varphi : A \to A'$ is an R-algebra map is equivalent to saying that φ is R-linear, $\varphi \circ u_A = u_{A'}$, and $\varphi \circ m_A = m_{A'} \circ (\varphi \otimes \varphi)$. An R-coalgebra map is defined as follows.

Definition 3.1.3 Let C and C' be R-coalgebras. Then we say that $\psi : C' \to C$ is an *R-coalgebra map* if ψ is R-linear, $\varepsilon_C \circ \psi = \varepsilon_{C'}$, and $\Delta_C \circ \psi = (\psi \otimes \psi) \circ \Delta_{C'}$.

Let B and B' be R-bialgebras. Then we say that $f : B \to B'$ is an *R-bialgebra map* if f is both an R-algebra map and an R-coalgebra map.

Example 3.1.4 Let B be an R-bialgebra, and $b \in B$. We say that b is a *group-like element* if $\varepsilon_B(b) = 1$ and $\Delta_B(b) = b \otimes b$ hold. Let G be a semigroup. Then the group algebra RG is an R-algebra. It has a unique R-bialgebra structure such that each $g \in G$ is group-like. Any semigroup homomorphism $G \to G'$ is uniquely extended to an R-bialgebra map $RG \to RG'$.

3.2 Hopf algebras

(3.2.1) We say that G is an *R-semigroup scheme* if G is an R-scheme, endowed with R-morphisms $e : \operatorname{Spec} R \to G$ and $\mu : G \times_R G \to G$ subject to the semigroup-law

$$\mu \circ (\mu \times 1_G) = \mu \circ (1_G \times \mu), \qquad \mu \circ (1_G \times e) = \rho_G, \qquad \mu \circ (e \times 1_G) = \lambda_G,$$

where $\rho_G : G \times \operatorname{Spec} R \cong G$ and $\lambda_G : \operatorname{Spec} R \times G \cong G$ are the canonical identifications. We call μ the product of G, and e the unit map of G. A homomorphism of R-semigroup schemes is an R-morphism which preserves μ and e.

The category of affine R-schemes is contravariantly equivalent to the category of commutative R-algebras. Hence, an *affine* R-semigroup scheme $G = \operatorname{Spec} B$, which is completely described in terms of objects and morphisms of the category of affine R-schemes, can be descried in terms of commutative R-algebras and R-algebra maps between them. In fact, B is a commutative R-bialgebra if and only if $G = \operatorname{Spec} B$ is an affine R-semigroup scheme. A homomorphism of affine R-semigroup schemes corresponds to an R-bialgebra map.

We say that G is an R-group scheme if G is an R-semigroup scheme such that there exists an R-morphism $\iota : G \to G$ such that

$$\mu_G \circ (1_G \times \iota) \circ \Delta_G = \mu_G \circ (\iota \times 1_G) \circ \Delta_G = e \circ u_G$$

holds, where $u_G : G \to \operatorname{Spec} R$ is the structure morphism of G, and $\Delta_G : G \to G \times_R G$ is the diagonalization. For an R-semigroup scheme G, such an ι is unique, if it exists, and it is called the inverse of G.

Considering the case that G is affine, translating the condition for ι into the context of commutative bialgebras, and generalizing it to the non-commutative case, we get the following definition.

Definition 3.2.2 Let B be an R-bialgebra. We say that $S : B \to B$ is an *antipode map* of B if S is R-linear, and the equality

$$m_B \circ (1_B \otimes S) \circ \Delta_B = m_B \circ (S \otimes 1_B) \circ \Delta_B = u_B \circ \varepsilon_B$$

holds. An R-bialgebra is called an *R-Hopf algebra* if it has an antipode map.

Remark 3.2.3 The following is known.

i If an antipode $S = S_B$ of B exists, then it is unique.

ii S is an anti-algebra anti-coalgebra map. Namely, the equalities

$$S(bb') = Sb' \cdot Sb, \qquad S(1) = 1, \qquad \tau \circ (S \otimes S) \circ \Delta = \Delta \circ S, \qquad \varepsilon \circ S = \varepsilon$$

hold, where $\tau : B \otimes B \to B \otimes B$ is the map given by $b \otimes b' \mapsto b' \otimes b$.

iii If B is commutative or cocommutative, then we have $S^2 = \mathrm{id}_B$.

iv If B and B' are R-Hopf algebras and $\varphi : B \to B'$ is an R-bialgebra map, then we have $\varphi \circ S_B = S_{B'} \circ \varphi$.

Exercise 3.2.4 In Example 3.1.4, G is a group if and only if RG is an R-Hopf algebra. If G is a group, then $S(g) = g^{-1}$ gives the antipode. Conversely, if S is an antipode of RG, then we have $g \cdot S(g) = S(g) \cdot g = 1$ by the definition of antipode, and it is easy to see that g is invertible in G. In particular, considering the case that G is the trivial group, R is an R-Hopf algebra.

In general, if B is an R-Hopf algebra and $b \in B$ is group-like, then we have $S(b) = b^{-1}$. Hence, the set of group-like elements $X(B)$ of B is a subgroup of $B^\times$. If G is a group and R has no non-trivial idempotents, then we have $X(RG) = G$, and we can recover G from the Hopf algebra RG.

Let A and A' be commutative R-algebras, C an A-coalgebra, and C' an A'-coalgebra. Then letting

$$C \otimes C' \xrightarrow{\Delta_C \otimes \Delta_{C'}} (C \otimes_A C) \otimes (C' \otimes_{A'} C') \cong (C \otimes C') \otimes_{A \otimes A'} (C \otimes C')$$

be the coproduct and $\varepsilon_C \otimes \varepsilon_{C'}$ be the counit map, $C \otimes C'$ is an $A \otimes A'$-coalgebra. If moreover C and C' are an A-bialgebra and an A'-bialgebra, respectively, then $C \otimes C'$ is an $A \otimes A'$-bialgebra. If moreover C and C' have the antipode S and S', respectively, then $S \otimes S'$ is an antipode map of $C \otimes C'$.

In particular, considering the case $A = A' = R$, a tensor product of R-coalgebras (resp. R-bialgebras, R-Hopf algebras) is again an R-coalgebra (resp. R-bialgebra, R-Hopf algebra). Considering the case $R = A$ and $C' = A'$, the base change $C \otimes A'$ is an A'-coalgebra (resp. A'-bialgebra, A'-Hopf algebra).

Exercise 3.2.5 Let B be an R-algebra R-coalgebra. Then B is an R-bialgebra (i.e., both ε_B and Δ_B are R-algebra maps) if and only if both u_B and m_B are R-coalgebra maps, where the coalgebra structure of the tensor product $B \otimes B$ is given as above.

3.3 Comodules

The notion of *comodule* is dual to that of module.

Definition 3.3.1 Let C be an R-coalgebra. We say that M is a right *C-comodule* if an R-linear map $\omega = \omega_M : M \to M \otimes C$ is given, and

$$\rho_M \circ (1_M \otimes \varepsilon_C) \circ \omega_M = \mathrm{id}_M, \qquad (\omega_M \otimes 1_C) \circ \omega_M = (1_M \otimes \Delta_C) \circ \omega_M$$

hold, where $\rho_M : M \otimes R \cong M$ is the canonical identification. We call ω_M the structure map of M.

Similarly, a left C-comodule is defined. In these notes, a C-comodule means a right C-comodule unless otherwise specified. Note that C itself is a C-comodule with the structure map Δ_C. Note also that Δ_C makes C a left C-comodule.

Definition 3.3.2 Let M and M' be C-comodules. We say that $f : M \to M'$ is a *C-comodule map* if f is R-linear and $\omega_{M'} \circ f = (f \otimes 1_C) \circ \omega_M$ holds.

It is easy to see that taking C-comodules as its objects and C-comodule maps as morphisms, we have an additive category. We denote this category by $\mathbb{M}^C$. Similarly, we have the category of left C-comodules, which we denote by ${}^C\mathbb{M}$.

Lemma 3.3.3 *If C is R-flat (as an R-module), then $\mathbb{M}^C$ is an R-linear abelian category which satisfies the* (AB5) *condition.*

Proof. It is easy to see that the set of C-comodule maps $\mathrm{Hom}_{\mathbb{M}^C}(M, M')$ is an R-submodule of $\mathrm{Hom}_R(M, M')$ for $M, M' \in \mathbb{M}^C$. As $? \otimes C$ preserves inductive limits (in particular, cokernels and direct sums) and kernels (as C is R-flat), inductive limits and kernels as R-modules are endowed with structures of C-comodules, and they are inductive limits and kernels in $\mathbb{M}^C$, respectively, in a natural way. The lemma follows. □

We denote $\mathrm{Hom}_{\mathbb{M}^C}(M, M')$ by $\mathrm{Hom}_C(M, M')$ for C-comodules M and M'.

Let C be an R-flat coalgebra. For $M \in \mathbb{M}^C$ and an R-submodule N of M, we say that N is a *C-subcomodule* of M if $\omega_M(N) \subset N \otimes C$ holds. Note that if N is a C-subcomodule of M, then N itself is a C-comodule with the structure map $\omega_M|_N$, and the inclusion $N \hookrightarrow M$ is a monomorphism of $\mathbb{M}^C$. Subobjects of M and subcomodules of M are in one-to-one correspondence, and we may identify them.

3.4 Sweedler's notation

Let C be an R-coalgebra, and $M \in \mathbb{M}^C$. For $n \geq 0$, we define $\omega_M^{(n)} : M \to M \otimes C^{\otimes n}$ by an inductive definition; $\omega_M^{(0)} := \mathrm{id}_M$ and $\omega_M^{(n)} := (\omega_M \otimes 1_{C^{\otimes(n-1)}}) \circ \omega_M^{(n-1)}$.

For $m \in M$, we express the element $\omega_M^{(n)}(m) \in M \otimes C^{\otimes n}$ as

$$\sum_{(m)} m_{(0)} \otimes m_{(1)} \otimes \cdots \otimes m_{(n)},$$

where we consider that $m_{(0)} \in M$, and $m_{(i)} \in C$ $(i = 1, \ldots, n)$. This notation is somewhat extraordinary, and we could write, for example,

$$\sum_i m_i^{(0)} \otimes m_i^{(1)} \otimes \cdots \otimes m_i^{(n)}.$$

However, the notation does not cause any confusion, and it is useful, because the summation itself tells us what the original element m is. This notation is called *Sweedler's notation*, and is accepted by many authors.

If $M = C$, then we sometimes express $\omega_C^{(n)}$ also as $\Delta_C^{(n+1)}$, and for $c \in C$, we denote $\Delta_C^{(n+1)}(c)$ by

$$\sum_{(c)} c_{(1)} \otimes c_{(2)} \otimes \cdots \otimes c_{(n+1)}.$$

This notation is also called Sweedler's notation.

For example, the counit law of C is equivalent to saying that the equality

$$\sum_{(c)} \varepsilon(c_{(2)})c_{(1)} = \sum_{(c)} \varepsilon(c_{(1)})c_{(2)} = c$$

holds for any $c \in C$. Messy commutative diagrams are sometimes expressed as a family of equalities of elements in a convenient way, using Sweedler's notation.

For an R-coalgebra C, we define $\Delta' : C \to C \otimes C$ by $\Delta'(c) := \sum_{(c)} c_{(2)} \otimes c_{(1)}$ for $c \in C$. It is easy to see that $(C, \Delta', \varepsilon)$ is an R-coalgebra. We call this R-coalgebra the *opposite coalgebra* of C, and denote it by C^{op}. Obviously, we have $(C^{\mathrm{op}})^{\mathrm{op}} = C$ as R-coalgebras. A right (resp. left) C-comodule is a left (resp. right) C^{op}-comodule in a natural way, and we have equivalences ${}^C\mathbb{M} \cong \mathbb{M}^{C^{\mathrm{op}}}$ and $\mathbb{M}^C \cong {}^{C^{\mathrm{op}}}\mathbb{M}$.

3.5 Bicomodules, Hom and ⊗

We say that M is a (C', C)-*bicomodule* if M is both a left C'-comodule and a right C-comodule, and $(1_{C'} \otimes \omega) \circ \omega' = (\omega' \otimes 1_C) \circ \omega$ holds in $\mathrm{Hom}(M, C' \otimes M \otimes C)$, where ω' (resp. ω) denotes the structure map of M as a left C'-comodule (resp. right C-comodule). Note that a (C', C)-bicomodule and a

$C \otimes (C')^{\mathrm{op}}$-comodule are the same thing. A (C', C)-bicomodule map is a map which is both a left comodule map and a right comodule map. It is the same as a $C \otimes (C')^{\mathrm{op}}$-comodule map. The category of (C', C)-bicomodules is denoted by ${}^{C'}\mathbb{M}^{C}$. Note that ${}^{C'}\mathbb{M}^{C}$ is equivalent to $\mathbb{M}^{C\otimes(C')^{\mathrm{op}}}$.

Let A and A' be commutative R-algebras, C an A-coalgebra, C' an A'-coalgebra, M a C-comodule, and N a C'-comodule. Then $M \otimes N$ is a $C \otimes C'$-comodule. In particular, the tensor product of a left C'-comodule and a right C-comodule is a (C', C)-bicomodule. If we consider the case $C' = A'$, V is an A'-module, and M is a C-comodule, then $M \otimes V$ is a $C \otimes A'$-comodule. In particular, if M is a C-comodule and V is an R-module, then $M \otimes V$ is a C-comodule.

For a C-comodule N which is also an A'-module, N is also a $C \otimes A'$-comodule, via the canonical isomorphism

$$N \otimes C \cong N \otimes_{A'} (C \otimes A'),$$

provided the R-module structures of N coming from the C-comodule structure of N and A'-module structure of N coincide (such a condition is not mentioned in the sequel, if there is no danger of misunderstandings).

Lemma 3.5.1 *Let $R \to R'$ be a homomorphism of commutative rings, C an R-coalgebra, M a C-comodule, and N a $C \otimes R'$-comodule. Then the canonical isomorphism*

$$\mathrm{Hom}(M, N) \cong \mathrm{Hom}_{R'}(M \otimes R', N)$$

induces an isomorphism

$$\mathrm{Hom}_C(M, N) \cong \mathrm{Hom}_{C'}(M \otimes R', N).$$

Let M and H be R-modules. Then the canonical map $\mathrm{Hom}(M, ?) \otimes H \cong \mathrm{Hom}(M, ? \otimes H)$ is an isomorphism in the following cases, see Lemma 2.1.7.

(3.5.2) H is R-flat, and M is a finitely presented R-module.

(3.5.3) M is R-projective, and H is finitely presented.

(3.5.4) H is R-finite projective.

Let C and C' be R-flat coalgebras, and V a C-comodule, and assume either that V is a finitely presented R-module, or that both C and C' are R-finite projective. Then for a C'-comodule V', we define

$${}_C\omega : \mathrm{Hom}(V, V') \to \mathrm{Hom}(V, C \otimes V') \cong C \otimes \mathrm{Hom}(V, V')$$

by ${}_C\omega(f)(v) := \sum_{(v)} v_1 \otimes f(v_0)$. We also define

$$\omega_{C'} : \mathrm{Hom}(V, V') \to \mathrm{Hom}(V, V' \otimes C') \cong \mathrm{Hom}(V, V') \otimes C'$$

by $\omega_{C'}(f) := \omega_{V'} \circ f$.

Lemma 3.5.5 *Let C, C', V and V' be as above. Then* $\mathrm{Hom}(V, V')$ *is a (C, C')-bicomodule with the structure maps ${}_C\omega$ and $\omega_{C'}$. If, moreover, C'' is an R-coalgebra, V is a (C'', C)-bicomodule, and V' is a (C'', C')-bicomodule, then* $\mathrm{Hom}_{C''}(V, V')$ *is a (C, C')-subbicomodule of* $\mathrm{Hom}(V, V')$. *The canonical map*

$$\mathrm{Hom}(V, R) \otimes V' \to \mathrm{Hom}(V, V')$$

is a (C, C')-bicomodule map.

Proof. We only prove the coassociativity with respect to ${}_C\omega$. By definition, if we set ${}_C\omega f := \sum_{(f)} f_1 \otimes f_0$, then the equality

$$\sum_{(f)} f_1 \otimes f_0 v = \sum_{(v)} v_1 \otimes f v_0$$

holds. As the isomorphism in Lemma 2.1.7 is natural on H, it suffices to prove

$$\sum_{(f)} \sum_{(f_0)} f_1 \otimes f_{01} \otimes f_{00} v = \sum_{(f)} \sum_{(f_1)} f_{11} \otimes f_{12} \otimes f_0 v.$$

Since we have

$$\sum_{(f)} \sum_{(f_0)} f_1 \otimes f_{01} \otimes f_{00} v = \sum_{(f)} \sum_{(v)} f_1 \otimes v_1 \otimes f_0 v_0$$
$$= \sum_{(v)} v_1 \otimes v_2 \otimes f v_0 = (\Delta \otimes 1) \sum_{(v)} v_1 \otimes f v_0 = (\Delta \otimes 1) \sum_{(f)} f_1 \otimes f_0 v,$$

the assertion follows. □

Lemma 3.5.6 *Let C and C' be R-coalgebras, M a (C', C)-bicomodule, and V a left C'-comodule. Then the map*

$$\Phi : \mathrm{Hom}_{C'}(M, V) \to \mathrm{Hom}_{(C', C)}(M, V \otimes C)$$

defined by $\Phi(f)(m) := \sum_{(m)} f m_{(0)} \otimes m_{(1)}$ is well-defined, and an isomorphism which is natural with respect to M and V, where $\sum_{(m)} m_0 \otimes m_1$ is the image of m by the structure map of M as a C-comodule.

Proof. It is easy to verify that $\Phi(f)$ is a (C', C)-bicomodule map, and hence Φ is well-defined. We define $\Psi : \mathrm{Hom}_{(C', C)}(M, V \otimes C) \to \mathrm{Hom}_{C'}(M, V)$ by $\Psi(g)(m) := Eg(m)$, where $E : V \otimes C \to V$ is defined by $E(v \otimes c) := \varepsilon(c)v$. It is also straightforward to check that $\Psi(g)$ is a C'-comodule map. When we set $gm = \sum_i v_i \otimes c_i$, then we have

$$\Phi\Psi(g)(m) = \sum_{(m)} E(gm_0) \otimes m_1 = E((gm)_0) \otimes (gm)_1$$
$$= \sum_i \sum_{(c_i)} E(v_i \otimes (c_i)_1) \otimes (c_i)_2 = \sum_i v_i \otimes c_i = gm.$$

This shows $\Phi\Psi = \mathrm{id}$. On the other hand, we have

$$\Psi\Phi(f)(m) = E(\Phi(f)(m)) = \varepsilon(m_1)f(m_0) = fm,$$

and hence $\Psi\Phi = \mathrm{id}$. Thus, Φ is an isomorphism. □

Let C be an R-coalgebra. Then the functor $F : \mathbb{M}^C \to {}_R\mathbb{M}$ has the right adjoint $G = ? \otimes C$ by Lemma 3.5.6. For $M \in \mathbb{M}^C$, we call the cobar resolution $\mathrm{Cobar}_F(M)$ (1.6) the cobar resolution of M, and denote it by $\mathrm{Cobar}_C(M)$. More explicitly, we have $\mathrm{Cobar}_C(M)^i = M \otimes C^{\otimes(i+1)}$, and the boundary map is given by

$$\partial^n = (-1)^{n+1}\omega_M \otimes 1_{C^{\otimes(n+1)}} + \sum_{i=0}^{n}(-1)^{n-i}1_M \otimes 1_{C^{\otimes i}} \otimes \Delta_C \otimes 1_{C^{\otimes(n-i)}}.$$

By definition, we have:

Lemma 3.5.7 *Let V be an R-module, and M a C-comodule. Then we have $V \otimes \mathrm{Cobar}_C(M) \cong \mathrm{Cobar}_C(V \otimes M)$ as complexes of C-comodules. Moreover, we have*

$$\mathrm{Cobar}_C(M)' \cong \mathrm{Cobar}_{C'}(M')$$

for any homomorphism of commutative rings $R \to R'$, where $(?)'$ denotes the functor $? \otimes R'$.

We also have the following, by the definition of the C-comodule structure of $\mathrm{Hom}(V, M)$.

Lemma 3.5.8 *Let C be an R-flat R-coalgebra, V a finitely presented R-module, and M a C-comodule. Then we have a canonical isomorphism*

$$\mathrm{Hom}(V, \mathrm{Cobar}_C(M)) \cong \mathrm{Cobar}_C(\mathrm{Hom}(V, M)).$$

Lemma 3.5.9 *Let C be an R-flat R-coalgebra. Then the set of C-comodules of the form $J \otimes C$ with J an injective R-module, is an injective cogenerator of $\mathbb{M}^C$. In particular, $\mathbb{M}^C$ has enough injectives. If, moreover, R is noetherian, then any C-injective comodule is R-injective.*

Proof. As $G = ? \otimes C$ has an exact left adjoint F, it preserves injectives. Hence, $G(J) = J \otimes C$ is injective as a C-comodule for any injective R-module J.

Let $M \in \mathbb{M}^C$, and take an injective hull of M as an R-module $i : M \hookrightarrow J$. Then $\Psi i : M \to J \otimes C$ is an injective C-comodule map, and the first assertion follows, where Ψ is the map in the proof of Lemma 3.5.6. The last assertion is obvious by Corollary 2.1.8. □

Corollary 3.5.10 *Let C be an R-flat R-coalgebra, V an R-finitely presented C-comodule, M a C-comodule, and R' an R-flat commutative R-algebra. Then the canonical map*

$$\mathrm{Hom}_C(V, M) \otimes R' \to \mathrm{Hom}_{C\otimes R'}(V \otimes R', M \otimes R')$$

is an isomorphism.

Proof. As both $\mathrm{Hom}_C(V, ?) \otimes R'$ and $\mathrm{Hom}_{C\otimes R'}(V \otimes R', ? \otimes R')$ are left exact, we may assume that M is of the form $J \otimes C$ with J an injective R-module by Lemma 3.5.9 and the five lemma. In this case, as the canonical map

$$\mathrm{Hom}(V, M) \otimes R' \to \mathrm{Hom}_{R'}(V \otimes R', M \otimes R')$$

is an isomorphism, the assertion follows from Lemma 3.5.6. □

Lemma 3.5.11 *Let C be an R-flat coalgebra. Then for any C-comodule M and any R-module V, we have*

$$\mathrm{Ext}^i_C(M, V \otimes C) \cong \mathrm{Ext}^i_R(M, V).$$

Proof. Obvious by Lemma 1.6.12. □

Lemma 3.5.12 *Let C be an R-flat R-coalgebra, V a finitely presented R-module, and M and N be C-comodules. Then the canonical map*

$$\mathrm{Hom}(V, N) \otimes V \to N \qquad (f \otimes v \mapsto fv)$$

is a C-comodule map. The isomorphism

$$\mathrm{Hom}(M \otimes V, N) \cong \mathrm{Hom}(M, \mathrm{Hom}(V, N))$$

induces an isomorphism

$$\mathrm{Hom}_C(M \otimes V, N) \cong \mathrm{Hom}_C(M, \mathrm{Hom}(V, N)).$$

Proof. Easy. □

If R is noetherian, then for an R-flat R-coalgebra C, any C-injective comodule is R-injective (Lemma 3.5.9). Hence, for any finite R-module V, $\mathrm{Ext}^i_R(V, ?)$ is the derived functor of $\mathrm{Hom}(V, ?)$ in the category $\mathbb{M}^C$. Hence, $\mathrm{Ext}^i_R(V, M)$ has a canonical C-comodule structure for a C-comodule M.

Proposition 3.5.13 *Let R be noetherian, C an R-flat R-coalgebra, V an R-finitely presented module, and M and N be C-comodules. If M or V is R-flat, then there is a spectral sequence*

$$E_2^{p,q} = \mathrm{Ext}^p_C(M, \mathrm{Ext}^q_R(V, N)) \Rightarrow \mathrm{Ext}^{p+q}_C(M \otimes V, N).$$

Proof. By Lemma 3.5.12, we have an isomorphism of left exact functors $\operatorname{Hom}_C(M \otimes V, ?) \cong \operatorname{Hom}_C(M, ?) \circ \operatorname{Hom}(V, ?)$. For an injective R-module J and $i > 0$, we have

$$\begin{aligned}\operatorname{Ext}^i_C(M, \operatorname{Hom}(V, J \otimes C)) &\cong \operatorname{Ext}^i_C(M, \operatorname{Hom}(V, J) \otimes C)\\ &\cong \operatorname{Ext}^i_R(M, \operatorname{Hom}(V, J)) = 0.\end{aligned}$$

Hence, $\operatorname{Hom}(V, I)$ is $\operatorname{Hom}_C(M, ?)$-acyclic for an injective C-comodule I, and the assertion follows from Grothendieck's spectral sequence theorem. □

3.6 The restriction and the induction

(3.6.1) Let B and C be R-coalgebras, and $\varphi : B \to C$ a coalgebra map. For $M \in \mathbb{M}^B$, letting

$$M \xrightarrow{\omega_M} M \otimes B \xrightarrow{1 \otimes \varphi} M \otimes C$$

be the structure map, we have $M \in \mathbb{M}^C$. It is also easy to see that this defines a functor $\operatorname{res}^B_C : \mathbb{M}^B \to \mathbb{M}^C$. Note that $\operatorname{res}^B_C M$ is nothing but M itself, as an R-module.

For a left C-comodule L and a right C-comodule N, we define the *cotensor product* of N and L, denoted by $N \boxtimes^C L$, by the exact sequence

$$0 \to N \boxtimes^C L \to N \otimes L \xrightarrow{\rho_{N,L}} N \otimes C \otimes L,$$

where $\rho_{N,L} := \omega_N \otimes 1_L - 1_N \otimes \omega_L$. If C' and C'' are R-flat R-coalgebras, N is a (C', C)-bicomodule, and L is a (C, C'')-bicomodule, then $N \otimes L$ and $N \otimes C \otimes L$ are (C', C'')-bicomodules in a natural way, and $\rho_{N,L}$ is a (C', C'')-bicomodule map. Hence, $N \boxtimes^C L$ has a (C', C'')-bicomodule structure. Note that the definition of cotensor product is dual to that of tensor product.

Almost by definition, the next lemma holds.

Lemma 3.6.2 *Let C be an R-coalgebra, L a left C-comodule, and N a right C-comodule. Then we have the following.*

1 *$? \boxtimes^C L$ and $N \boxtimes^C ?$ preserve any* filtered *inductive limits.*

2 *Let $F : \mathbb{M}^C \to {}_R\mathbb{M}$ be the forgetful functor. Then $? \boxtimes^C L$ is F-left exact.*

3 *If C and L are R-flat, then $? \boxtimes^C L$ is a left exact functor from $\mathbb{M}^C$ to ${}_R\mathbb{M}$.*

Lemma 3.6.3 *Let V be an R-module, N a right C-comodule, and L a left C-comodule. Then there are isomorphisms of R-modules $N \boxtimes^C (C \otimes V) \cong N \otimes V$ and $(V \otimes C) \boxtimes^C L \cong V \otimes L$, which are natural with respect to V, N and L. If C is R-flat, then these isomorphisms are isomorphisms of C-comodules.*

Proof. The cobar resolution

$$\mathrm{Cobar}_C(N) : 0 \to N \xrightarrow{\omega} N \otimes C \xrightarrow{d^0} N \otimes C \otimes C \to \cdots$$

of N is a split exact sequence of R-modules (if, moreover, C is R-flat, then it is an exact sequence of C-comodules which R-splits). As we have $d^0 \otimes 1_V = -\rho_{N,C\otimes V}$ and $\mathrm{Cobar}_C(N) \otimes V$ is still exact, the first isomorphism follows. The second isomorphism is obtained in a similar way. □

(3.6.4) Let $\varphi : B \to C$ be as in (3.6.1). Assume that B is R-flat. As B is a (C, B)-bicomodule, $\mathrm{ind}_C^B(N) := N \boxtimes^C B$ is a B-comodule, and we have a functor $\mathrm{ind}_C^B : \mathbb{M}^C \to \mathbb{M}^B$. Note that ind_C^B is additive, and preserves filtered inductive limits.

Lemma 3.6.5 *Let $f : B \to C$ be an R-coalgebra map. If B and C are R-flat, then ind_C^B is right adjoint to res_C^B.*

Proof. First, note that ind_C^B is left exact by Lemma 3.6.2, **3**.

It is easy to see that the composite map

$$\mathrm{Hom}_C(\mathrm{res}_C^B(M), V) \subset \mathrm{Hom}(M, V)$$
$$\cong \mathrm{Hom}_B(M, V \otimes B) \xrightarrow{\rho_{V,B}} \mathrm{Hom}_B(M, V \otimes C \otimes B)$$

is 0 for $M \in \mathbb{M}^B$ and $V \in \mathbb{M}^C$. As the sequence

$$0 \to \mathrm{Hom}_B(M, \mathrm{ind}_C^B(V)) \xrightarrow{u} \mathrm{Hom}_B(M, V \otimes B) \xrightarrow{\rho_{V,B}} \mathrm{Hom}_B(M, V \otimes C \otimes B)$$

is exact, we get a map

$$q : \mathrm{Hom}_C(\mathrm{res}_C^B(M), V) \to \mathrm{Hom}_B(M, \mathrm{ind}_C^B(V)),$$

which is natural with respect to M and V.

To prove that q is an isomorphism, we may assume that V is of the form $V = V_0 \otimes C$ with V_0 an R-module by the five lemma, as the sequence

$$0 \to V \to V \otimes C \to V \otimes C \otimes C$$

is exact. In this case, we may identify $\mathrm{ind}_C^B(V) = V_0 \otimes B$, while $u : \mathrm{ind}_C^B(V) \to V \otimes B$ is identified with

$$V_0 \otimes B \xrightarrow{1\otimes((f\otimes 1)\circ\Delta_B)} V_0 \otimes C \otimes B.$$

Hence, as we have $\sum_{(b)} \varepsilon_C(fb_1)b_2 = b$ for $b \in B$, q is given by

$$q(\varphi)(m) = \sum_{(m)} (1_{V_0} \otimes \varepsilon_C)(\varphi m_0) \otimes m_1,$$

and hence q is identified with the composite map

$$\mathrm{Hom}_C(\mathrm{res}_C^B(M), V_0 \otimes C) \cong \mathrm{Hom}(M, V_0) \cong \mathrm{Hom}_B(M, V_0 \otimes B)$$

(see the proof of Lemma 3.5.6), and is an isomorphism. □

Example 3.6.6 Let B be an R-flat R-coalgebra. Set $C := R$, and consider the R-coalgebra map $\varepsilon : B \to C = R$. Then we have $\operatorname{ind}_C^B V = V \otimes B$ for an R-module V, where $V \otimes B$ is endowed with a B-comodule structure by $\omega(v \otimes b) = v \otimes \Delta_B(b)$. The counit of adjunction $M \to M \otimes B$ is nothing but ω_M, and hence ω_M is a B-comodule map.

Definition 3.6.7 Let C be an R-coalgebra. We say that B is an *R-subcoalgebra* of C if B is a pure R-submodule of C, and $\Delta_C(B) \subset B \otimes B$ holds. A subalgebra subcoalgebra of a bialgebra is called a *subbialgebra*.

(3.6.8) Assume that C is R-flat. Note that $B \subset C$ is an R-subcoalgebra of C if and only if it is a (C, C)-subbicomodule of C, which is also a pure R-submodule. If this is the case, B is also R-flat. If B is an R-subcoalgebra of C, then $\operatorname{ind}_C^B(N)$ is identified with the C-subcomodule

$$\{n \in N \mid \omega_N(n) \in N \otimes B(\subset N \otimes C)\}$$

of N for $N \in \mathbb{M}^C$, via the injective map $\operatorname{ind}_C^B(N) = N \boxtimes^C B \subset N \boxtimes^C C \cong N$.

Remark 3.6.9 In particular, if C is an R-flat R-coalgebra and $B \subset C$ is an R-subcoalgebra of C, then the counit of adjunction $(\operatorname{ind}_C^B \circ \operatorname{res}_C^B)(M) \to M$ is an isomorphism. Hence, res_C^B is fully faithful (and is obviously exact, as $\operatorname{res}_C^B(M) = M$). Thus, $\mathbb{M}^B$ is identified with the full subcategory of $\mathbb{M}^C$ consisting of C-comodules N such that $\omega_N(N) \subset N \otimes B$. Note that $\mathbb{M}^B$ is closed under subobjects, quotients, and direct sums.

The converse holds in the following weak form.

Lemma 3.6.10 *Let k be a field, C a k-coalgebra, and $\mathcal{B}$ a full subcategory of $\mathbb{M}^C$ closed under subobjects, quotients, and direct sums. We denote the inclusion $\mathcal{B} \hookrightarrow \mathbb{M}^C$ by i, and denote its right adjoint (it does exist, see* Lemma 1.10.2) *by j. Then there is a unique k-subcoalgebra B of C such that any B-comodule is in $\mathcal{B}$, and $\operatorname{res}_C^B : \mathbb{M}^B \to \mathcal{B}$ is an equivalence. In fact, B is given as $j(C)$.*

Proof. Let V be a k-vector space, and $M \in \mathbb{M}^C$. Then the canonical map $V \otimes jM \to j(V \otimes M)$ is an isomorphism. This is trivial when $V = k$, and hence also for the case $\dim V < \infty$, as j is additive. As j is also compatible with filtered inductive limits by Lemma 1.10.3, the general case follows.

Now let $N \in \mathcal{B}$. Then the coaction $\omega_N : N \to N \otimes C$ is a C-comodule map, and hence it factors through $j(N \otimes C) \cong N \otimes j(C)$. Considering the case $N = j(C)$, we have that $j(C)$ is a k-subcoalgebra of C. This also shows that any object of $\mathcal{B}$ is a $j(C)$-comodule. Conversely, as $j(C) \in \mathcal{B}$, any $j(C)$ comodule N is in $\mathcal{B}$, as N is a subobject of $N \otimes j(C)$. Hence, the existence of B is proved.

We show the uniqueness. As $B \in \mathcal{B}$, we have $B \subset j(C)$ by the definition of j. We show $j(C) \subset B$. As $\mathbb{M}^B$ is closed under isomorphisms in $\mathbb{M}^C$, we have that any object of $\mathcal{B}$ is a B-comodule. In particular, $j(C)$ is a B-comodule. For $c \in j(C)$, we have $c = \sum_{(c)} \varepsilon_C(c_{(1)})c_{(2)} \in B$, and the uniqueness is proved. □

Lemma 3.6.11 *Let C be an R-coalgebra, C' and C'' be R-flat coalgebras, P an R-finite R-projective (C, C')-bicomodule, N a (C'', C)-bicomodule. Then there is a (C'', C')-bicomodule isomorphism*

$$N \boxtimes^C P \cong \operatorname{Hom}_C(P^*, N)$$

which is natural with respect to P and N.

Proof. It is easy to see that the kernel of the map $N \otimes P \to N \otimes C \otimes P$ agrees with the image of $\operatorname{Hom}_C(P^*, N) \hookrightarrow \operatorname{Hom}(P^*, N) \cong N \otimes P^{**} = N \otimes P$. □

Corollary 3.6.12 *Let $B \to C$ be an R-coalgebra map. If B is R-finite projective, then there is an isomorphism of functors $\operatorname{ind}_C^B \cong \operatorname{Hom}_C(B^*, ?)$.*

(3.6.13) From now on, until the end of this subsection, we assume that C is an *R-flat* R-coalgebra. Note that $\mathbb{M}^C$ is abelian, and the forgetful functor $F : \mathbb{M}^C \to {}_R\mathbb{M}$ is faithful exact, and its right adjoint $G = ? \otimes C$ is also exact.

Definition 3.6.14 For $M \in \mathbb{M}^C$ and a left C-comodule N, we denote $R^i_F(? \boxtimes^C N)(M)$ by $\operatorname{Cotor}^i_C(M, N)$, and call it the ith *cotorsion module* of M and N.

By definition, we have

$$\operatorname{Cotor}^i_C(M, N) = H^i(\operatorname{Cobar}_C(M) \boxtimes^C N).$$

As $\operatorname{Cobar}_C(M) \boxtimes^C N$ is the complex of the form

$$0 \to M \otimes N \to M \otimes C \otimes N \to M \otimes C^{\otimes 2} \otimes N \to \cdots,$$

it is symmetric with respect to M and N, up to sign change. We denote it by $\operatorname{Cobar}_C(M, N)$. We have $\operatorname{Cobar}_C(M, N) \cong \operatorname{Cobar}_{C^{\mathrm{op}}}(N, M)$. Hence, we have

$$\operatorname{Cotor}^i_C(M, N) \cong H^i(\operatorname{Cobar}_C(M, N)) \cong H^i(\operatorname{Cobar}_{C^{\mathrm{op}}}(N, M)) \cong \operatorname{Cotor}^i_{C^{\mathrm{op}}}(N, M).$$

Lemma 3.6.15 *For a homomorphism of commutative rings $R \to R'$, we have*

$$\mathrm{Cobar}_C(M,N)' \cong \mathrm{Cobar}_{C'}(M',N')$$

in a natural way, where $(?)' = ? \otimes R'$.

Proof. Obvious.

By Lemma 3.6.11 and Lemma 1.6.12, we have the following.

Lemma 3.6.16 *Let M be an R-finite R-projective C-comodule, and N a C-comodule. Then we have natural isomorphisms*

$$\begin{aligned}\mathrm{Ext}^i_C(M,N) &\cong H^i(\mathrm{Hom}_C(M,\mathrm{Cobar}_C(N))) \\ &\cong H^i(\mathrm{Cobar}_C(N,M^*)) \cong \mathrm{Cotor}^i_C(N,M^*).\end{aligned}$$

The isomorphisms are natural with respect to M and N.

Lemma 3.6.17 *Let N be a left C-comodule, and (M_λ) a filtered inductive system of C-comodules. Then*

$$\varinjlim \mathrm{Cotor}^i_C(M_\lambda,N) \cong \mathrm{Cotor}^i_C(\varinjlim M_\lambda,N)$$

for $i \geq 0$.

Proof. Easy. □

Lemma 3.6.18 *Let I be an F-injective C-comodule, and N an R-flat left C-comodule. Then I is $? \boxtimes^C N$-acyclic as an object of $\mathbb{M}^C$. In particular, if N is R-flat, then we have*

$$\mathrm{Cotor}^i_C(M,N) \cong R^i(? \boxtimes^C N)(M)$$

for any C-comodule M.

Proof. We may assume that I is F-cofree, that is to say, $I = V \otimes C$ for some R-module V. We take an R-injective resolution $J^\bullet$ of V. Then we have

$$R^i(? \boxtimes^C N)(V) = H^i((J^\bullet \otimes C) \boxtimes^C N) \cong H^i(J^\bullet \otimes N).$$

As N is R-flat, we have that $J^\bullet \otimes N$ is quasi-isomorphic to $V \otimes N$, and the assertion follows. □

Corollary 3.6.19 *Let B be an R-flat coalgebra, and $B \to C$ an R-coalgebra map. Then we have*

$$R^i \mathrm{ind}^B_C \cong H^i(\mathrm{Cobar}_C(?,B)) = \mathrm{Cotor}^i_C(?,B).$$

We have the following.

Proposition 3.6.20 *Let $R \to R'$ be a flat morphism of commutative rings. We denote the functor $? \otimes R'$ by $(?)'$. For $i \geq 0$, we have*

1 *For $M \in \mathbb{M}^C$ and $N \in {}^C\mathbb{M}$, we have*

$$\mathrm{Cotor}^i_C(M, N)' \cong \mathrm{Cotor}^i_{C'}(M', N').$$

2 *Let M be a C-comodule, and $B \to C$ an R-coalgebra map with B R-flat. Then we have*

$$R^i \operatorname{ind}^B_C(M)' \cong R^i \operatorname{ind}^{B'}_{C'}(M').$$

3 *Assume that R is noetherian. Then for any R-finite C-comodule V and any C-comodule M, we have*

$$\mathrm{Ext}^i_C(V, M) \otimes R' \cong \mathrm{Ext}^i_{C'}(V', M').$$

4 *Assume that R is noetherian. For any R-finite C-comodule V and any C'-comodule N, we have*

$$\mathrm{Ext}^i_C(V, N) \cong \mathrm{Ext}^i_{C'}(V', N).$$

The proof is straightforward. We give a remark on **4**. If $I^\bullet$ is a C-injective resolution of M, then $I^\bullet \otimes R'$ is a $\mathrm{Hom}_{C'}(V', ?)$-acyclic resolution of M', but it is not a C'-injective resolution in general.

3.7 Locally noetherian property

Throughout this subsection, we assume that C is an R-flat R-coalgebra.

Lemma 3.7.1 *Let C be an R-flat R-coalgebra. Let $M \in \mathbb{M}^C$, and X be an R-finite submodule of M. Then there exist some C-subcomodule L of M and an R-finite R-submodule N of M such that $X \subset L \subset N \subset M$.*

Proof. Take a generator $x_1, \ldots, x_n$ of X as an R-module. We express $\omega(x_i) = \sum_j m_{ij} \otimes c_{ij}$, and define N to be the R-finite R-submodule of M generated by all m_{ij}'s. The inverse image of $N \otimes C$ by the C-comodule map

$$\omega_M : M \to M \otimes C = \operatorname{ind}^C_R \operatorname{res}^C_R(M),$$

say $L := \omega_M^{-1}(N \otimes C)$, is a C-subcomodule of M, and it contains X. On the other hand, as we have

$$L = \mathrm{id}_M(L) = (\mathrm{id}_M \otimes \varepsilon_C)\omega_M(L) \subset (\mathrm{id}_M \otimes \varepsilon_C)(N \otimes C) \subset N,$$

it follows that $L \subset N$. □

Corollary 3.7.2 *In the lemma, if moreover R is noetherian, then there exists an R-finite C-subcomodule of M which contains X. In particular, M is the filtered inductive limit of its R-finite C-subcomodules.*

We have already seen that $\mathbb{M}^C$ has enough injectives. By the following corollary, we see that $\mathbb{M}^C$ has injective envelopes by Theorem 1.7.6.

Corollary 3.7.3 *Let C be an R-flat R-coalgebra. Then the abelian category $\mathbb{M}^C$ is Grothendieck. If moreover R is noetherian, then $\mathbb{M}^C$ is locally noetherian, and $\mathbb{M}^C_f$ consists of all R-finite C-comodules.*

Proof. We already know by Lemma 3.3.3 that $\mathbb{M}^C$ is an abelian category which satisfies (AB5). We show that $\mathbb{M}^C$ has a small set of G-generators. The isomorphism classes of C-comodules which are R-submodules of R-finite modules are parameterized by a small set I. Taking representatives from these isomorphism classes, and collecting them, we have a small set of G-generators $(U_i)_{i\in I}$ of $\mathbb{M}^C$ by the lemma. This shows $\mathbb{M}^C$ is Grothendieck.

Now we assume that R is noetherian. Obviously, any R-finite C-comodule is noetherian. Hence, $\mathbb{M}^C$ is locally noetherian, as each U_i in the first paragraph is R-finite. As any object in $\mathbb{M}^C_f$ is a homomorphic image of a finite direct sum of copies of U_i's, it is R-finite. □

Proposition 3.7.4 *Let R be noetherian, C a flat R-coalgebra, and V an R-finite C-comodule. Then the canonical map*

$$\varinjlim \operatorname{Ext}^i_C(V, M_\lambda) \to \operatorname{Ext}^i_C(V, \varinjlim M_\lambda)$$

is an isomorphism for any filtered inductive system (M_λ) of C-comodules and any $i \geq 0$.

Proof. Follows immediately from Corollary 3.7.3 and Lemma 1.9.5. □

3.8 The dual algebra of a coalgebra

(3.8.1) For R-modules V and W, we define $\rho : V^* \otimes W^* \to (V \otimes W)^*$ by $\rho(\varphi \otimes \psi)(v \otimes w) = \varphi(v) \cdot \psi(w)$. If V and W are R-projective (e.g., R is a field), then ρ is injective, and we may assume $V^* \otimes W^* \subset (V \otimes W)^*$. In case either V or W is finite projective, then the inclusion becomes an identification, as ρ is an isomorphism in this case.

Let C be an R-coalgebra. Then C^* is an R-algebra with the product $m_{C^*} := \Delta_C^* : C^* \otimes C^* \to (C \otimes C)^* \to C^*$. The unit map is ε_C^*. Using Sweedler's notation, we can write

$$(b^*c^*)(c) = \sum_{(c)} b^*c_{(1)} \cdot c^*c_{(2)} \qquad (b^*, c^* \in C^*,\ c \in C).$$

We call the R-algebra C^* the *dual algebra* of C.

Definition 3.8.2 Let R be a field. Let V be an R-space, and W a subspace of V^*. We say that W is a *dense subspace* of V^*, if $V \to W^*$ $(v \mapsto (w \mapsto \langle w, v\rangle))$ is injective. Let R be a general commutative ring, and V and W be R-modules. We say that an R-linear map $f : W \to V^*$ is *universally dense*, if the R-linear map

$$\theta_U : U \otimes V \to \mathrm{Hom}(W, U) \qquad u \otimes v \mapsto (w \mapsto \langle w, v\rangle u)$$

is injective for any R-module U.

In the definition above, the pairing $W \otimes V \to R$ which corresponds to f by the isomorphism

$$\mathrm{Hom}(W, V^*) \cong \mathrm{Hom}(W \otimes V, R)$$

is denoted by $\langle -, -\rangle$. For a homomorphism $R \to R'$ of commutative rings, we still denote the pairing $W' \otimes_{R'} V' \to R'$ obtained by the base change by the same symbol $\langle -, -\rangle$. From the pairing $\langle -, -\rangle$,

$$f' : \mathrm{Hom}_{R'}(W', \mathrm{Hom}_{R'}(V', R'))$$

is induced. Note that the associated map $V' \to \mathrm{Hom}_{R'}(W', R')$ $(v' \mapsto (w' \mapsto \langle w', v'\rangle))$ is nothing but the composite map

$$V' = R' \otimes V \xrightarrow{\theta_{R'}} \mathrm{Hom}(W, R') \cong \mathrm{Hom}_{R'}(W', R').$$

Note also that θ_U in the definition above is natural with respect to U.

Lemma 3.8.3 *Let R be a field, V an R-vector space, and $f : W \to V^*$ an R-linear map. Then the following are equivalent.*

1 *f is universally dense.*

2 *$\theta_R : V \to W^*$ is injective.*

3 *$\mathrm{Im}\, f$ is a dense subspace of V^*.*

Proof. **1⇒2** is obvious. As $\theta_R : V \to W^*$ is the composite of $V \to (\mathrm{Im}\, f)^*$ and the injective map $(\mathrm{Im}\, f)^* \to W^*$, **2⇔3** is obvious. We show **2⇒1**. It is obvious that θ_U is injective for finite dimensional U. Now consider the case $\dim_R U = \infty$, and assume that θ_U is not injective. As there is a finite dimensional R-subspace U_0 of U such that $\mathrm{Ker}\, \theta_U \cap (U_0 \otimes V) \neq 0$ and θ is natural, θ_{U_0} is not injective, and this is a contradiction. □

Lemma 3.8.4 *Let $f : W \to V^*$ be an R-linear map. Consider the following two conditions.*

1 *f is universally dense.*

2 *V is R-flat, and the induced map*

$$f\langle\mathfrak{p}\rangle : W(\mathfrak{p}) \to \operatorname{Hom}_{\kappa(\mathfrak{p})}(V(\mathfrak{p}), \kappa(\mathfrak{p}))$$

is universally dense for $\mathfrak{p} \in \operatorname{Spec} R$, *where* $?(\mathfrak{p})$ *denotes the functor*

$$\kappa(\mathfrak{p})\otimes ? : {}_R\mathbb{M} \to {}_{\kappa(\mathfrak{p})}\mathbb{M}.$$

In general, we have **1**$\Rightarrow$**2**. *If, moreover, R is noetherian, then we have* **2**$\Rightarrow$**1**.

Proof. We prove **1**$\Rightarrow$**2**. For any injective R-linear map $g : U \to U'$, we prove that $g\otimes 1_V : U\otimes V \to U'\otimes V$ is injective. As $g_* : \operatorname{Hom}(W, U) \to \operatorname{Hom}(W, U')$ is injective, θ_U is injective, and θ is natural so that $\theta_{U'} \circ (g \otimes 1) = g_* \circ \theta_U$ holds, we have that $g \otimes 1$ is injective. This shows that V is R-flat. As we have

$$\theta_{\kappa(\mathfrak{p})} : V(\mathfrak{p}) \to \operatorname{Hom}(W, \kappa(\mathfrak{p})) \cong \operatorname{Hom}_{\kappa(\mathfrak{p})}(W(\mathfrak{p}), \kappa(\mathfrak{p}))$$

is injective, we have that $f\langle\mathfrak{p}\rangle$ is universally dense by Lemma 3.8.3.

Next, assuming that R is noetherian, we prove **2**$\Rightarrow$**1**. Assume the contrary. Then we have an R-module U such that θ_U is not injective. Take a non-zero element $\sum_i u_i \otimes v_i$ in $\operatorname{Ker}\theta_U$, and set U_0 to be the R-submodule of U generated by all u_i's. Then θ_{U_0} is not injective, and we may assume that U is R-finite, replacing U by U_0. As V is R-flat, if

$$0 \to U_1 \to U_2 \to U_3$$

is an exact sequence of R-modules and θ_{U_1} and θ_{U_3} are injective, then θ_{U_2} is also injective by the five lemma. As any finitely generated module over R has a finite filtration whose successive subquotients are of the form $R/\mathfrak{p}$ with $\mathfrak{p} \in \operatorname{Spec} R$, there exists some $\mathfrak{p} \in \operatorname{Spec} R$ such that $\theta_{R/\mathfrak{p}}$ is not injective. So we may assume that $U = R/\mathfrak{p}$. Replacing $R/\mathfrak{p}$ by R, we may assume that R is an integral domain, and $\theta_R : V \to W^*$ is not injective. We set $K = \kappa(0)$ to be the quotient field of R. As $r : R \hookrightarrow K$ is injective and V is R-flat, $r \otimes 1_V : V \to K \otimes V$ is injective. On the other hand, $\theta_K : K \otimes V \to \operatorname{Hom}_K(K \otimes W, K)$ is injective. As the injective map $\theta_K \circ (r \otimes 1_V)$ agrees with the composite map $V \to W^* \to \operatorname{Hom}_K(K \otimes W, K)$, we have that θ_R is injective, and this is a contradiction. □

Lemma 3.8.5 *Let* $W \to V^*$ *be a universally dense R-linear map. Then for any R-module U, we have that* $\theta_U : U \otimes V \to \operatorname{Hom}(W, U)$ *is R-pure. In particular,* $\theta_R : V \to W^*$ *is R-pure.*

Proof. Let X be an R-module. Then the composite map

$$X \otimes U \otimes V \xrightarrow{1_X \otimes \theta_U} X \otimes \mathrm{Hom}(W, U) \to \mathrm{Hom}(W, X \otimes U)$$

agrees with $\theta_{X \otimes U}$, which is injective. Hence, $1_X \otimes \theta_U$ is also injective for any X, and this shows that θ_U is R-pure. □

Exercise 3.8.6 Prove that if V and W are R-projective modules, then $V^* \otimes W^* \to (V \otimes W)^*$ is universally dense. Considering the case $W = R$, $\mathrm{id}_{V^*} : V^* \to V^*$ is universally dense, if V is R-projective. Utilizing this, prove that if C is an R-projective R-coalgebra, then C is cocommutative if and only if C^* is commutative.

Let M be a right C-comodule. Then M is a (left) C^*-module with the action

$$c^* m := \sum_{(m)} (c^* m_{(1)}) m_{(0)} \qquad (c^* \in C^*,\ m \in M).$$

Any C-comodule map is a C^*-linear map.

Exercise 3.8.7 Prove the assertions above.

Thus, for an R-algebra map $A \to C^*$, the exact functor $\Phi : \mathbb{M}^C \to {}_{C^*}\mathbb{M} \to {}_A\mathbb{M}$ is defined in an obvious way.

3.9 The dual coalgebra of an algebra

Let A be an R-algebra. In general, A^* does not have a canonical R-coalgebra structure even if R is a field. This is because even if $f \in A^*$, $m_A^*(f) \in (A \otimes A)^*$ does not belong to $A^* \otimes A^*$ in general. However, if R is a field, then there is a dense subspace of A^* which is endowed with a canonical coalgebra structure.

In this subsection, we assume $R = k$ is a field. We set

$$A^\circ := \{ f \in A^* \mid m_A^*(f) \in A^* \otimes A^* \}.$$

Lemma 3.9.1 *For $f \in A^*$, the following conditions are equivalent.*

i *$f \in A^\circ$, that is, $m_A^*(f) \in A^* \otimes A^*$*

ii $m_A^*(f) \in A^\circ \otimes A^\circ$

iii *There exists some ideal I of A such that* $\dim_k A/I < \infty$ *and* $f(I) = 0$.

Hence, we have $m_A^*(A^\circ) \subset A^\circ \otimes A^\circ$, and A° is a k-coalgebra with the coproduct m_A^*. We call A° the *dual coalgebra* of A.

If A and B are k-algebras, then we have $A^\circ \otimes B^\circ = (A \otimes B)^\circ$. If $\dim_k A < \infty$, then we have $A^* = A^\circ$, and $A \cong A^{**}$. In this case, $\Phi : \mathbb{M}^{A^*} \to {}_A\mathbb{M}$ is an equivalence.

3.10 Rational modules

In this subsection, R denotes a general commutative ring again. Let C be an R-coalgebra, and $A \to C^*$ a universally dense R-algebra map (hence, C is R-flat).

We denote the canonical functor $\mathbb{M}^C \to {}_A\mathbb{M}$ by Φ. As an R-module, we have $\Phi(M) = M$.

Let V be an A-module. We denote the A-action $A \otimes V \to V$ by a_V. The canonical isomorphism

$$\mathrm{Hom}(A \otimes V, V) \cong \mathrm{Hom}(V, \mathrm{Hom}(A, V))$$

is denoted by ρ_V. By the universal density assumption,

$$\theta_V : V \otimes C \to \mathrm{Hom}(A, V)$$

is injective.

We define the *rational part* of V to be the pull-back $(\rho_V a_V)^{-1}(\mathrm{Im}(\theta_V))$ of $\mathrm{Im}(\theta_V)$ by the map $\rho_V a_V : V \to \mathrm{Hom}(A, V)$, and denote it by V_{rat}.

Lemma 3.10.1 *With the notation above, V_{rat} is an A-submodule of V. Moreover, $V_{\mathrm{rat}} \hookrightarrow V \to \mathrm{Hom}(A, V)$ factors through*

$$V_{\mathrm{rat}} \otimes C \xrightarrow{\theta_{V_{\mathrm{rat}}}} \mathrm{Hom}(A, V_{\mathrm{rat}}) \hookrightarrow \mathrm{Hom}(A, V),$$

the coaction $\omega_{V_{\mathrm{rat}}} : V_{\mathrm{rat}} \to V_{\mathrm{rat}} \otimes C$ is canonically defined, and V_{rat} has a C-comodule structure. The A-module structure of $\Phi(V_{\mathrm{rat}})$ agrees with that of V_{rat} as an A-submodule of V.

Proof. For $v \in V_{\mathrm{rat}}$, we may write $(\rho_V a_V)(v) = \sum_i \theta_V(v_i \otimes c_i)$. For $a, a' \in A$, as we have

$$\begin{aligned}((\rho_V a_V)(av))(a') &= a'(av) = (a'a)v \\ &= \sum_i \langle a'a, c_i\rangle v_i = \sum_i \sum_{(c_i)} \langle a', (c_i)_{(1)}\rangle \langle a, (c_i)_{(2)}\rangle v_i,\end{aligned}$$

we have

$$(\rho_V a_V)(av) = \sum_i \sum_{(c_i)} \langle a, (c_i)_{(2)}\rangle \theta_V(v_i \otimes (c_i)_{(1)}),$$

and it follows that $av \in V_{\mathrm{rat}}$. Hence, V_{rat} is an A-submodule of V.

This shows that $(\rho_V a_V)(v)$ $(v \in V_{\mathrm{rat}})$ is contained in the image of $\mathrm{Hom}(A, V_{\mathrm{rat}}) \hookrightarrow \mathrm{Hom}(A, V)$, and hence it is zero in $\mathrm{Hom}(A, V/V_{\mathrm{rat}})$. As A is universally dense,

$$\theta_{V/V_{\mathrm{rat}}} : V/V_{\mathrm{rat}} \otimes C \to \mathrm{Hom}(A, V/V_{\mathrm{rat}})$$

is injective, and hence $\sum_i v_i \otimes c_i$ is zero in $V/V_{\mathrm{rat}} \otimes C$ by naturality of θ. As C is R-flat, we have $\sum_i v_i \otimes c_i \in V_{\mathrm{rat}} \otimes C$. So we have the definition of $\omega_{V_{\mathrm{rat}}} : V_{\mathrm{rat}} \to V_{\mathrm{rat}} \otimes C$.

We show that V_{rat} is a C-comodule with the structure map $\omega_{V_{\mathrm{rat}}}$. Let $v \in V_{\mathrm{rat}}$, and set $\omega_{V_{\mathrm{rat}}} v = \sum_i v_i \otimes c_i$. As we have

$$\sum_i \varepsilon_C(c_i) v_i = \sum_i \langle 1_A, c_i \rangle v_i = 1_A v = v,$$

the counit law is satisfied.

Next, the map

$$P : V \otimes C \otimes C \to \mathrm{Hom}(A \otimes A, V) \qquad (v \otimes c \otimes c' \mapsto (a \otimes a' \mapsto \langle c, a \rangle \langle c', a' \rangle v))$$

is injective. In fact, P is the composite of the injective maps

$$V \otimes C \otimes C \xrightarrow{\theta_V \otimes 1_C} \mathrm{Hom}(A, V) \otimes C \xrightarrow{\theta_{\mathrm{Hom}(A,V)}} \mathrm{Hom}(A, \mathrm{Hom}(A, V)) \cong \mathrm{Hom}(A \otimes A, V).$$

For any $a, a' \in A$, we have

$$P(\sum_i \omega_{V_{\mathrm{rat}}}(v_i) \otimes c_i)(a \otimes a') = \sum_i \langle a', c_i \rangle (a v_i) = a \sum_i \langle a', c_i \rangle v_i = a(a'v) = (aa')v$$
$$= \sum_i \langle aa', c_i \rangle v_i = \sum_i \sum_{(c_i)} \langle a, (c_i)_{(1)} \rangle \langle a', (c_i)_{(2)} \rangle v_i = P(\sum_i v_i \otimes \Delta_C(c_i))(a \otimes a').$$

As P is injective, the coassociativity law also holds.

By the definition of $\omega_{V_{\mathrm{rat}}}$, the A-module structure of $\Phi(V_{\mathrm{rat}})$ agrees with that of V_{rat} as an A-submodule of V. □

Proposition 3.10.2 *Let $V \in \mathbb{M}^C$, $W \in {}_A\mathbb{M}$, and $f \in \mathrm{Hom}_A(\Phi V, W)$. Then we have $f(V) \subset W_{\mathrm{rat}}$, and an isomorphism*

$$\mathrm{Hom}_A(\Phi V, W) \cong \mathrm{Hom}_C(V, W_{\mathrm{rat}})$$

is induced. In particular, if $W' \in {}_A\mathbb{M}$ and $g \in \mathrm{Hom}_A(W', W)$, then we have $g(W'_{\mathrm{rat}}) \subset W_{\mathrm{rat}}$, and letting $g_{\mathrm{rat}} = g|_{W'_{\mathrm{rat}}} \in \mathrm{Hom}_C(W'_{\mathrm{rat}}, W_{\mathrm{rat}})$, $(?)_{\mathrm{rat}}$ is a functor. Moreover, $(?)_{\mathrm{rat}}$ is right adjoint to Φ, $\mathrm{id}_V : V \cong V = (\Phi V)_{\mathrm{rat}}$ is the unit, and $\Phi(W_{\mathrm{rat}}) = W_{\mathrm{rat}} \hookrightarrow W$ is the counit of adjunction.

Proof. Easy.

Corollary 3.10.3 *The functor $(?)_{\mathrm{rat}}$ is left exact, and preserves injective objects. Φ is fully faithful.*

Proof. As $(?)_{\mathrm{rat}}$ has an exact left adjoint Φ, it is left exact and preserves injective objects. The unit of adjunction $(\Phi V)_{\mathrm{rat}} \to V$ is clearly an isomorphism, hence Φ is fully faithful. □

If V is an A-module such that $V_{\mathrm{rat}} = V$, then we say that V is *rational.* For a C-comodule M, we have that $M = \Phi M$ is rational. Conversely, if V is rational, then $V = V_{\mathrm{rat}}$ is a C-comodule. So we identify rational A-modules with C-comodules.

Exercise 3.10.4 Rational A-modules are closed under submodules, factor modules, and inductive limits in ${}_A\mathbb{M}$.

Exercise 3.10.5 If M is a flat R-module and V is an A-module, then the canonical map $V_{\mathrm{rat}} \otimes M \to (V \otimes M)_{\mathrm{rat}}$ is an isomorphism.

Exercise 3.10.6 If there is a universally dense R-algebra map $A \to C^*$, then the intersection of (infinitely many) C-subcomodules of a C-comodule is again a C-subcomodule.

3.11 FPCP coalgebras and IFP coalgebras

Throughout this subsection, R denotes a noetherian commutative ring. Let C be an R-coalgebra.

Definition 3.11.1 We say that C is *ind-finite projective* (*IFP*, for short) if for any R-finite R-submodule M of C, there exists some R-finite R-projective R-subcoalgebra of C containing M.

By definition, we have

Lemma 3.11.2 *If C is an IFP R-coalgebra, then a base change $R' \otimes C$ is an IFP R'-coalgebra for any commutative noetherian R-algebra R' of R.*

Lemma 3.11.3 *If C is IFP, then C is R-Mittag-Leffler. In particular, C is R-flat. If R is noetherian and C is R-countable, then C is R-projective.*

Proof. Follows immediately from Lemma 2.2.11 and Proposition 2.2.6. □

Now we assume that C is R-flat.

Definition 3.11.4 We say that C satisfies the *finite projective cover property* (resp. projective cover property), FPCP (resp. PCP) for short, if for any R-finite C-comodule (resp. any C-comodule) M, there exists some surjective C-comodule map $P \to M$ with P R-finite projective (resp. R-projective).

Lemma 3.11.5 *If C satisfies FPCP, then C satisfies PCP. Conversely, if C satisfies PCP and* gl.dim $R \leq 2$, *then C satisfies FPCP.*

Proof. Assume that C satisfies FPCP and M is a C-comodule. By Corollary 3.7.2, M is the inductive limit $\varinjlim M_\lambda$ of the inductive system (M_λ) of R-finite C-subcomodules of M. As C satisfies FPCP, we can take a surjective C-comodule map $f_\lambda : P_\lambda \to M_\lambda$ with P_λ R-finite projective for each λ. Then the composite map

$$\bigoplus_\lambda P_\lambda \to \bigoplus_\lambda M_\lambda \to \varinjlim M_\lambda = M$$

is a surjective C-comodule map, and $\bigoplus_\lambda P_\lambda$ is R-projective. This shows that C satisfies PCP.

Assume that C satisfies PCP, $\mathrm{gl.dim}\, R \leq 2$, and M is an R-finite C-comodule. As C satisfies PCP, there exists some surjective C-comodule map $f : P \to M$ with P R-projective. We can take an R-finite C-subcomodule N of P such that $f(N) = M$, by Lemma 3.7.1. We can also take an R-free module F with a basis B which contains P as its direct summand. Then we may consider $N \subset P \subset F$, and there exists some finite subset B_0 of B such that $F_0 := R \cdot B_0$ contains N. Now we define Q to be the kernel of the composite of the C-comodule maps

$$\rho : P \xrightarrow{\omega_P} P \otimes C \hookrightarrow F \otimes C \to F/F_0 \otimes C.$$

For $q \in Q$, we have $q = \sum_{(q)} \varepsilon_C(c_{(1)})c_{(0)} \in F_0$, and hence $Q \subset F_0$ is R-finite. As both P and $F/F_0 \otimes C$ are R-flat and $\mathrm{gl.dim}\, R \leq 2$, we have that Q is R-flat, and hence is R-finite projective. As we have $\rho(N) = 0$, it follows that $f(Q) \supset f(N) = M$, and hence the restriction $f|_Q : Q \to M$ is a surjective C-comodule map. Hence, C satisfies FPCP.

Lemma 3.11.6 *If C is IFP, then C satisfies FPCP.*

Proof. Let M be an R-finite C-comodule. As the image of $\omega_M : M \to M \otimes C$ is R-finite, there exists an R-finite R-projective R-subcoalgebra D of C such that $\mathrm{Im}\, \omega_M \subset M \otimes D$, by assumption. This shows that M is a D-comodule, and hence is a D^*-module. As D^* is R-finite R-projective, there exists some surjective D^*-linear map $f : P \to M$ such that P is R-finite R-projective. As a D^*-module is always a D-comodule and a D^*-linear map is a D-comodule map, we have that C satisfies FPCP. □

Lemma 3.11.7 *Let R be a noetherian ring, $R \to K$ an injective homomorphism of commutative rings, P an R-projective module, and M_K a finite K-submodule of $P_K := K \otimes P$. Then $M_K \cap P$ is R-finite.*

Proof. Replacing P by an R-free module which contains P as its direct summand, we may assume that P is an R-free module with a basis B. As M_K is K-finite, there exists some finite subset B_0 of B such that M_K is contained in the K-span $K \cdot B_0$ of B_0. Replacing P by $R \cdot B_0$, we may assume that P is R-finite, which case is trivial. □

Corollary 3.11.8 *Let R be a reduced noetherian ring, P an R-projective module, and M an R-submodule of P. If $M_{\mathfrak{p}}$ is a finite dimensional $R_{\mathfrak{p}}$-vector space for any $\mathfrak{p} \in \operatorname{Min}(R)$, then M is R-finite.*

Proof. Set K to be the total quotient ring $\prod_{\mathfrak{p}\in\operatorname{Min}(R)} R_{\mathfrak{p}}$ of R. By the lemma, $M_K \cap P$ is R-finite, where $M_K := M \otimes K$. As $M \subset M_K \cap P$, M is also R-finite. □

Lemma 3.11.9 *Let R be a hereditary (i.e., $\operatorname{gl.dim} R \leq 1$) noetherian ring, and C an R-projective R-coalgebra. Then C is IFP. That is to say, for any R-finite R-submodule M of C, there exists some R-finite projective R-subcoalgebra D of C which contains M.*

Proof. First, we consider the case $\operatorname{gl.dim} R = 0$. As the image of M by $\Delta^{(2)} : C \to C\otimes C\otimes C$ $(c \mapsto \sum_{(c)} c_{(1)}\otimes c_{(2)}\otimes c_{(3)})$ is R-finite, there exists some R-finite R-submodule N of C such that $\Delta^{(2)}(M)$ is contained in $C \otimes N \otimes C$. If we set $D := (\Delta^{(2)})^{-1}(C \otimes N \otimes C)$, then we have $M \subset D$. It is easy to check that $\Delta(D) \subset D \otimes D$. As $\operatorname{gl.dim} R = 0$, D is an R-subcoalgebra of C. If $d \in D$, then $d = \sum_{(d)} \varepsilon(d_1)\varepsilon(d_3)d_2 \in N$, and hence $D \subset N$. This shows that D is R-finite. It is R-projective, as $\operatorname{gl.dim} R = 0$.

Next, we consider the case $\operatorname{gl.dim} R = 1$. Let K be the total quotient ring of R. As $\operatorname{gl.dim} K = 0$, there is a K-finite K-subcoalgebra D_K of $K\otimes C$ which contains $K \otimes M$. Set $D := D_K \cap C$. By Lemma 3.11.7, D is R-finite. As C/D is torsion-free, it is R-flat, since $\operatorname{gl.dim} R = 1$. This shows that D is a pure submodule of C, and D is R-finite projective. Moreover, the composite map

$$D \hookrightarrow C \xrightarrow{\Delta} C \otimes C \to K \otimes (C/D \otimes C \oplus C \otimes C/D)$$

is zero by the choice of D_K. Hence, $\Delta(D) \subset D \otimes D$, and D is an R-subcoalgebra of C. □

3.12 ⊗ and Hom of modules and comodules over a Hopf algebra

In this subsection, R denotes an arbitrary commutative ring again.

(3.12.1) Let U be an R-Hopf algebra. For $V, W \in {}_U\mathbb{M}$, we define the U-module structure of $V \otimes W$ by

$$u(v \otimes w) = \sum_{(u)} u_{(1)}v \otimes u_{(2)}w \quad (u \in U,\ v \in V,\ w \in W),$$

and the U-module structure of $\operatorname{Hom}(V,W)$ by

$$(uf)(v) = \sum_{(u)} u_{(1)}(f((Su_{(2)})v)) \quad (u \in U,\, f \in \operatorname{Hom}_R(V,W),\, v \in V),$$

where $S = S_U$ denotes the antipode of U.

Exercise 3.12.2 Check that the definitions above do give U-modules.

(3.12.3) An R-module M endowed with the U-module structure given by $um := \varepsilon(u)m$ $(u \in U,\ m \in M)$ is called a *trivial* U-module. If we want to emphasize the trivial U-structure, we denote it by M^{triv}. However, by the U-module R we always mean the trivial module R^{triv}, unless otherwise specified.

Lemma 3.12.4 *Let V, W, and X be U-modules. The standard R-linear maps*

(3.12.5) $\operatorname{Hom}(V,W) \otimes \operatorname{Hom}(X,V) \to \operatorname{Hom}(X,W) \quad (f \otimes g \mapsto f \circ g)$

(3.12.6) $\operatorname{Hom}(V \otimes X, W) \cong \operatorname{Hom}(V, \operatorname{Hom}(X,W))$

(3.12.7) $V \otimes R \cong V \cong R \otimes V \quad (v \otimes 1 \mapsto v \mapsto 1 \otimes v)$

(3.12.8) $(V \otimes W) \otimes X \cong V \otimes (W \otimes X) \quad ((v \otimes w) \otimes x \mapsto v \otimes (w \otimes x))$

(3.12.9) $X \otimes \operatorname{Hom}(V,W) \to \operatorname{Hom}(V, X \otimes W) \quad (x \otimes f \mapsto (v \mapsto x \otimes fv))$

are U-linear, and natural with respect to V, W, and X, where the map in (3.12.6) *is given by*

$$f \mapsto (v \mapsto (x \mapsto f(v \otimes x))).$$

Proof. Straightforward. □

(3.12.10) See Lemma 2.1.7 for sufficient conditions for the map (3.12.9) to be an isomorphism. If $W = R$, then (3.12.9) is nothing but the map

$$X \otimes V^* \to \operatorname{Hom}(V,X) \qquad (x \otimes \varphi \mapsto (v \mapsto \langle \varphi, v\rangle x)).$$

If U is cocommutative, then $\tau : V \otimes W \cong W \otimes V$ $(\tau(v \otimes w) = w \otimes v)$ is also U-linear and natural with respect to V and W.

(3.12.11) We denote the functor $\operatorname{Ext}^i_U(R, ?)$ by $H^i(U, ?)$. For U-modules V and W, we have

$$\operatorname{Hom}_U(V,W) \cong \operatorname{Hom}_U(R, \operatorname{Hom}_R(V,W)) = H^0(U, \operatorname{Hom}_R(V,W)).$$

Hence, taking $H^0(U, ?)$ of both sides of (3.12.6), we have

(3.12.12) $\operatorname{Hom}_U(V \otimes X, W) \cong \operatorname{Hom}_U(V, \operatorname{Hom}_R(X,W)).$

This shows that $? \otimes X$ is left adjoint to $\mathrm{Hom}_R(X, ?)$. In particular, considering the case that V is U-projective and X is R-projective, as the left-hand side is an exact functor on W, we have that $V \otimes X$ is U-projective.

Considering the case $V = \mathrm{Hom}_R(X, W)$ in (3.12.12), the element of the left-hand side corresponding to id_V is nothing but the evaluation map

$$\mathrm{ev} : \mathrm{Hom}_R(X, W) \otimes X \to W \qquad (f \otimes x \mapsto fx).$$

Hence, ev is U-linear.

If U is cocommutative moreover, then the U-linear map which corresponds to $\mathrm{ev} \circ \tau$ by the isomorphism

$$\mathrm{Hom}_U(X \otimes \mathrm{Hom}_R(X, W), W) \cong \mathrm{Hom}_U(X, \mathrm{Hom}_R(\mathrm{Hom}_R(X, W), W))$$

is nothing but the duality map $x \mapsto (f \mapsto fx)$. Hence, the duality map is U-linear, if U is cocommutative.

(3.12.13) Let H be an R-Hopf algebra. For H-comodules M and N, we define the H-comodule structure of $M \otimes N$ by

$$M \otimes N \to M \otimes N \otimes H \qquad (m \otimes n \mapsto \sum_{(m),(n)} m_{(0)} \otimes n_{(0)} \otimes m_{(1)} n_{(1)}).$$

(3.12.14) We have seen that in the following cases, $\mathrm{Hom}(M, ?) \otimes H \cong \mathrm{Hom}(M, ? \otimes H)$ is an isomorphism for R-modules M and H, see Lemma 2.1.7.

(3.12.15) H is R-flat and M is of finite presentation.

(3.12.16) M is R-projective and H is of finite presentation.

(3.12.17) H is R-finite projective.

Assume one of the conditions above is satisfied. Let N be an H-comodule. Then we define the H-comodule structure of $\mathrm{Hom}(M, N)$, defining $\omega(f) \in \mathrm{Hom}(M, N \otimes H)$ by

$$\omega(f)(m) := \sum_{(m)} \sum_{(fm_{(0)})} (fm_{(0)})_{(0)} \otimes (fm_{(0)})_{(1)} S_H(m_{(1)})$$

for $f \in \mathrm{Hom}(M, N)$. It is left to interested readers to check that these definitions do give H-comodules.

(3.12.18) For an R-module M, the H-comodule M with the structure map $\omega(m) = m \otimes 1$ is again denoted by M. If necessary, it is denoted by M^{triv}. The functor $(?)^{\mathrm{triv}}$ is nothing but the restriction via the R-coalgebra map $u_H : R \to H$, and its right adjoint is the induction $? \otimes H$. The H-comodule R means R^{triv}, unless otherwise specified.

We denote the functor $\mathrm{Ext}^i_{\mathbb{M}^H}(R, ?)$ by $H^i(\mathbb{M}^H, ?)$. Note that the functor $H^0(\mathbb{M}^H, ?) = \mathrm{Hom}_{\mathbb{M}^H}(R, ?)$ is nothing but the induction via u_H, and it is called the *H-invariance*.

3.13 The dual Hopf algebra

(3.13.1) Let R be a commutative ring, H an R-bialgebra, and U an R-bialgebra, and $\langle -, - \rangle : U \otimes H \to R$ an R-linear map. We assume the following conditions.

1 $\langle -, - \rangle$ is a pairing of R-bialgebras. Namely, the induced map $U \to H^*$ is an R-algebra map, and $H \to U^*$ is also an R-algebra map.

2 $U \to H^*$ is a universally dense injective map.

We call such a pair U and $\langle -, - \rangle$ a *generalized hyperalgebra* of H.

(3.13.2) When is there a generalized hyperalgebra of H? As a necessary condition, H must be R-flat by Lemma 3.8.4.

In the following two cases, a generalized hyperalgebra of H exists.

Example 3.13.3 Let R be a field, and H an R-Hopf algebra. As H is an R-algebra, $U := H^\circ$ is an R-coalgebra. It is easy to verify that H° is an R-subalgebra of H^*, and is an R-Hopf algebra. We call H° the *dual Hopf algebra* of H. The canonical map $H \to U^*$ is an algebra map. If H is commutative, then H° is cocommutative [1, Corollary 2.3.17]. If H is commutative and of finite type over R, then the inclusion $H^\circ \to H^*$ is universally dense, and hence H° is a generalized hyperalgebra of H.

We prove the last assertion. By Lemma 3.8.3, it suffices to show that the canonical map $\theta : H \to (H^\circ)^*$ $(\theta(h)(h^*) = \langle h^*, h \rangle)$ is injective. Assume that $h \in \operatorname{Ker} \theta$. For any maximal ideal $\mathfrak{m}$ of H and $n \geq 1$, as $H/\mathfrak{m}^n$ is a finite dimensional R-space by the Hilbert Nullstellensatz (see [110, Theorem 5.3]), the image of h in $H/\mathfrak{m}^n = (H/\mathfrak{m}^n)^{**}$ is 0 by Lemma 3.9.1. Hence, we have $\mathfrak{m} \notin \operatorname{supp} Hh$. As $\mathfrak{m}$ is arbitrary, we have $Hh = 0$. □

Example 3.13.4 Assume that H is R-finite projective. Then as we have $H^* \otimes H^* \cong (H \otimes H)^*$, $U = H^*$ (and id_U) is a generalized hyperalgebra of H. If H is commutative, then U is cocommutative.

(3.13.5) Let U be a generalized hyperalgebra of H. If M is an H-comodule, then M is an H^*-module, hence is a U-module, and an exact functor $\Phi : \mathbb{M}^H \to {}_U\mathbb{M}$ is induced. As $U \to H^*$ is a universally dense algebra map, there is a right adjoint $(?)_{\mathrm{rat}} : {}_U\mathbb{M} \to \mathbb{M}^H$ of Φ, and Φ is fully faithful (Corollary 3.10.3).

In this situation, more is true. The functor Φ also preserves tensor products. That is, for $M, N \in \mathbb{M}^H$, the identity map $\Phi M \otimes \Phi N \cong M \otimes N \cong \Phi(M \otimes N)$ is a U-isomorphism. Moreover, Φ preserves Hom. That is, if

$M, N \in \mathbb{M}^H$ and one of (3.12.15–3.12.17) is satisfied, then the identity map $\operatorname{Hom}(\Phi(M), \Phi(N)) \cong \Phi(\operatorname{Hom}(M, N))$ is a U-isomorphism.

Moreover, the functor Φ preserves trivial representations. That is to say, $\Phi(M^{\text{triv}}) = M^{\text{triv}}$ for any R-module M.

Hence, rational U-modules are closed under tensor products, and if moreover one of the conditions (3.12.15–3.12.17) is satisfied, then they are also closed under Hom.

3.14 Module algebras and comodule algebras

The reference for this subsection is [114].

(3.14.1) Let R be a commutative ring, and U an R-Hopf algebra. We say that A is a *U-module R-algebra* if A is an R-algebra and a U-module, and the product $m_A : A \otimes A \to A$ is U-linear. In this case, $u_A : R \to A$ is also U-linear. For U-module R-algebras A and B, we say that $\varphi : A \to B$ is a U-module R-algebra map if φ is U-linear and is an R-algebra map.

(3.14.2) Let U be an R-Hopf algebra, and A a U-module algebra. We say that M is a (U, A)-module if M is both a U-module and is an A-module, and the A-action $A \otimes M \to M$ is U-linear. For (U, A)-modules M and N, we say that $f : M \to N$ is (U, A)-linear if f is both U-linear and A-linear. We denote the category of (U, A)-modules and (U, A)-linear maps by ${}_{U,A}\mathbb{M}$. Note that the U-module R-algebra R is a U-module algebra.

(3.14.3) Let A and U be as in (3.14.2). We define the *smash product* $A\#U$ of A and U as follows. As an R-module, $A\#U$ is $A \otimes U$. The product of $A\#U$ is given by

$$(a \otimes u)(b \otimes v) = \sum_{(u)} a(u_{(1)}b) \otimes u_{(2)}v \qquad (a, b \in A, \;\; u, v \in U).$$

It is easy to see that $A\#U$ is an R-algebra. Note that both $A \to A\#U$ ($a \mapsto a \otimes 1$) and $U \to A\#U$ ($u \mapsto 1 \otimes u$) are R-algebra maps. So any $A\#U$-module is in a natural way a U-module A-module, which is also a (U, A)-module. Conversely, if M is a (U, A)-module, then defining $(a \otimes u)(m) = a(um)$ for $a \otimes u \in A\#U$ and $m \in M$, M is an $A\#U$-module. These correspondences are quasi-inverse to each other, and an $A\#U$-module and a (U, A)-module are one and the same thing. Thus, we have that ${}_{A\#U}\mathbb{M}$ and ${}_{U,A}\mathbb{M}$ are equivalent. We always identify an $A\#U$-module and a (U, A)-module. As $R\#U \cong U$, we have ${}_{U,R}\mathbb{M} \cong {}_U\mathbb{M}$. Note that $H^0(U, A)$ is an R-subalgebra of A, and $H^0(U, ?)$ is a left exact functor from ${}_{U,A}\mathbb{M}$ to ${}_{H^0(U,A)}\mathbb{M}$.

(3.14.4) Let H be an R-flat Hopf algebra. We say that B is an *H-comodule R-algebra* if B is an R-algebra H-comodule, and the product

$m_B : B \otimes B \to B$ is an H-comodule map. By an H-comodule R-algebra map we mean an H-comodule map which is also an R-algebra map. We say that M is an (H, B)-*Hopf module* if M is an H-comodule B-module, and the action $B \otimes M \to M$ is an H-comodule map. A B-linear H-comodule map between (H, B)-Hopf modules is called an (H, B)-linear map. The category of (H, B)-Hopf modules and (H, B)-linear maps is abelian, and we denote it by ${}_B\mathbb{M}^H$. Note that ${}_B\mathbb{M}^H$ satisfies the (AB5) condition. Note also that R is an H-comodule algebra, and we have ${}_R\mathbb{M}^H = \mathbb{M}^H$. We have that $B^H = H^0(\mathbb{M}^H, B)$ is an R-subalgebra of B, and $H^0(\mathbb{M}^H, ?)$ is a left exact functor from ${}_B\mathbb{M}^H$ to ${}_{B^H}\mathbb{M}$.

(3.14.5) Let U be a generalized hyperalgebra of H. If B is an H-comodule algebra, then $\Phi B = B$ is a U-module algebra in a natural way. If M is an (H, B)-Hopf module, then M is a (U, B)-module. Thus, we have an exact functor $\Phi : {}_B\mathbb{M}^H \to {}_{U\#B}\mathbb{M}$.

Conversely, if A is a U-module algebra, then A_{rat} is an R-subalgebra of A, and A_{rat} is an H-comodule algebra. If M is a (U, A)-module, then M_{rat} is an (H, A_{rat})-Hopf module, and we obtain a functor $(?)_{\text{rat}} : {}_{U,A}\mathbb{M} \to {}_{A_{\text{rat}}}\mathbb{M}^H$.

If B is an H-comodule algebra (and hence $B = B_{\text{rat}}$), then $(?)_{\text{rat}} : {}_{U\#B}\mathbb{M} \to {}_B\mathbb{M}^H$ is right adjoint to Φ, and preserves injective objects. Note also that Φ is fully faithful in this case.

3.15 Coalgebras and comodules over a scheme

Let X be a scheme. We say that $\mathcal{C}$ is an $\mathcal{O}_X$-coalgebra if $\mathcal{C}$ is a quasi-coherent $\mathcal{O}_X$-module, $\mathcal{O}_X$-module maps $\varepsilon : \mathcal{C} \to \mathcal{O}_X$ and $\omega : \mathcal{C} \to \mathcal{C} \otimes_{\mathcal{O}_X} \mathcal{C}$ are given, and the coassociativity and the counit laws are satisfied. Similarly, $\mathcal{O}_X$-algebra, $\mathcal{O}_X$-bialgebra, and $\mathcal{O}_X$-Hopf algebra are defined, replacing an R-module by a quasi-coherent $\mathcal{O}_X$-module. In [71], an $\mathcal{O}_X$-algebra is called a quasi-coherent $\mathcal{O}_X$-algebra.

Notes and References. For basics on Hopf algebra theory, see [140], [1], [93], and [114]. As the base ring R in our text is not restricted to a field, we have discussed some difficulties arising from this point. In particular, the notion of universal density and related results on rational modules and generalized hyperalgebras, and the notion of IFP, FPCP, and PCP are new here. Some of the important properties of a flat coalgebra and its comodules in this section are proved in [145].

4 From representation theory

4.1 Group schemes as faisceaux

(4.1.1) Let X be a scheme. The category of X-schemes Sch/X with the fppf topology is a site, and a sheaf with respect to the Grothendieck topology is called an X-faisceau, see Example 1.8.14. For an X-scheme Y, $y(Y) = \mathrm{Hom}_{\mathrm{Sch}/X}(?, Y)$ is a set-valued X-faisceau. An X-faisceau is called *representable* if it is isomorphic to $y(Y)$ for some $Y \in \mathrm{Sch}/X$, see (1.1.7).

(4.1.2) We denote the full subcategory of the category of X-schemes Sch/X consisting of affine X-schemes by X-aff. Note that X-aff is also a site with the fppf topology. If F is an X-faisceau, then F is completely determined by its restriction to X-aff. Hence, F can be viewed as a faisceau over X-aff, and is a covariant functor from the category of X-algebras X-alg to Set, where an X-algebra is a commutative ring A together with a morphism $\operatorname{Spec} A \to X$ (note that X-alg is contravariantly equivalent to X-aff). We call a covariant functor on X-alg an X-functor. The functor which maps an X-faisceau F to the X-functor F has a left adjoint $(\tilde{?})$ by (1.8.8). For an X-functor P, the sheafification $\tilde{P}$ is called the *associated faisceau* of P. For more, see [43]. We remark the following.

Lemma 4.1.3 *Let F be a subfunctor of an X-faisceau G. Then $\tilde{F}$ is the subfunctor of G given by*

$$\tilde{F}(A) = \{x \in G(A) \mid \exists \text{ fppf } A\text{-algebra } B \text{ such that } x \in F(B)\}$$

for $A \in X$-alg, where fppf means faithfully flat of finite presentation.

Definition 4.1.4 A group (semigroup)-valued X-functor G is called an X-group scheme if G, viewed as a set-valued functor (with composing the forgetful functor), is a representable X-faisceau.

By Yoneda's lemma (Lemma 1.1.6), to say that an X-scheme G is a semigroup-valued functor is the same as to say that X-morphisms $\mu_G : G \times_X G \to G$ and $e : X \to G$ are given, and the semigroup laws $\mu_G \circ (1_G \times \mu_G) = \mu_G(\mu_G \times 1_G)$, $\mu_G \circ (e \times 1_G) \circ \lambda_G^{-1} = 1_G = \mu_G \circ (1_G \times e) \circ \rho_G^{-1}$ are satisfied, where $\lambda_G : X \times_X G \to G$ and $\rho_G : G \times_X X \to G$ are the canonical identifications. Further, G is group-valued if and only if there is an X-morphism $\iota_G : G \to G$ such that

$$\mu_G \circ (1_G \times \iota_G) \circ \Delta_G = e \circ u_G = \mu_G \circ (\iota_G \times 1_G) \circ \Delta_G$$

is satisfied, where $u_G : G \to X$ is the structure map, and $\Delta_G : G \to G \times_X G$ is the diagonalization. Thus, we see that the definition above agrees with that in (3.2.1), when X is affine. In particular, if both $X = \operatorname{Spec} R$ and $G = \operatorname{Spec} H$ are affine, then to give an X-group scheme (resp. X-semigroup scheme) structure to G is the same as to give an R-Hopf algebra (resp. R-bialgebra) structure to the commutative R-algebra H.

(4.1.5) We say that a semigroup X-scheme G acts on an X-scheme Y (from the right) if the X-functor G acts on Y from the right. Translating this situation in terms of Yoneda's lemma, an action of G on Y is nothing but an X-morphism $a : Y \times_X G \to Y$ such that $a \circ (a \times 1_G) = a \circ (1_Y \times \mu_G)$ and the unit element acts as the identity morphism. The left action is defined similarly. Unless otherwise specified, an action of a semigroup scheme on a scheme is a right action. However, if G is a group scheme, then a right action $a : Y \times_X G \to Y$ is sometimes viewed as the left action $G \times_X Y \to Y$ given by $g \cdot y := yg^{-1}$.

The quotient of Y by G, denoted by Y/G, is the associated faisceau of the X-functor F defined by $F(A) := X(A)/G(A)$. We say that a subscheme Z of Y is *G-stable* if $Z(A)$ is a G-stable subset of $Y(A)$ for any X-algebra A.

4.2 Rational representations of an algebraic group

(4.2.1) Let R be a commutative ring.

Definition 4.2.2 Let G be an affine R-semigroup scheme with $H = R[G]$. We call an H-comodule a *G-module* or a *rational G-module*.

There is an alternative definition, which is more natural. Let X be a scheme, and G an X-semigroup scheme. For a quasi-coherent $\mathcal{O}_X$-module $\mathcal{M}$, we define an X-semigroup functor $\operatorname{End}(\mathcal{M})$ (resp. X-group functor $GL(\mathcal{M})$) by

$$\operatorname{End}(\mathcal{M})(Y) := \operatorname{End}_{\mathcal{O}_Y}(f^*\mathcal{M}) \qquad (\text{resp. } GL(\mathcal{M})(Y) := \operatorname{End}_{\mathcal{O}_Y}(f^*\mathcal{M})^\times)$$

for each X-scheme $f : Y \to X$.

We say that $\mathcal{M}$ is a G-module, if $\mathcal{M}$ is a quasi-coherent $\mathcal{O}_X$-module, equipped with a morphism $G \to \operatorname{End}(\mathcal{M})$ of X-semigroup functors. This definition looks more like that of group representation.

If G is an X-group scheme, then the representation $G \to \operatorname{End}(\mathcal{M})$ factors through $GL(\mathcal{M})$. If, moreover, $\mathcal{M}$ is locally free coherent, then we have

$$\operatorname{End}(\mathcal{M}) = \underline{\operatorname{Spec}}(\operatorname{Sym} \underline{\operatorname{Hom}}_{\mathcal{O}_X}(\mathcal{M}, \mathcal{M})^\vee),$$

and both $\operatorname{End}(\mathcal{M})$ and $GL(\mathcal{M})$ are representable. In this case, the representation $G \to \operatorname{End}(\mathcal{M})$ or $G \to GL(\mathcal{M})$ is a morphism of X-schemes, by Yoneda's lemma.

(4.2.3) We briefly review the correspondence between the two definitions provided in the last paragraph in the case where both $X = \operatorname{Spec} R$ and $G = \operatorname{Spec} H$ are affine [90]. If M is an H-comodule, then we have a morphism $G \to GL(M)$ given by

$$g \mapsto (a \otimes m \mapsto \sum_{(m)} ag(m_{(1)}) \otimes m_{(0)})$$

for each A and $g \in G(A) = \operatorname{Hom}_{R\text{-alg}}(H, A)$.

Conversely, assume that a morphism $h : G \to GL(M)$ of R-group functors is given. Then as $\mathrm{id}_H \in G(H) = \operatorname{Hom}_{R\text{-alg}}(H, H)$, we have $h_H(\mathrm{id}_H) \in \operatorname{End}_H(M \otimes H)$. It is easy to see that M is an H-comodule, letting the composite map

$$M \hookrightarrow M \otimes H \xrightarrow{h_H(\mathrm{id}_H)} M \otimes H$$

be its coaction.

A little more generally, if $G = \underline{\operatorname{Spec}}\,\mathcal{H}$ is affine over X, then $\mathcal{H}$ is an $\mathcal{O}_X$-Hopf algebra, and a G-module and an $\mathcal{H}$-comodule are the same.

In the sequel, we only consider group schemes $G = \underline{\operatorname{Spec}}\,\mathcal{H}$ affine over the base scheme X.

(4.2.4) Let X be a scheme, and G an X-group scheme affine over X. Let $\mathcal{M}$ and $\mathcal{M}'$ be G-modules. We say that $\varphi : \mathcal{M} \to \mathcal{M}'$ is a *G-linear map* if φ is an $\mathcal{O}_X$-module map, and for any morphism $f : Y \to X$,

$$\Gamma(Y, f^*\varphi) : \Gamma(Y, f^*\mathcal{M}) \to \Gamma(Y, f^*\mathcal{M}')$$

is $G(Y)$-linear. We denote the category of G-modules and G-linear maps by ${}_G\mathbb{M}$. Note that ${}_G\mathbb{M}$ is equivalent to the category $\mathbb{M}^{\mathcal{H}}$.

(4.2.5) For G-modules $\mathcal{M}$ and $\mathcal{M}'$, we define the G-module structure of $\mathcal{M} \otimes_{\mathcal{O}_X} \mathcal{M}'$. For $f : Y \to X$ with Y affine and $g \in G(Y)$, g acts on

$$f^*(\mathcal{M} \otimes_{\mathcal{O}_X} \mathcal{M}') \cong f^*\mathcal{M} \otimes_{\mathcal{O}_Y} f^*\mathcal{M}'$$

so that the action on the right hand side is given by $g \otimes g$. This definition agrees with the tensor product of $\mathcal{H}$-comodules.

Lemma 4.2.6 *Let $\mathcal{A}$ be both a G-module and an $\mathcal{O}_X$-algebra. Then the following are equivalent.*

1 *The coaction $\omega_{\mathcal{A}} : \mathcal{A} \to \mathcal{A} \otimes_{\mathcal{O}_X} \mathcal{H}$ is an $\mathcal{O}_X$-algebra map.*

2 *The product map $\mathcal{A} \otimes_{\mathcal{O}_X} \mathcal{A} \to \mathcal{A}$ is G-linear.*

(4.2.7) If the conditions above are satisfied, then we say that $\mathcal{A}$ is an $\mathcal{H}$-*comodule algebra* or a *G-algebra.* Applying the functor $\underline{\mathrm{Spec}}$ to the coaction $\omega_{\mathcal{A}}$, we get a right action $a_Z : Z \times_X G \to Z$, where $Z = \underline{\mathrm{Spec}}\,\mathcal{A}$. Conversely, if an affine morphism $f : Z \to X$ and a right action $a_Z : Z \times_X G \to Z$ are given, then $f_*\mathcal{O}_Z$ is a G-algebra in a natural way. Thus, a G-algebra and a right G-action affine over X are one and the same thing.

Lemma 4.2.8 *Let R be a noetherian commutative ring, and $G = \mathrm{Spec}\, H$ an affine R-group scheme of finite type. Assume that H is IFP. Then the coordinate ring H of G is R-projective. Moreover, there exists some n such that G is a closed subgroup of $GL_n(R)$.*

Proof. Note that H is countably generated as an R-module. The first assertion is obvious by Lemma 3.11.3.

As $H = k[G]$ is of finite type over R, there exists some R-finite projective R-subcoalgebra D of H, which generates H as an R-algebra, by the definition of IFP group. Note that the dual algebra D^* is an R-pure subalgebra of $(\mathrm{End}\, D^*)^{\mathrm{op}} = \mathrm{End}\, D$, via the right multiplication. This is trivial when R is a field, and the general case follows easily from Lemma 2.1.4. This shows that there is a surjective coalgebra map $(\mathrm{End}\, D)^* \to D$. The composite coalgebra map $(\mathrm{End}\, D)^* \to D \hookrightarrow H$ is uniquely extended to a k-algebra map $\mathrm{Sym}(\mathrm{End}\, D)^* \to H$, which is obviously an R-bialgebra map. This map is surjective, because the image of this map contains D, which generates H as an R-algebra. Taking the corresponding geometric morphism, we have a closed immersion R-semigroup homomorphism $G \to \mathrm{End}\, D$. As D is a direct summand of R^n for some n, there is a closed immersion R-semigroup homomorphism $G \to \mathrm{End}\, R^n$. Because G is a group, this map factors through $G \hookrightarrow GL_n$, which is also a closed immersion. □

(4.2.9) Let $\psi : H \to G$ be a homomorphism of R-flat affine semigroup schemes. Then we have a bialgebra map $R[G] \to R[H]$. The restriction $\mathrm{res}^{R[G]}_{R[H]}$ and induction $\mathrm{ind}^{R[G]}_{R[H]}$ are respectively denoted by res^G_H and ind^G_H, and called the *restriction* functor and the *induction* functor. However, note that this induction is not the same as induction in the context of representations of finite groups or representations of (the enveloping algebras of) Lie algebras, defined via the tensor product. The terms 'induction' and 'coinduction' are sometimes interchanged depending on the context. Inductions via tensor products are right exact, while our induction via cotensor products is left exact but not necessarily right exact.

(4.2.10) Let G be an affine flat R-group scheme of finite type. Then G acts on G itself via the adjoint action $((x, y) \mapsto y^{-1}xy)$. Thus, $H = R[G]$ is a G-module. As the unit element e is fixed by the action, the defining ideal $I := \mathrm{Ker}\, \varepsilon_H$ of e is a G-submodule of H. As the product of H is

G-linear, I/I^2 is also a G-module. The Zariski tangent space $(I/I^2)^*$ of the unit element is denoted by $\mathrm{Lie}(G)$, and called the *Lie algebra* of G (it is an R-Lie algebra, in fact). The G-module $\mathrm{Lie}(G)$, as $(I/I^2)^*$, is called the *adjoint representation* of G.

4.3 Algebraic tori

(4.3.1) For a positive integer n, the X-group scheme $GL\mathcal{O}_X^{\oplus n}$ is denoted by $GL(n, X)$ or $GL_n(X)$. We denote $\mathcal{O}_X^\times = GL(1, X)$ by $\mathbb{G}_{m,X}$ or $\mathbb{G}_m$. The direct product $\mathbb{G}_m^n$ of $\mathbb{G}_m$ is called the n-fold *split torus*. An X-group scheme which is isomorphic to the n-fold split torus for some n is also called a split torus. An X-group scheme T is called an *n-torus* if T is affine flat of finite type over X, with its all geometric fibers n-fold split tori. An n-fold split torus is an n-torus, but the converse is not true.

(4.3.2) Consider the case $X = \operatorname{Spec} R$ is affine. Then we can express $R[\mathbb{G}_m] = R[t, t^{-1}]$, with t group-like. Hence, if G is an affine R-group scheme, then to give a rank-one R-free representation of G is the same as to give a homomorphism of R-group schemes $G \to \mathbb{G}_m$, and it is the same as to give a bialgebra map $R[t, t^{-1}] \to R[G]$, which is given by a group-like element of $R[G]$. Thus, the set of isomorphism classes of rank-one R-free representations $X(G)$ of G, and the set of group-like elements $X(R[G])$ of $R[G]$ are identified. Moreover, the canonical bijection $X(G) \to X(R[G])$ is an isomorphism of abelian groups, where the product of $X(G)$ is given by tensor products, and the product of $X(R[G])$ is the product of $R[G]$. However, it is common to view $X(G)$ as an additive group, and express its product by '+'. The group $X(G)$ or $X(R[G])$ is called the *character group* of G.

(4.3.3) As the coordinate ring of $T := \mathbb{G}_{m,R}^n$ is expressed as

$$H := R[T] = k[t_1^{\pm 1}, \ldots, t_n^{\pm n}],$$

with each t_i group-like, it is easy to see that $X(H) = \{t^\lambda \mid \lambda \in \mathbb{Z}^n\}$, where as usual $t^\lambda := t_1^{\lambda_1} \cdots t_n^{\lambda_n}$ for $\lambda = (\lambda_1, \ldots, \lambda_n) \in \mathbb{Z}^n$. By the map given by $t^\lambda \mapsto \lambda$, we have an isomorphism of additive groups $X(H) \cong \mathbb{Z}^n$. When T is an R-split torus, we call $n = \operatorname{rank}_{\mathbb{Z}} X(T)$ the *rank* of T. Note that $X(H)$ above is an R-basis of H, and H is a direct sum of rank-one R-free R-subcoalgebras.

(4.3.4) Let $T = \operatorname{Spec} H$ be as in (4.3.3). If V is a T-module, then we have a direct sum decomposition $V = \bigoplus_{\lambda \in X(H)} V_\lambda$, where

$$V_\lambda = \{v \in V \mid \omega_V(v) = v \otimes \lambda\}. \qquad (4.3.5)$$

If $V_\lambda \neq 0$, then λ is called a *weight* of V. If $f : V \to V'$ is a T-homomorphism, then obviously we have $f(V_\lambda) \subset (V'_\lambda)$. Thus, we have a canonical functor from ${}_T\mathbb{M}$ to the category of $X(T)$-graded R-modules. Conversely, letting (4.3.5) be the definition, we have its quasi-inverse, and we see that a T-module is nothing but an $X(T)$-graded module.

Even if the base scheme X is not affine, for a split torus $T = \mathbb{G}_m^n$ we have $X(T) \cong \mathbb{Z}^n$ by the same reasoning, and $X(T)$-graded quasi-coherent $\mathcal{O}_X$-modules and T-modules are the same thing.

4.4 Maximal tori, Borel subgroups, and reductive groups

(4.4.1) Let k be an algebraically closed field, and G a reduced affine algebraic k-group scheme. In this situation, G and the group $G(k)$ are sometimes identified. Any Zariski-closed subset of $G(k)$ is considered as a reduced closed subscheme of G.

G has a maximum connected normal solvable subgroup, which is a closed subgroup of G, called the *radical* of G. We say that G is *reductive* if G is connected and non-trivial, and the radical of G is a torus. If G is reductive, then the connected component $Z(G)^\circ$ of the center $Z(G)$ of G containing the unit element agrees with the radical of G. For example, a torus, $GL(n,k)$, $SL(n,k)$, $SO(n,k)$, $Sp(n,k)$, and their direct products are reductive. If G is reductive, then the derived subgroup $[G,G]$ and $G/Z(G)$ are also reductive.

(4.4.2) Let k and G be as in (4.4.1). A maximal connected solvable subgroup of G is called a *Borel subgroup* of G. A Borel subgroup is Zariski closed. It is not unique, but is unique up to conjugacy. For a closed subgroup P of G, P contains some Borel subgroup of G if and only if G/P is a k-projective variety. If the equivalent conditions are satisfied, then P is called a *parabolic subgroup* of G. If, moreover, there is no closed subgroup Q of G such that $P \subsetneq Q \subsetneq G$, then P is called *maximal*.

A subgroup which is maximal among closed subgroups which are tori is called a *maximal torus*. Note that maximal tori are conjugate to one another, hence their ranks are equal. We call the rank of a maximal torus of G the rank of G.

By Lemma 4.2.8, G is a closed subgroup of some $GL(n,k)$. We say that $x \in GL(n,k)$ is *unipotent* if its eigenvalues are 1 only. As this is equivalent to $(x-1)^n = 0$, the set of unipotent matrices in $GL(n,k)$ is Zariski closed. Hence, the set of unipotent elements G_u in $G \subset GL(n,k)$ is a closed subset of G. Note that G_u is independent of the embedding $G \hookrightarrow GL(n,k)$. If $G = G_u$, then we say that G is unipotent. Let $\mathcal{B}$ be a Borel subgroup of G. Then $\mathcal{B}_u$ is a normal subgroup of $\mathcal{B}$, $\mathcal{B}$ contains a maximal torus T of G, and $\mathcal{B}$ is a semidirect product of $\mathcal{B}_u$ and T.

4.5 Split reductive groups

(4.5.1) Let R be a noetherian commutative ring. We say that G is a *reductive* R-group scheme if G is an affine flat R-group scheme of finite type, and all geometric fibers of G are reductive in the sense of (4.4.1). By definition, a reductive R-group scheme is R-smooth with connected geometric fibers. Any base change of a reductive group scheme is again reductive.

Let G be an R-smooth group scheme with connected geometric fibers. A closed subgroup H of G is called a *maximal torus* of G if H is a torus, and all geometric fibers $H \otimes k$ are maximal tori of $G \otimes k$ in the sense of (4.4.2). A maximal torus which is a split torus is called a *split maximal torus.*

We say that an R-group scheme G is *split reductive* if G is reductive, and there exists some split maximal torus T such that the following conditions are satisfied. As the adjoint representation $\operatorname{Lie} G$ of G is a G-module, it is a T-module. As T is a split torus, we can decompose $\operatorname{Lie} G$ as in (4.3.4). What we require here is that, for any $\lambda \in X(T)$, $(\operatorname{Lie} G)_\lambda$ is R-free.

As can be seen easily, the base change of the adjoint representation of a smooth R-group scheme is again the adjoint representation, and a base change of a split reductive group scheme is again split reductive. Note that a reductive group with a split maximal torus splits Zariski-locally.

Theorem 4.5.2 *Any split reductive group scheme is a base change of a split reductive group scheme over* $\mathbb{Z}$*. For any reductive* R*-group scheme* G *and any geometric point* x *of* $\operatorname{Spec} R$*, there exists some affine étale neighborhood* V *of* x *such that* $G \times_X V$ *is a split reductive group over* V*. In particular, any reductive group over a strictly Henselian local ring is split.*

(4.5.3) Let R be a noetherian commutative ring, G a split reductive group scheme over R, and T its split maximal torus such that $T \hookrightarrow G$ is defined over $\mathbb{Z}$. In particular, $(\operatorname{Lie} G)_\lambda$ is R-free for any $\lambda \in X(T)$. For a G-module V, a weight of V as a T-module is called a weight of the G-module V. A non-zero weight of the adjoint representation $\operatorname{Lie} G$ is called a root of G. The set of roots of G is denoted by $\Sigma_G = \Sigma$. Note that we have $(\operatorname{Lie} G)_0 = \operatorname{Lie} T$, and we have a direct sum decomposition

$$\operatorname{Lie} G = \operatorname{Lie} T \oplus \bigoplus_{\alpha \in \Sigma} (\operatorname{Lie} G)_\alpha.$$

If $\alpha \in \Sigma$, then $(\operatorname{Lie} G)_\alpha$ is rank-one R-free. If $\alpha \in \Sigma$, then $-\alpha \in \Sigma$.

For any R-scheme X, $\operatorname{Hom}_{R\text{-sch}}(X, \mathbb{A}^1_R) = \Gamma(X, \mathcal{O}_X)$ as an additive group forms a representable R-group. Thus, $\mathbb{A}^1_R$ can be viewed as a commutative R-group scheme (by addition), which we denote by $\mathbb{G}_a$.

For $\alpha \in \Sigma$, there exists some R-group homomorphism $x_\alpha : \mathbb{G}_a \to G$ such that the conditions

1 For any commutative R-algebra A, any $t \in T(A)$ and any $a \in A = \mathbb{G}_a(A)$,

$$t x_\alpha(a) t^{-1} = x_\alpha(\alpha(t)a).$$

2 The tangent map dx_α is an isomorphism $\mathrm{Lie}\,\mathbb{G}_a \cong (\mathrm{Lie}\,G)_\alpha$.

are satisfied. Note that x_α is unique up to isomorphisms of $\mathbb{G}_a$ by the action of $R^\times$ by multiplications. We always assume that x_α is the base change of an x_α defined over $\mathbb{Z}$ (and hence is uniquely determined up to sign change). Note that x_α is a closed immersion. We denote the scheme-theoretic image $x_\alpha(\mathbb{G}_a)$ by U_α, and call it the root subgroup of G with respect to α. The group functor U_α represents $A \mapsto \mathrm{Im}(x_\alpha(A))$.

(4.5.4) Let R, G, T and Σ be as in (4.5.3). The set

$$Y(T) := \mathrm{Hom}_{R\text{-group}}(\mathbb{G}_m, T)$$

is an abelian group. If $T \cong \mathbb{G}_m^n$, then $Y(T) \cong \mathbb{Z}^n$. For $f, g \in Y(T)$, the sum $f + g$ is nothing but the composite

$$\mathbb{G}_m \xrightarrow{\Delta_{\mathbb{G}_m}} \mathbb{G}_m \times \mathbb{G}_m \xrightarrow{(f,g)} T \times T \xrightarrow{\mu_T} T.$$

For $\lambda \in X(T)$ and $f \in Y(T)$, the integer which corresponds to $\lambda \circ f$ by the isomorphism

$$\mathrm{Hom}_{R\text{-group}}(\mathbb{G}_m, \mathbb{G}_m) \cong \mathbb{Z} \qquad ((t \mapsto t^r) \mapsto r)$$

is denoted by $\langle \lambda, f \rangle$. Note that $\langle ?, - \rangle : X(T) \times Y(T) \to \mathbb{Z}$ is bilinear, and induces an isomorphism $Y(T) \cong \mathrm{Hom}_{\mathbb{Z}}(X(T), \mathbb{Z})$.

(4.5.5) Let R, G, T and Σ be as in (4.5.3). For $\alpha \in \Sigma$, there exists a unique homomorphism of R-group schemes $\varphi_\alpha : SL_2 \to G$ such that for any commutative R-algebra A and $a \in A$,

$$\varphi_\alpha \begin{bmatrix} 1 & a \\ 0 & 1 \end{bmatrix} = x_\alpha(a) \quad \text{and} \quad \varphi_\alpha \begin{bmatrix} 1 & 0 \\ a & 1 \end{bmatrix} = x_{-\alpha}(\pm a)$$

hold. For any A and $a \in A^\times$, we have

$$\alpha^\vee(a) := \varphi_\alpha \begin{bmatrix} a & 0 \\ 0 & a^{-1} \end{bmatrix} \in T(A),$$

and thus we have a definition of $\alpha^\vee \in Y(T)$. Note that $\alpha^\vee$ is independent of the choice of x_α, and is determined only by α.

We set $X(T)_{\mathbb{R}} := X(T) \otimes_{\mathbb{Z}} \mathbb{R}$ and $Y(T)_{\mathbb{R}} := Y(T) \otimes_{\mathbb{Z}} \mathbb{R}$. Then the pairing

$$\langle -, ? \rangle : X(T) \otimes_{\mathbb{Z}} Y(T) \to \mathbb{Z}$$

is extended to the pairing $\langle -, ? \rangle : X(T)_{\mathbb{R}} \otimes_{\mathbb{R}} Y(T)_{\mathbb{R}} \to \mathbb{R}$, and we may identify $Y(T)_{\mathbb{R}} \cong \mathrm{Hom}_{\mathbb{R}}(X(T), \mathbb{R})$ in a natural way. For $\alpha \in \Sigma$, we define $s_\alpha : X(T)_{\mathbb{R}} \to X(T)_{\mathbb{R}}$ by

$$s_\alpha \lambda := \lambda - \langle \lambda, \alpha^\vee \rangle \alpha \qquad (\lambda \in X(T)_{\mathbb{R}}).$$

Note that s_α is an $\mathbb{R}$-linear involution of $X(T)_{\mathbb{R}}$. We call the subgroup of $GL(X(T)_{\mathbb{R}})$ generated by $\{s_\alpha \mid \alpha \in \Sigma\}$ the *Weyl group* of G, and denote it by $W = W(G)$. Obviously, W maps $X(T)$ to $X(T)$.

(4.5.6) Let R, G, T and Σ be as in (4.5.3). The set of roots Σ of G is an abstract root system [87, p. 229] of the $\mathbb{R}$-span $\mathbb{R} \cdot \Sigma$ of Σ in $X(T)_{\mathbb{R}}$. Note that the restriction of s_α to $\mathbb{R} \cdot \Sigma$ is the reflection which corresponds to α. For any R-algebra A which is an integral domain, we have

$$W(G) \cong N_G(T)(A)/T(A) \cong (N_G(T)/T)(A),$$

where in the first isomorphism, for any $\alpha \in \Sigma$, $s_\alpha \in W(G)$ corresponds to the image of

$$\varphi_\alpha \begin{bmatrix} 0 & 1 \\ -1 & 0 \end{bmatrix} \in N_G(T)(A)$$

in $N_G(T)(A)/T(A)$. This map is always defined regardless of what A is, and it is injective unless A is the null ring. Hence, we have a canonical map $W \to N_G(T)/T$, and W acts on T via the adjoint action of $N_G(T)/T$ on T. Hence, W acts on both $X(T)$ and $Y(T)$, and the pairing $\langle -, ? \rangle$ is W-invariant (note that the action of W on $X(T)$ so obtained agrees with the action of W on $X(T)$ given in (4.5.5)).

As Σ is a root system, we can take a base $\Delta = \Delta_G$. We say that Δ is a *base* of Σ, if $\Delta = \{\alpha_1, \ldots, \alpha_l\}$ is a basis of $\mathbb{R} \cdot \Sigma$, and for any $\alpha \in \Sigma$, when we uniquely express α as the linear combination $\alpha = \sum_i c_i \alpha_i$ of this basis, c_i are non-zero integers of like sign. If the c_i's are all positive (resp. negative), we say that $\alpha \in \Sigma$ is a *positive root* (resp. *negative root*). The set of all positive roots (resp. negative roots) is denoted by $\Sigma^+ = \Sigma_G^+$ (resp. $\Sigma^- = \Sigma_G^-$). Note that the positivity/negativity of roots is a notion which depends on the choice of Δ. By the definition of a base, we have $\Sigma = \Sigma^+ \coprod \Sigma^-$.

In the rest of these notes, if a split reductive R-group scheme G is given, then we take it for granted that a split maximal torus T defined over $\mathbb{Z}$ and a base $\Delta = \{\alpha_1, \ldots, \alpha_l\}$ of Σ_G are fixed.

Note that W is a finite group generated by $\{s_\alpha \mid \alpha \in \Delta\}$. For $w \in W$, the minimum h such that there exists an expression $w = s_{\alpha_{i(1)}} \cdots s_{\alpha_{i(h)}}$ $(1 \leq i(1), \ldots, i(h) \leq l)$ is called the *length* of w, and is denoted by $l(w)$. There is a unique element of maximum length in W. We call it the *longest element* of W, and denote it by w_0. An element w in W equals w_0 if and only if $w(\Sigma^+) = \Sigma^-$.

(4.5.7) Let R be a noetherian commutative ring, and G a split reductive R-group scheme. We denote by U the closed subgroup of G generated by all U_α with $\alpha \in \Sigma^-$. By the product map (in any order), we have an isomorphism of R-schemes

$$\prod_{\alpha \in \Sigma^-} U_\alpha \cong U.$$

Hence, as an R-scheme, we have $U \cong \mathbb{A}_R^{\#\Sigma^-}$. Note that the maximal torus T normalizes U. The semidirect product $B = TU = T \ltimes U$ is called the *negative Borel subgroup* of G. If $R = k$ is an algebraically closed field, then B is a Borel subgroup of G in the sense of (4.4.2).

(4.5.8) Let $I \subset \Delta$. We set $\Sigma_I := \Sigma \cap \mathbb{Z}I$, and we define L_I to be the closed subgroup of G generated by T and all U_α with $\alpha \in \Sigma_I$. We have that L_I is again a split reductive R-group scheme, and its Weyl group is $W_I := \langle s_\alpha \mid \alpha \in I \rangle$. We call a closed subgroup of the form L_I for some I a (standard) *Levi subgroup* of G.

(4.5.9) For $\lambda, \mu \in X(T)$, we say that $\lambda \leq \mu$ if we have an expression $\mu - \lambda = \sum_{i=1}^{l} c_i \alpha_i$ with $c_i \in \mathbb{N}_0$. This defines an ordering of $X(T)$. We call this ordering the *dominant order* of $X(T)$.

We say that $\lambda \in X(T)$ is dominant if for any $\alpha \in \Delta$, we have $\langle \lambda, \alpha^\vee \rangle \geq 0$. We denote the set of dominant weights by X_G^+.

Lemma 4.5.10 *We have that X_G^+, as a subset of $X(T)$, is a countable ordered set. For any $\lambda \in X_G^+$, there are only finitely many $\mu \in X_G^+$ such that $\mu \leq \lambda$.*

For the proof, see [88, 13.2, Lemma B].

For $\lambda \in X(T)$, $\lambda \in X_G^+$ if and only if $-w_0\lambda \in X_G^+$. For $\lambda \in X(T)$, we denote $-w_0\lambda$ by λ^*. We have $\lambda^{**} = \lambda$, and $\lambda \geq \mu$ if and only if $\lambda^* \geq \mu^*$.

4.6 General linear groups

Taking $G = GL(V) = GL(n, R)$ as an example, we review the basics on reductive groups explained in (4.5) more explicitly.

(4.6.1) Let R be a noetherian commutative ring, and $G = GL(V) = GL(n, R)$. We choose T as the subgroup of invertible diagonal matrices.

(4.6.2) In this case, any rank-one R-free representation of T is of the form

$$\begin{bmatrix} t_1 & & & \\ & t_2 & & \\ & & \ddots & \\ & & & t_n \end{bmatrix} \mapsto t_1^{\lambda_1} t_2^{\lambda_2} \cdots t_n^{\lambda_n}$$

($\lambda = (\lambda_1, \lambda_2, \ldots, \lambda_n) \in \mathbb{Z}^n$). Thus, we have $X(T) \cong \mathbb{Z}^n$ in a natural way.

(4.6.3) We set $\varepsilon_i = (0, 0, \ldots, 0, 1, 0, \ldots, 0) \in \mathbb{Z} \cong X(T)$ (the jth entry is δ_{ij}, where δ_{ij} denotes Kronecker's delta). Then

$$\Sigma = \{\varepsilon_i - \varepsilon_j \mid 1 \leq i, j \leq n \quad i \neq j\}$$

is the set of roots of G (with respect to our choice of T). When we set $\alpha_i := \varepsilon_i - \varepsilon_{i+1}$, then $\Delta = \{\alpha_1, \ldots, \alpha_{n-1}\}$ is a base of the root system Σ.

Now we fix the choice of the base Δ of Σ as above. Then the set of positive roots is

$$\Sigma^+ = \{\varepsilon_i - \varepsilon_j \mid 1 \leq i < j \leq n\}.$$

(4.6.4) Let T and Δ be as above. Then the negative Borel subgroup B is nothing but the subgroup of invertible lower triangular matrices. Note that U consists of all elements of B whose diagonal entries are all 1.

(4.6.5) In the situation above, we have $W(G) \cong \mathfrak{S}_n$, and W acts on T as the adjoint action of the group of permutation matrices. The action of W on $X(T) = \mathbb{Z}^n$ is given by the permutation of entries. The longest element w_0 is the permutation given by $w_0(i) = n + 1 - i$.

(4.6.6) We have $\langle \lambda, (\varepsilon_i - \varepsilon_j)^\vee \rangle = \lambda_i - \lambda_j$ for $\lambda = (\lambda_1, \ldots, \lambda_n) \in \mathbb{Z}^n = X(T)$ and $1 \leq i \neq j \leq n$. Thus, λ is dominant if and only if $\lambda_1 \geq \lambda_2 \geq \cdots \geq \lambda_n$.

4.7 Representations of reductive groups over an algebraically closed field

Let R be a noetherian commutative ring, G an R-split reductive group, T a split maximal torus of G defined over $\mathbb{Z}$, and let Δ be a base of the root system Σ of G. We recall that U is a normal subgroup of the negative Borel subgroup B of G, and B is a semidirect product of U and T, with U normal. Let $\lambda \in X(T)$, namely, λ is a rank-one R-free T-module. Then letting U act on λ trivially, λ is a rank-one R-free B-module, whose restriction to T is the original λ. We denote this rank-one R-free B-module by R_λ.

Definition 4.7.1 For $\lambda \in X^+$, we denote $\operatorname{ind}_B^G(R_\lambda)$ by $\nabla(\lambda) = \nabla_G(\lambda)$, and call it the *induced module* of highest weight λ. The G-module $\nabla(\lambda^*)^*$ is denoted by $\Delta(\lambda) = \Delta_G(\lambda)$, and called the *Weyl module* of highest weight λ.

(4.7.2) From now on, $R = k$ denotes an algebraically closed field. We denote R_λ by k_λ. The following is well-known.

(4.7.3) The set $\{k_\lambda \mid \lambda \in X(T)\}$ is a complete set of representatives of the isomorphism classes of simple B-modules. Any simple U-module is trivial.

(4.7.4) If M is a finite dimensional B-module, then $R^i\mathrm{ind}_B^G(M)$ is also finite dimensional. If $i > \dim G/B$, then we have $R^i\mathrm{ind}_B^G(M) = 0$ (this vanishing holds also for infinite dimensional B-modules, see Lemma 3.6.17). In particular, for $\lambda \in X^+$, we have that $\nabla_G(\lambda)$ and $\Delta_G(\lambda)$ are finite dimensional.

(4.7.5) For $\lambda \in X(T)$, we have

$$\mathrm{ind}_B^G(k_\lambda) \neq 0 \iff \lambda \in X_G^+.$$

In the representation theory of the reductive group G, induced modules and Weyl modules play important roles.

Theorem 4.7.6 (Kempf's vanishing) *If $\lambda \in X^+$, then for $i > 0$ we have $R^i\mathrm{ind}_B^G(k_\lambda) = 0$.*

(4.7.7) For $\lambda \in X^+$, we have $\mathrm{soc}(\nabla(\lambda)) \cong \mathrm{top}(\Delta(\lambda))$, and they are simple. We denote this simple G-module by $L(\lambda) = L_G(\lambda)$, and call it the simple G-module of highest weight λ. As a result, we have that $\{L(\lambda) \mid \lambda \in X^+\}$ is a complete set of representatives of isomorphism classes of simple G-modules. Moreover, if $\lambda, \mu \in X^+$ and $L(\mu)$ is a subquotient of $\mathrm{rad}\,\Delta(\lambda) \oplus \nabla(\lambda)/\mathrm{soc}\,L(\lambda)$, then $\mu < \lambda$.

Theorem 4.7.8 (Cline–Parshall–Scott–van der Kallen, [39]) *For any dominant weights $\lambda, \mu \in X_G^+$, we have*

$$\mathrm{Ext}_G^i(\Delta(\lambda), \nabla(\mu)) \cong \begin{cases} k & (i = 0,\ \lambda = \mu) \\ 0 & (\textit{otherwise}) \end{cases}.$$

(4.7.9) Let V be a finite dimensional G-module. Then it is a T-module by restriction, hence we have a decomposition $V = \bigoplus_{\lambda \in X(T)} V_\lambda$. The element

$$\mathrm{ch}(V) := \sum_{\lambda \in X(T)} (\dim_k V_\lambda)\lambda \in \mathbb{Z}X(T) = \mathbb{Z}[T_\mathbb{Z}]$$

is called the *formal character* of V. It is easy to see that $\mathrm{ch}(V) \in (\mathbb{Z}X(T))^W$, where W is the Weyl group of G.

As a consequence of Weyl's character formula [90, p. 250], we have that for any $\lambda \in X_G^+$, $\mathrm{ch}(\Delta_G(\lambda)) = \mathrm{ch}(\nabla_G(\lambda))$, and they are determined only by λ, and independent of k. In particular, we have $\dim_k \Delta_G(\lambda) = \dim_k \nabla_G(\lambda)$ are independent of k.

(4.7.10) Another important property of weights of induced (or Weyl) modules is: if $\nabla_G(\lambda)_\mu \neq 0$, then we have $w_0\lambda \leq \mu \leq \lambda$. Moreover, $\nabla_G(\lambda)_\lambda = 1$. Similarly for $\Delta_G(\lambda)$ and $L_G(\lambda)$. The name 'highest weight' comes from this fact. It follows that, if V is a G-module, $\lambda \in X_G^+$, $\dim_k V_\lambda = 1$, and $\dim_k V_\mu = 0$ for $\mu \in X_G^+ \setminus \{\lambda\}$, then $V \cong L_G(\lambda)$.

(4.7.11) If k is of characteristic 0, then we have $\Delta(\lambda) \cong \nabla(\lambda)$ for $\lambda \in X^+$.

For the results above, we refer the reader to [90].

4.8 Universal module functors

(4.8.1) Let X be a scheme. We say that $\mathcal{U} = (U_A)$ is a *universal family* over X if for any X-algebra A (i.e., a morphism $\operatorname{Spec} A \to X$), a full subcategory U_A of ${}_A\mathbb{M}$ closed under isomorphisms corresponds, and for any X-algebra map $A \to B$ (i.e., a morphism $\operatorname{Spec} B \to \operatorname{Spec} A$ of X-schemes), $M \in U_A$ implies $M \otimes_A B \in U_B$. For example, the family $\mathcal{P}_X = (P_A)$, where P_A is the full subcategory of ${}_A\mathbb{M}$ consisting of finite projective A-modules, is a universal family.

Definition 4.8.2 Let $s, t \geq 0$, and $\mathcal{U}_1, \ldots, \mathcal{U}_{s+t}$, and $\mathcal{V}$ be universal families. We say that

$$\mathcal{M} = ((M_A), (\rho_f)) : \mathcal{U}_1 \times \cdots \times \mathcal{U}_s \times \mathcal{U}_{s+1}^{\mathrm{op}} \times \cdots \times \mathcal{U}_{s+t}^{\mathrm{op}} \to \mathcal{V}$$

is a *universal functor* of type (r, s), if for each commutative X-algebra A,

$$M_A : (U_1)_A \times \cdots \times (U_s)_A \times (U_{s+1})_A^{\mathrm{op}} \times \cdots \times (U_{s+t})_A^{\mathrm{op}} \to V_A$$

is a functor, for each X-algebra map $f : A \to B$,

$$\rho_f : (B\otimes ?) \circ M_A \to M_B \circ ((B\otimes ?)^s \times ((B\otimes ?)^{\mathrm{op}})^t)$$

is a natural isomorphism, and for any composable X-algebra maps

$$A \xrightarrow{f} B \xrightarrow{g} C,$$

the diagram

$$\begin{array}{ccc}
C\otimes_B B\otimes_A M_A & \xrightarrow{C\otimes_B \rho_f} & C\otimes_B M_B((B\otimes_A ?)^r, (B\otimes_A ?)^s) \\
 & & \downarrow \rho_g((B\otimes_A ?)^r, (B\otimes_A ?)^s) \\
\downarrow \alpha_{f,g} M_A & & M_C((C\otimes_B (B\otimes_A ?))^r, (C\otimes_B (B\otimes_A ?))^s) \\
 & & \downarrow M_C(\alpha^r, \alpha^s) \\
C\otimes_A M_A & \xrightarrow{\rho_{gf}} & M_C((C\otimes_A ?)^r, (C\otimes_A ?)^s)
\end{array}$$

is commutative, where $\alpha : C\otimes_B (B\otimes_A ?) \to C\otimes_A ?$ is the usual identification.

If $\mathcal{U}_1 = \cdots = \mathcal{U}_{s+t} = \mathcal{P}_X$ and $\mathcal{V} = ({}_A\mathbb{M})$, then we say that $\mathcal{M}$ is a *universal module functor* of type (r, s). If $\mathcal{U}_1 = \cdots = \mathcal{U}_{s+t} = \mathcal{V} = \mathcal{P}_X$, then we say that $\mathcal{M}$ is a *universally projective functor* of type (r, s). If it happens that $X = \operatorname{Spec} R$ with R a PID, then a universally projective functor is sometimes referred as a *universally free functor.*

This definition could be made as a special case of part of the theory of fibered categories and pseudo-functors [66, VI]. In the sequel, we only treat universal module functors, for simplicity.

Definition 4.8.3 Let $\mathcal{M} = ((M_A), (\rho_f))$ and $\mathcal{N} = ((N_A), (\rho'_f))$ be universal module functors of type (r, s) over X. We say that $\varphi = (\varphi_A) : \mathcal{M} \to \mathcal{N}$ is a *universal map* if for any X-algebra A, $\varphi_A : M_A \to N_A$ is a natural transformation, and for any X-algebra map $f : A \to B$,

$$\rho'_f \circ ((B \otimes_A ?)\varphi_A) = \varphi_B((B \otimes_A ?)^s, ((B \otimes_A ?)^{\mathrm{op}})^t) \circ \rho_f$$

holds.

We have a category of universal module functors of type (r, s) over X. We denote this category by $\mathrm{UMF}(r, s; X)$. Note that $\mathrm{UMF}(r, s; X)$ is abelian.

Exercise 4.8.4 The functor $(\mathrm{Hom}_A(?, -))$ is a universally free functor of type $(1, 1)$ over $\operatorname{Spec} \mathbb{Z}$. $(\rho_f)(M, N) : B \otimes \mathrm{Hom}_A(M, N) \to \mathrm{Hom}_B(B \otimes_A M, B \otimes_A N)$ is given by $b \otimes h \mapsto (b' \otimes m \mapsto bb' \otimes h(m))$. The tensor product $(V_1, \ldots, V_r) \mapsto V_1 \otimes_A \cdots \otimes_A V_r$ is a universally free functor of type $(r, 0)$.

$$\rho_f(V_1, \ldots, V_r) : B \otimes_A (V_1 \otimes_A \cdots \otimes_A V_r) \to (B \otimes_A V_1) \otimes_B \cdots \otimes_B (B \otimes V_r)$$

is given by

$$b \otimes (v_1 \otimes \cdots \otimes v_r) \mapsto b(v_1 \otimes 1) \otimes \cdots \otimes (v_r \otimes 1).$$

Similarly, the exterior power $V \mapsto \bigwedge^i V$, the symmetric power $V \mapsto S_i V$, and the divided power $V \mapsto D_i V = (S_i V^*)^*$ are universally free of type $(1, 0)$ for $i \geq 0$.

Exercise 4.8.5 Note that a universal module functor (resp. universally projective functor) of type $(0, 0)$ is nothing but a quasi-coherent $\mathcal{O}_X$-module (resp. locally free coherent $\mathcal{O}_X$-module).

Lemma 4.8.6 *Let $r, s \geq 0$, and $\mathcal{M}_1, \ldots, \mathcal{M}_{r+s}$ be universally projective functors of type*

$$(r_1, s_1), \ldots, (r_{r+s}, s_{r+s})$$

over X, respectively. If $\mathcal{N}$ is a universal module functor (resp. universally projective functor) of type (r, s), then the 'composite'

$$\mathcal{L} := \mathcal{N} \circ (\mathcal{M}_1, \ldots, \mathcal{M}_{r+s})$$

defined by

$$L_A := N_A((M_1)_A(?), \ldots, (M_{r+s})_A(?))$$

is a universal module functor (resp. universally projective functor) of type

$$(\sum_{i=1}^{r} r_i + \sum_{i=r+1}^{s+1} s_i, \sum_{i=1}^{r} s_i + \sum_{i=r+1}^{s+1} r_i).$$

(4.8.7) Let X be an R-scheme. We denote the category of locally free coherent $\mathcal{O}_X$-modules by P_X. Let $\mathcal{M} = ((M_A), (\rho_f))$ be a universal module functor of type (r, s), $\mathcal{V}_1, \dots, \mathcal{V}_{r+s} \in P_X$, and $r \geq 1$. As an extreme case of Lemma 4.8.6, we have that $\mathcal{M}(\mathcal{V}_1, \dots, \mathcal{V}_{r+s})$ is a universal module functor of type $(0, 0)$, i.e., a quasi-coherent $\mathcal{O}_X$-module.

Lemma 4.8.8 *Let $\mathcal{M} = ((M_A), (\rho_f))$ be a universal module functor of type (r, s) over X, and $\mathcal{V}_1, \dots, \mathcal{V}_{r+s} \in P_X$. Then the quasi-coherent sheaf $\mathcal{N} := \mathcal{M}(\mathcal{V}_1, \dots, \mathcal{V}_{r+s})$ is a module over*

$$G := GL(\mathcal{V}_1) \times \cdots \times GL(\mathcal{V}_{r+s})$$

with the action $\varphi : G \to GL(\mathcal{N})$ such that, for any commutative X-algebra $f : \operatorname{Spec} A \to X$,

$$\varphi_A(g_1, \dots, g_s, g_{s+1}, \dots, g_{s+r}) = M_A(g_1, \dots, g_s, g_{s+1}^{-1}, \dots, g_{s+r}^{-1}).$$

4.9 Tilting modules

Let A be a ring.

Definition 4.9.1 We say that $T \in {}_A\mathbb{M}$ is a *tilting A-module* if the following conditions are satisfied.

(4.9.2) $T \in \widehat{\operatorname{add}_A A}$

(4.9.3) ${}_A A \in (\operatorname{add} T)^\vee$

(4.9.4) $\operatorname{Ext}^i_A(T, T) = 0 \ (i > 0)$.

(4.9.5) Let A be a ring, and T a tilting A-module. Set $B := \operatorname{End}_A(T)^{\mathrm{op}}$. Then T is an (A, B)-bimodule in a natural way. As a right B-module, T is a tilting module, and we have $\operatorname{End}_B(T) \cong A$. In this sense, the axioms for tilting modules are symmetric with respect to B and A. Hence, we sometimes say that T is a tilting (A, B)-bimodule.

(4.9.6) Let A, T and B be as above. For $e \geq 0$, we define $\mathrm{KE}_e(T)$ to be the full subcategory of ${}_A\mathbb{M}$ consisting of A-modules M such that $\operatorname{Ext}^i_A(T, M) = 0$ for $i \neq e$. We define $\mathrm{KT}_e(T)$ to be the full subcategory of ${}_B\mathbb{M}$ consisting of B-modules N such that $\operatorname{Tor}^B_i(T, N) = 0$ for $i \neq e$. Then $\mathrm{KE}_e(T)$ and $\mathrm{KT}_e(T)$ are equivalent. In fact, $\operatorname{Ext}^e_A(T, ?) : \mathrm{KE}_e(T) \to \mathrm{KT}_e(T)$ and $\operatorname{Tor}^B_e(T, ?) : \mathrm{KT}_e(T) \to \mathrm{KE}_e(T)$ are the quasi-inverse of each other.

(4.9.7) We have $\operatorname{left.gl.dim} A < \infty \iff \operatorname{left.gl.dim} B < \infty$.

(4.9.8) Let A and B be artinian algebras (i.e., module finite algebras over artinian commutative centers). Let us denote by $G(A)$ and $G(B)$ the Grothendieck groups of $({}_A\mathbb{M})_f$ and $({}_B\mathbb{M})_f$, respectively. Then the maps

$$\mathrm{Ext} : G(A) \to G(B), \;\; [X] \mapsto \sum_{i \geq 0} (-1)^i [\mathrm{Ext}^i_A(T, X)]$$

and

$$\mathrm{Tor} : G(B) \to G(A), \;\; [Y] \mapsto \sum_{i \geq 0} (-1)^i [\mathrm{Tor}^B_i(T, Y)]$$

are inverse to each other.

(4.9.9) Let A and B be as in (4.9.8). Then the number of isomorphism classes of indecomposable direct summands of T, the number of isomorphism classes of simple A-modules, and the number of isomorphism classes of simple B-modules are all equal.

For the details of the results summarized above, see [113]. See also [67], [68] and [35].

Let T be a tilting A-module. When we set $F := \mathrm{Hom}_A(T, ?)$, then $F : \mathrm{KE}_0(T) \to \mathrm{KT}_0(T)$ is an equivalence by (4.9.6), and $\mathrm{add}(T)$ corresponds to $\mathrm{add}(B)$ by F.

4.10 Cotilting modules

Let A be a left noetherian ring, and B a right noetherian ring.

Definition 4.10.1 ([112, p. 586]) Let U be an (A, B)-bimodule. We say that U is a *cotilting* (A, B)*-bimodule* if the following conditions are satisfied.

(4.10.2) ${}_AU \in {}_A\mathbb{M}_f$, $U_B \in \mathbb{M}_{Bf}$

(4.10.3) $\mathrm{inj.dim}\, {}_AU < \infty$, $\mathrm{inj.dim}\, U_B < \infty$

(4.10.4) $\mathrm{Ext}^i_A(U, U) = 0$ $(i > 0)$, $\mathrm{Ext}^i_B(U, U) = 0$ $(i > 0)$

(4.10.5) The canonical maps $B^{\mathrm{op}} \to \mathrm{End}_A U$ and $A \to \mathrm{End}_B U$ are isomorphisms.

By definition, the condition for being a cotilting module is left–right symmetric. In other words, an (A, B)-bimodule U is cotilting if and only if it is cotilting as a $(B^{\mathrm{op}}, A^{\mathrm{op}})$-bimodule.

By condition (4.10.5), a usage such as '${}_AU$ is a cotilting A-module' or 'U_B is a cotilting right B-module' makes sense. Assuming that A is both left and right noetherian, a cotilting (A, A)-bimodule is called a *dualizing bimodule* of A.

(4.10.6) Let k be a field, and A and B be finite dimensional algebras over k. Then T is a tilting (A, B)-bimodule if and only if T^* is a cotilting (B, A)-bimodule.

(4.10.7) Cotilting modules are deeply related to Auslander–Buchweitz contexts (1.12) in ${}_A\mathbb{M}_f$.

Theorem 4.10.8 **([112])** *Let A be a left noetherian ring, B a right noetherian ring, and U a cotilting (A, B)-bimodule. When we set $\mathcal{A} := {}_A\mathbb{M}_f$, $\mathcal{X} := {}^\perp U$, $\omega :=$ add U, and $\mathcal{Y} := \hat{\omega}$, then $(\mathcal{X}, \mathcal{Y}, \omega)$ is an Auslander–Buchweitz context in $\mathcal{A}$. The functor*

$$\mathrm{Hom}_A(?, {}_AU_B) : \mathcal{X}_A = {}^\perp_A U \to {}^\perp U_B$$

is a contravariant equivalence of exact categories, with $\mathrm{Hom}_B(?, {}_AU_B)$ *its quasi-inverse.*

By Lemma 2.11.10, we have the following.

Lemma 4.10.9 *If A is a noetherian commutative ring with a dualizing A-module U_A, then A is Cohen–Macaulay, and the minimal injective resolution of U_A is a dualizing complex of A. Conversely, if A is a Cohen–Macaulay ring with a dualizing complex $I^\bullet$, then the A-module* $\Gamma(\mathrm{Spec}\, A, \omega_{\mathrm{Spec}\, A})$ *is dualizing, where $\omega_{\mathrm{Spec}\, A}$ is the canonical sheaf* (2.11.9) *of* Spec A. *In this case, A is of finite Krull dimension.*

(4.10.10) If the conditions in the lemma are satisfied, then the assumption of Theorem 4.10.8 is satisfied, and we have an Auslander–Buchweitz context $(\mathcal{X}_A, \mathcal{Y}_A, \omega_A)$. More generally, even if A is not finite dimensional, we can construct an Auslander–Buchweitz context.

Let A be a commutative noetherian ring. An A-module M is called *pointwise dualizing* if M is A-finite, and for any $\mathfrak{p} \in \mathrm{Spec}\, A$, the $A_\mathfrak{p}$-module $M_\mathfrak{p}$ is dualizing. By Theorem 2.5.6, a commutative noetherian ring with a pointwise dualizing module is Cohen–Macaulay.

An A-module M is called a *maximal Cohen–Macaulay* module if M is A-finite, and for any $\mathfrak{p} \in \mathrm{Spec}\, A$, $M_\mathfrak{p}$ is a maximal Cohen–Macaulay $A_\mathfrak{p}$-module in the sense of (2.5.1).

Proposition 4.10.11 *Let A be a commutative noetherian ring, and K_A a pointwise dualizing module of A. Set $\mathcal{A}_A := {}_A\mathbb{M}_f$, $\omega_A :=$ add K_A, $\mathcal{X}_A$ to be the full subcategory of $\mathcal{A}_A$ consisting of maximal Cohen–Macaulay A-modules, and $\mathcal{Y}_A$ to be the full subcategory of $\mathcal{A}_A$ consisting of $M \in \mathcal{A}_A$ such that for any $\mathfrak{p} \in \mathrm{Spec}\, A$, $\mathrm{inj.dim}_{A_\mathfrak{p}} M_\mathfrak{p} < \infty$. We define $\mathcal{P}_A :=$ add A. Then the following hold.*

(4.10.12) $\mathcal{X}_A = {}^{\perp}K_A$

(4.10.13) $\mathcal{Y}_A = \hat{\omega}_A$

(4.10.14) $(\mathcal{X}_A, \mathcal{Y}_A, \omega_A)$ *is an Auslander–Buchweitz context.*

(4.10.15) $\mathrm{Hom}_A(?, K_A)$ *is a contravariant equivalence of exact categories from* $\mathcal{X}_A$ *to itself, with* $\mathrm{Hom}_A(?, K_A)$ *itself its quasi-inverse.*

(4.10.16) $\mathrm{Hom}_A(K_A, ?)$ *is an equivalence of exact categories from* $\mathcal{Y}_A$ *to* $\hat{\mathcal{P}}_A$, *with* $K_A \otimes_A ?$ *its quasi-inverse. In fact,*

$$K_A \otimes_A \mathrm{Hom}_A(K_A, N) \to N$$

given by $q \otimes f \mapsto fq$ *is an isomorphism for* $N \in \mathcal{Y}_A$, *and*

$$M \to \mathrm{Hom}_A(K_A, K_A \otimes M)$$

given by $m \mapsto (q \mapsto q \otimes m)$ *is an isomorphism for* $M \in \hat{\mathcal{P}}_A$.

Proof. For $M \in \mathcal{A}_A$ and $\mathfrak{p} \in \mathrm{Spec}\, A$, we set $c_M(\mathfrak{p}) := \dim A_{\mathfrak{p}} - \operatorname{depth} M_{\mathfrak{p}}$. Let

$$\mathbb{F} : \cdots \to F_2 \xrightarrow{d_2} F_1 \xrightarrow{d_1} F_0 \xrightarrow{d_0} M \to 0$$

be a free resolution of M with each F_i A-finite. We set $\Omega_i M := \mathrm{Im}\, d_i$. Then it is easy to see that $(\Omega_i M)_{\mathfrak{p}}$ is a maximal Cohen–Macaulay $A_{\mathfrak{p}}$-module if and only if $c_M(\mathfrak{p}) \leq i$. By Corollary 2.12.3, c_M is an upper semicontinuous function over $\mathrm{Spec}\, A$. A similar argument applied to $p_M(\mathfrak{p}) := \mathrm{proj.dim}_{A_{\mathfrak{p}}} M_{\mathfrak{p}}$ instead of c_M yields that p_M is also upper semicontinuous.

As $\mathrm{Spec}\, A$ is quasi-compact, we have $d := \max c_M(\mathfrak{p}) < \infty$, and by definition, $\Omega_d M$ is maximal Cohen–Macaulay. Hence, we have $M \in \hat{\mathcal{X}}_A$, and $\mathcal{A}_A = \hat{\mathcal{X}}_A$. Similarly, an object M of $\mathcal{A}_A$ belongs to $\mathcal{P}_A$ if and only if $\mathrm{proj.dim}_{A_{\mathfrak{p}}} M_{\mathfrak{p}} < \infty$ for any $\mathfrak{p} \in \mathrm{Spec}\, A$.

Next, we show that $\mathcal{X}_A = {}^{\perp}K_A$. To prove this, we may assume that A is local. In this case, as K_A is a dualizing complex of A, the assertion follows easily from the local duality (Theorem 2.10.7).

Also, for any $M \in \mathcal{X}_A$, we have that $\mathrm{Hom}_A(M, K_A) \in \mathcal{X}_A$ and

$$M \to \mathrm{Hom}_A(\mathrm{Hom}_A(M, K_A), K_A)$$

is an isomorphism. This is also checked after localization. Hence, (4.10.15) holds.

Next, we show that ω_A is an injective cogenerator of $\mathcal{X}_A$. We already know that $\omega_A = \mathrm{add}\, K_A$ is $\mathcal{X}$-injective, and $\omega_A \subset \mathcal{X}_A$. When we take an exact sequence

$$0 \to N \to F \to \mathrm{Hom}_A(M, K_A) \to 0$$

with F A-finite free, then the sequence is an exact sequence in $\mathcal{X}_A$. Applying the functor $\mathrm{Hom}_A(?, K_A)$ to the exact sequence, we have an exact sequence

$$0 \to M \to \mathrm{Hom}_A(F, K_A) \to \mathrm{Hom}_A(N, K_A) \to 0$$

in $\mathcal{X}_A$ again. As $\mathrm{Hom}_A(F, K_A) \in \omega_A$, we have that ω_A is an injective cogenerator of $\mathcal{X}_A$.

Next, we show that $\mathcal{X}_A \cap \mathcal{Y}_A \subset \omega_A$. To verify this, we take $M \in \mathcal{X}_A \cap \mathcal{Y}_A$, and it suffices to show that $\mathrm{Hom}_A(M, K_A) \in \mathcal{P}_A$. Hence, we may assume that $(A, \mathfrak{m})$ is local. As $\mathrm{Hom}_A(M, K_A)$ is maximal Cohen–Macaulay, it suffices to show that

$$\mathrm{proj.dim}_A \mathrm{Hom}_A(M, K_A) < \infty$$

by Theorem 2.5.3. Let $\mathbb{F}$ be the minimal free resolution of $A/\mathfrak{m}$, and $I^\bullet$ the minimal injective resolution of K_A. Then as we know that $M \in {}^\perp K_A$, we have

$$R\,\mathrm{Hom}^\bullet_A(\mathbb{F}, M) \cong R\,\mathrm{Hom}^\bullet_A(\mathbb{F}, R\,\mathrm{Hom}^\bullet_A(R\,\mathrm{Hom}^\bullet_A(M, I^\bullet), I^\bullet))$$
$$\cong R\,\mathrm{Hom}^\bullet_A(\mathbb{F} \otimes^L_A \mathrm{Hom}^\bullet_A(M, I^\bullet), I^\bullet) \cong R\,\mathrm{Hom}^\bullet_A(\mathbb{F} \otimes^L_A \mathrm{Hom}_A(M, K_A), I^\bullet).$$

Note that $R\,\mathrm{Hom}^\bullet_A(\mathbb{F}, M)$ has bounded homology groups, since $M \in \mathcal{Y}_A$. Hence,

$$\mathbb{F} \otimes^L_A \mathrm{Hom}_A(M, K_A) \cong R\,\mathrm{Hom}^\bullet_A(R\,\mathrm{Hom}^\bullet_A(\mathbb{F} \otimes^L_A \mathrm{Hom}_A(M, K_A), I^\bullet), I^\bullet)$$

also has bounded homologies. This shows $\mathrm{proj.dim}_A \mathrm{Hom}_A(M, K_A) < \infty$. Hence, we have that (4.10.14) holds. The assertions (4.10.12) and (4.10.13) are consequences of Theorem 1.12.10.

The assertion (4.10.16) is well-known. It was first proved by Sharp [136], and is generalized by Avramov and Foxby [18, Corollary 3.6].

We prove the first assertion of (4.10.16). As $N \in \mathcal{Y}_A = \hat{\omega}_A$, there is a finite ω_A-resolution $\mathbb{W}$ of N. Since $K_A \otimes_A \mathrm{Hom}_A(K_A, K_A) \to K_A$ is an isomorphism, we have that $K_A \otimes_A \mathrm{Hom}_A(K_A, \mathbb{W}) \to \mathbb{W}$ is an isomorphism of complexes. As the augmented complex $\mathbb{W} \to N \to 0$ is a bounded exact complex consisting of objects of $\mathcal{Y}_A \subset K_A^\perp$, we have that $\mathrm{Hom}_A(K_A, \mathbb{W})$ is a resolution of $\mathrm{Hom}_A(K_A, N)$. By the five lemma, $K_A \otimes_A \mathrm{Hom}_A(K_A, N) \to N$ is also an isomorphism. The second assertion is proved similarly, utilizing Theorem 4.10.19 below. □

As in the proof of the theorem, the following is easy to prove.

Lemma 4.10.17 *Let R be a regular ring, and $V \in {}_R\mathbb{M}_f$. Then we have* $\mathrm{proj.dim}_R V < \infty$.

(4.10.18) Let R be a noetherian commutative ring. We say that an R-module N is locally of finite flat dimension if $\text{flat.dim}_{R_\mathfrak{p}} N_\mathfrak{p} < \infty$ for $\mathfrak{p} \in \operatorname{Spec} R$. Note that if N is of finite flat dimension, then it is locally of finite flat dimension. The converse is true if the Krull dimension of R is finite, see [55, Corollary 3.4]. Related to the proof of (4.10.16), the following holds [18, Corollary 3.6].

Theorem 4.10.19 *Let R be a Cohen–Macaulay ring, and M a maximal Cohen–Macaulay R-module. If N is an R-module locally of finite flat dimension, then we have $\operatorname{Tor}_i^R(M, N) = 0$ for $i > 0$.*

Note that the proof is easily reduced to the case that $(R, \mathfrak{m})$ is a complete local ring, and we may also assume that $M = \bar{\omega}_R$ by (4.10.14).

Corollary 4.10.20 *Let R and M be as in the theorem, and let N be a perfect R-module of codimension h. Then we have $\operatorname{Ext}_R^i(N, M) = 0$ $(i \neq h)$.*

By Lemma 2.9.1, we have $\text{proj.dim}_R \operatorname{Ext}_R^h(N, R) < \infty$, and

$$\operatorname{Ext}_R^i(N, M) \cong \operatorname{Tor}_{h-i}^R(\operatorname{Ext}_R^h(N, R), M).$$

By the theorem, the assertion follows. □

Proposition 4.10.11 is well-known as the *Cohen–Macaulay approximation*. The name 'approximation' comes from the fact that $\mathcal{X}_A$ is contravariantly finite (see Theorem 1.12.10). In Theorem 4.10.8, we have that any object of $\mathcal{Y}$ is of finite injective dimension. However, in the situation of Proposition 4.10.11, this is not true any more if $\dim A = \infty$.

(4.10.21) For a finite dimensional algebra A over a field, Auslander–Buchweitz contexts in $({}_A\mathbb{M})_f$ and basic cotilting modules of A are in one-to-one correspondence. This beautiful result was proved by Auslander–Reiten [11].

Let A be a ring, and M an A-module. We say that M is *basic* if M does not have any direct summand of the form $N \oplus N$, where N is a non-zero A-module.

Theorem 4.10.22 *Let k be a field, and A a finite dimensional k-algebra. We set $\mathcal{A}_A := ({}_A\mathbb{M})_f$. Consider the following.*

a *A full subcategory ω in $\mathcal{A}_A$ such that $\omega = \operatorname{add} \omega \subset \omega^\perp$ and $\hat{\mathcal{X}}_\omega = \mathcal{A}_A$, where $\mathcal{X}_\omega$ is the full subcategory of $\mathcal{A}_A$ consisting of $X \in {}^\perp\omega$ such that there exists an exact sequence*

$$0 \to X \to T^0 \xrightarrow{f^0} T^1 \to \cdots \to T^n \xrightarrow{f^n} T^{n+1} \to \cdots$$

such that $T^i \in \omega$ and $\operatorname{Im} f^i \in {}^\perp\omega$.

b *A full subcategory $\mathcal{X}$ of $\mathcal{A}_A$ which is closed under extensions, epikernels, and direct summands, and has an injective cogenerator, such that $\hat{\mathcal{X}} = \mathcal{A}_A$.*

c *A covariantly finite full subcategory $\mathcal{Y}$ of $\mathcal{A}_A$ closed under extensions, monocokernels, and direct summands such that ${}_AA^* \in \mathcal{Y}$ and any object in $\mathcal{Y}$ is of finite injective dimension.*

d *An Auslander–Buchweitz context $(\mathcal{X}, \mathcal{Y}, \omega)$ in $\mathcal{A}_A$.*

e *The isomorphism class of a basic cotilting A-module T.*

The objects **a–e** *above are in one-to-one correspondence by the following correspondences.*

a⇒b *ω to $\mathcal{X}_\omega$.*

b⇒c *$\mathcal{X}$ to $\mathcal{X}^\perp$.*

c⇒a *$\mathcal{Y}$ to ${}^\perp\mathcal{Y} \cap \mathcal{Y}$.*

a,b,c⇒d *Obvious correspondence.*

a,b,c,d⇒e *As we have ${}_AA^* \in \mathcal{X}^\perp = \hat{\omega}$, there is an exact sequence*

$$0 \to \omega_n \to \cdots \to \omega_1 \to \omega_0 \to {}_AA^* \to 0$$

with $\omega_i \in \omega$. Starting with the Krull–Schmidt decomposition of $\omega_0 \oplus \cdots \oplus \omega_n$, we get a basic module T, removing N whenever we find $N \oplus N$ in the decomposition. The isomorphism class of T is uniquely determined by $(\mathcal{X}, \mathcal{Y}, \omega)$, and T is the corresponding cotilting module.

e⇒a *T to* $\operatorname{add} T$.

Corollary 4.10.23 *Let A be a finite dimensional algebra over a field k, and $(\mathcal{X}, \mathcal{Y}, \omega)$ an Auslander–Buchweitz context in ${}_A\mathbb{M}_f$. Then the number of isomorphism classes of indecomposable objects in ω is equal to the number of isomorphism classes of simple A-modules.*

Proof. By the theorem, the number of isomorphism classes of indecomposable objects in ω agrees with the number of indecomposable direct summands of T in the theorem, which agrees with the number of indecomposable direct summands of T^*. As T^* is a tilting A^{op}-module, these numbers agree with the number of simples of A^{op}, which is equal to the number of simples of A by (4.9.9). □

Notes and References. There is no new result at all in this section. In this section, we listed basic results in the representation theory of algebraic groups and algebras. For more, we refer the reader to [87], [90], [42] for algebraic groups, and to [12], [11], [112], [113] for algebras. Although we assume that a reductive group (over an algebraically closed field) is connected, note that it is not always assumed in the literature.

5 Basics on equivariant modules

5.1 Cocommutative Hopf algebra actions

(5.1.1) Let R be a commutative ring, U a cocommutative R-Hopf algebra, and A a commutative U-module R-algebra.

Let M be an $A\#U$-module, and N a U-module. The R-module $M \otimes N$ is an $A\#U$-module by

$$(a \otimes u)(m \otimes n) = \sum_{(u)} a(u_{(1)}m) \otimes u_{(2)}n \quad (a \in A,\, u \in U,\, m \in M,\, n \in N),$$

where $A\#U$ denotes the smash product (3.14.3). If M is a U-module and N an $A\#U$-module, then letting A act on N we get an $A\#U$-module $M \otimes N$. If both M and N are $A\#U$-modules, then there are two different ways to see that $M \otimes N$ is an $A\#U$-module. Unless otherwise specified, we understand that A acts on M (so we take the former definition).

Let M and N be $A\#U$-modules. When we define an R-linear map

$$d : M \otimes (A \otimes N) \to M \otimes N$$

by $d(m \otimes a \otimes n) = am \otimes n - m \otimes an$, then it is an $A\#U$-linear map, and $M \otimes_A N = \operatorname{Coker} d$ also has an $A\#U$-module structure.

(5.1.2) Let M be an $A\#U$-module, and N a U-module. Then $\operatorname{Hom}(M, N)$ is an $A\#U$-module in a natural way. The action of U is given as in (3.12.1), and the action of A is that on M. If M is a U-module and N is an $A\#U$-module, then letting A act on N, $\operatorname{Hom}(M, N)$ is an $A\#U$-module. If both M and N are $A\#U$-modules, then there are two different ways to see $\operatorname{Hom}(M, N)$ as an $A\#U$-module. We give priority to the definition in which A acts on N.

If both M and N are $A\#U$-modules, then $\operatorname{Hom}_A(M, N)$ is an $A\#U$-submodule of $\operatorname{Hom}(M, N)$. Note that we have

$$\operatorname{Hom}_{A\#U}(M, N) = \operatorname{Hom}_A(M, N) \cap \operatorname{Hom}_U(M, N) = H^0(U, \operatorname{Hom}_A(M, N)).$$

(5.1.3) Let R and U be as in (5.1.1). Let A and B be commutative U-module algebras, and $\varphi : A \to B$ a U-module algebra map. Then we define

$$\varphi\#U : A\#U = A \otimes U \to B \otimes U = B\#U$$

to be $\varphi \otimes \mathrm{id}_U$. It is easy to see that $\varphi\#U$ is an R-algebra map. In particular, any $B\#U$-module is an $A\#U$-module by restriction.

Let M be a $B\#U$-module, and V an $A\#U$-module. Then $M \otimes_A V$, $V \otimes_A M$, $\operatorname{Hom}_A(M, V)$ and $\operatorname{Hom}_A(V, M)$ are $B\#U$-modules in a natural way, and as $A\#U$-modules they agree with the ones defined in (5.1.1) and (5.1.2).

(5.1.4) Let R, U, and $\varphi : A \to B$ be as in (5.1.3). Let M and N be $B\#U$-modules, and V and W be $A\#U$-modules. Then the standard maps

$$
\begin{array}{lll}
(5.1.5) & & M \to \operatorname{Hom}_A(A, M) \qquad (m \mapsto (a \mapsto am)) \\
(5.1.6) & \circ : & \operatorname{Hom}_A(V, W) \otimes_A \operatorname{Hom}_A(M, V) \to \operatorname{Hom}_A(M, W) \\
(5.1.7) & \circ : & \operatorname{Hom}_A(W, M) \otimes_A \operatorname{Hom}_A(V, W) \to \operatorname{Hom}_A(V, M) \\
(5.1.8) & \Psi : & \operatorname{Hom}_A(V \otimes_A W, M) \cong \operatorname{Hom}_A(V, \operatorname{Hom}_A(W, M)) \\
(5.1.9) & \Psi : & \operatorname{Hom}_A(M \otimes_A V, W) \cong \operatorname{Hom}_A(M, \operatorname{Hom}_A(V, W)) \\
(5.1.10) & \Psi : & \operatorname{Hom}_A(V \otimes_A M, W) \cong \operatorname{Hom}_A(V, \operatorname{Hom}_A(M, W)) \\
(5.1.11) & \Psi : & \operatorname{Hom}_B(V \otimes_A M, N) \cong \operatorname{Hom}_A(V, \operatorname{Hom}_B(M, N)) \\
(5.1.12) & \Psi : & \operatorname{Hom}_B(M \otimes_A V, N) \cong \operatorname{Hom}_B(M, \operatorname{Hom}_A(V, N)) \\
(5.1.13) & \Psi : & \operatorname{Hom}_A(M \otimes_B N, V) \cong \operatorname{Hom}_B(M, \operatorname{Hom}_A(N, V)) \\
(5.1.14) & & M \otimes_A A \cong M \cong A \otimes_A M \\
(5.1.15) & & (M \otimes_A V) \otimes_A W \cong M \otimes_A (V \otimes_A W) \\
(5.1.16) & \tau : & M \otimes_A V \cong V \otimes_A M \\
(5.1.17) & & M \to \operatorname{Hom}_A(\operatorname{Hom}_A(M, V), V) \qquad \text{(the duality map)} \\
(5.1.18) & & M \otimes_A \operatorname{Hom}_A(V, W) \to \operatorname{Hom}_A(V, M \otimes_A W) \\
(5.1.19) & & M \otimes_B \operatorname{Hom}_A(V, N) \to \operatorname{Hom}_A(V, M \otimes_B N)
\end{array}
$$

are $B\#U$-linear maps, which are natural with respect to M, N, V, and W. As a special case, we will use the case $A = R$ or $A = B$ frequently.

Taking the invariances $H^0(U, ?)$ of both sides of (5.1.11), (5.1.12), and (5.1.13), we get natural isomorphisms

$$
\begin{array}{lll}
(5.1.20)\ \Psi : \operatorname{Hom}_{B\#U}(V \otimes_A M, N) & \cong & \operatorname{Hom}_{A\#U}(V, \operatorname{Hom}_B(M, N)) \\
(5.1.21)\ \Psi : \operatorname{Hom}_{B\#U}(M \otimes_A V, N) & \cong & \operatorname{Hom}_{B\#U}(M, \operatorname{Hom}_A(V, N)) \\
(5.1.22)\ \Psi : \operatorname{Hom}_{A\#U}(M \otimes_B N, V) & \cong & \operatorname{Hom}_{B\#U}(M, \operatorname{Hom}_A(N, V)),
\end{array}
$$

respectively. In particular, we have

Lemma 5.1.23 *Let R, U, $f : A \to B$, M, N, and V be as in* (5.1.4). *Then the following hold.*

1 *If N is $B\#U$-injective and M is A-flat, then* $\operatorname{Hom}_B(M, N)$ *is $A\#U$-injective.*

2 *If N is $B\#U$-injective and V is A-flat, then* $\operatorname{Hom}_A(V, N)$ *is $B\#U$-injective.*

3 *If N is B-flat and V is $A\#U$-injective, then* $\operatorname{Hom}_A(N, V)$ *is $B\#U$-injective.*

4 *If M is B-projective and V is $A\#U$-projective, then $V \otimes_A M$ is $B\#U$-projective.*

5 *If M is $B\#U$-projective and V is A-projective, then $M \otimes_A V$ is $B\#U$-projective.*

6 *If N is A-projective and M is $B\#U$-projective, then $M \otimes_B N$ is $A\#U$-projective.*

In particular, if V is an $A\#U$-projective module, then $V \otimes_A B$ is $B\#U$-projective.

Moreover, since $B\#U \cong B \otimes_A (A\#U) \cong (A\#U) \otimes_A B$ as $A\#U$-modules, we have

Corollary 5.1.24 *If B is A-projective, then any $B\#U$-projective module is $A\#U$-projective.*

When we consider the case $M = B$ in Lemma 5.1.23, **1**, we have the following.

Corollary 5.1.25 *If B is A-flat, then any $B\#U$-injective module is $A\#U$-injective.*

5.2 Tor^A and Ext_A as $A\#U$-modules

(5.2.1) Let R be a commutative ring, U a cocommutative R-Hopf algebra, A and B commutative U-module algebras, and $\varphi : A \to B$ a U-module algebra map. We assume that U is R-projective.

Lemma 5.2.2 *Any $A\#U$-projective module is A-projective. Any $A\#U$-injective module is A-injective.*

Proof. We prove the first assertion. It suffices to show that $A\#U$ is A-projective. As the action of A on $A\#U$ is given by $a(b \otimes u) = ab \otimes u$ and U is R-projective, $A\#U$ is A-projective.

We prove the second assertion. It is enough to show that any injective $A\#U$-module of the form $\mathrm{Hom}(A\#U, I)$, with I an injective R-module, is A-injective. Hence, it suffices to show that the A-module induced by $A\#U_{A\#U}$ is A-projective (as A is not contained in the center of $A\#U$, this is not completely the same thing as the first paragraph). To verify this, we show that $f : {}_A(A\#U) \to (A\#U)_A$ defined by $f(a \otimes u) = \sum_{(u)} u_{(1)}a \otimes u_{(2)}$ is an A-linear isomorphism. This is A-linear, as A is commutative and we have

$$(f(a \otimes u))b = \sum_{(u)} (u_{(1)}a)(u_{(2)}b) \otimes u_{(3)} = \sum_{(u)} u_{(1)}(ab) \otimes u_{(2)} = f(b(a \otimes u)).$$

Since U is cocommutative, the inverse of f is given by $a \otimes u \mapsto \sum_{(u)} (Su_{(1)})a \otimes u_{(2)}$, where S is the antipode of U. Hence, f is an A-isomorphism. □

(5.2.3) Let R, U, and $\varphi : A \to B$ be as in (5.2.1). Let M be a $B\#U$-module, and V an $A\#U$-module. Then the $B\#U$-module $L_i(M\otimes_A ?)(V)$ is isomorphic to $\operatorname{Tor}_i^A(M,V)$ as an A-module, since any $A\#U$-projective module is A-flat. So we denote the $B\#U$-module $L_i(M\otimes_A ?)(V)$ by $\operatorname{Tor}_i^A(M,V)$. If $\mathbb{F}$ is an A-flat resolution of M in ${}_{B\#U}\mathbb{M}$, then we have $H_i(\mathbb{F}\otimes_A V) \cong \operatorname{Tor}_i^A(M,V)$ in ${}_{B\#U}\mathbb{M}$. In particular, if B is A-flat, then we have

$$L_i(? \otimes_A V)(M) \cong \operatorname{Tor}_i^A(M,V),$$

since any $B\#U$-projective module is A-flat in this case.

Similarly, we define

$$\begin{aligned}
\operatorname{Tor}_i^A(V,M) &:= L_i(? \otimes_A M)(V);\\
\operatorname{Ext}_A^i(M,V) &:= R^i(\operatorname{Hom}_A(M,?))(V);\\
\operatorname{Ext}_A^i(V,M) &:= R^i(\operatorname{Hom}_A(?,M))(V),
\end{aligned}$$

and these are $B\#U$-modules. We have $L_i(V\otimes_A ?)(M) \cong \operatorname{Tor}_i^A(V,M)$ in ${}_{B\#U}\mathbb{M}$, if B is A-flat. We have $R^i(\operatorname{Hom}_A(?,V))(M) \cong \operatorname{Ext}_A^i(M,V)$ in ${}_{B\#U}\mathbb{M}$, if B is A-projective. We have $R^i(\operatorname{Hom}_A(V,?))(M) \cong \operatorname{Ext}_A^i(V,M)$ in ${}_{B\#U}\mathbb{M}$, if B is A-flat, by Corollary 5.1.25.

Proposition 5.2.4 *Let M and N be $B\#U$-modules, and V an $A\#U$-module. If M is A-flat, then we have spectral sequences*

$$(5.2.5)\qquad E_2^{p,q} = \operatorname{Ext}_A^p(V, \operatorname{Ext}_B^q(M,N)) \Rightarrow \operatorname{Ext}_B^{p+q}(V\otimes_A M, N)$$

$$(5.2.6)\qquad E_2^{p,q} = \operatorname{Ext}_{A\#U}^p(V, \operatorname{Ext}_B^q(M,N)) \Rightarrow \operatorname{Ext}_{B\#U}^{p+q}(V\otimes_A M, N)$$

in the categories ${}_{B\#U}\mathbb{M}$ and ${}_{H^0(U,B)}\mathbb{M}$, respectively. If, moreover, M is B-projective, then we have isomorphisms

$$(5.2.7)\qquad \operatorname{Ext}_A^i(V, \operatorname{Hom}_B(M,N)) \cong \operatorname{Ext}_B^i(V\otimes_A M, N)$$

$$(5.2.8)\qquad \operatorname{Ext}_{A\#U}^i(V, \operatorname{Hom}_B(M,N)) \cong \operatorname{Ext}_{B\#U}^i(V\otimes_A M, N)$$

of ${}_{B\#U}\mathbb{M}$ and ${}_{H^0(U,B)}\mathbb{M}$, respectively.

Proof. We prove the existence of a spectral sequence (5.2.5). Let $\mathbb{F}$ be an $A\#U$-projective resolution of V, and $\mathbb{I}$ a $B\#U$-injective resolution of N. Then we have quasi-isomorphisms of complexes

$$\operatorname{Hom}_A^\bullet(\mathbb{F}, \operatorname{Hom}_B^\bullet(M,\mathbb{I})) \cong \operatorname{Hom}_B^\bullet(\mathbb{F}\otimes_A M, \mathbb{I}) \leftarrow \operatorname{Hom}_B^\bullet(V\otimes_A M, \mathbb{I}),$$

by (5.1.11) and the A-flatness of M. As $\mathbb{F}$ is an A-projective complex, the assertion follows immediately.

The rest of the proposition is proved similarly, using (5.1.20). □

(5.2.9) For later use, we need to study unbounded derived functors. Let R and U be as in (5.2.1), and A a commutative U-module algebra.

For $\mathbb{F}, \mathbb{G} \in C({}_{A\#U}\mathbb{M})$, the complex $\mathrm{Hom}_A^\bullet(\mathbb{F}, \mathbb{G})$ is defined as in (1.4.2), and it is again a complex of $A\#U$-modules. It is easy to see that the construction induces a bifunctor

$$\mathrm{Hom}_A^\bullet : K({}_{A\#U}\mathbb{M})^{\mathrm{op}} \times K({}_{A\#U}\mathbb{M}) \to K({}_{A\#U}\mathbb{M}).$$

Lemma 5.2.10 *If $\mathbb{F}$ is an exact A-complex and $\mathbb{G}$ is a K-injective $A\#U$-complex, then $\mathrm{Hom}_A^\bullet(\mathbb{F}, \mathbb{G})$ is an exact A-complex.*

Proof. We have a standard isomorphism

$$\mathrm{Hom}_A^\bullet(\mathbb{F}, \mathbb{G}) \cong \mathrm{Hom}_{A\#U}^\bullet((A\#U) \otimes_A \mathbb{F}, \mathbb{G}).$$

By the proof of Lemma 5.2.2, the A-module $A\#U$ induced by the right regular action $(A\#U)_{A\#U}$ is A-projective. Hence, $(A\#U) \otimes_A \mathbb{F}$ is still exact. As any $A\#U$-linear chain map $(A\#U) \otimes_A \mathbb{F}[-n] \to \mathbb{G}$ is null homotopic by the K-injectivity of $\mathbb{G}$, the complex $\mathrm{Hom}_{A\#U}^\bullet((A\#U) \otimes_A \mathbb{F}, \mathbb{G})$ is exact, and hence so is $\mathrm{Hom}_A^\bullet(\mathbb{F}, \mathbb{G})$. □

By the lemma and Theorem 1.4.12, there is a derived functor

$$\underline{R}\,\mathrm{Hom}_A^\bullet : D({}_{A\#U}\mathbb{M})^{\mathrm{op}} \times D({}_{A\#U}\mathbb{M}) \to D({}_{A\#U}\mathbb{M}).$$

For $\mathbb{F}, \mathbb{G} \in K({}_{A\#U}\mathbb{M})$, $\underline{R}\,\mathrm{Hom}_A^\bullet(\mathbb{F}, \mathbb{G})$ is $\mathrm{Hom}_A^\bullet(\mathbb{F}, \mathbb{I})$, where $\mathbb{I}$ is the K-injective resolution of $\mathbb{G}$. By the lemma, we also have that the construction of $\underline{R}\,\mathrm{Hom}_A^\bullet$ is compatible with the forgetful functor $D({}_{A\#U}\mathbb{M}) \to D({}_A\mathbb{M})$.

(5.2.11) For $\mathbb{A} \in C({}_{A\#U}\mathbb{M})$, there is a quasi-isomorphism $\mathbb{F} \to \mathbb{A}$ in $C({}_{A\#U}\mathbb{M})$ such that $\mathbb{F}$ is the inductive limit of an inductive system indexed by $\mathbb{N}$, of $A\#U$-projective complexes bounded above [137], [25]. This shows that the bifunctor $\otimes_A^\bullet$ induces a bifunctor

$$\otimes_A^L : D({}_{A\#U}\mathbb{M}) \times D({}_{A\#U}\mathbb{M}) \to D({}_{A\#U}\mathbb{M}),$$

and the construction of $\otimes_A^L$ is compatible with the forgetful functor

$$D({}_{A\#U}\mathbb{M}) \to D({}_A\mathbb{M}).$$

For $\mathbb{F} \in D({}_{A\#}\mathbb{M})$, the composition induces

$$\underline{R}\,\mathrm{Hom}_A^\bullet(\mathbb{F}, \mathbb{G}) \otimes_A^L \underline{R}\,\mathrm{Hom}_A^\bullet(\mathbb{E}, \mathbb{F}) \to \underline{R}\,\mathrm{Hom}_A^\bullet(\mathbb{E}, \mathbb{G}),$$

which is natural with respect to $\mathbb{E} \in D({}_{A\#U}\mathbb{M})^{\mathrm{op}}$ and $\mathbb{G} \in D({}_{A\#U}\mathbb{M})$. This is calculated as the composite

$$y : \mathbb{Q} \otimes_A^\bullet \mathrm{Hom}_A^\bullet(\mathbb{E}, \mathbb{F}') \to \mathrm{Hom}_A^\bullet(\mathbb{F}', \mathbb{G}') \otimes_A^\bullet \mathrm{Hom}_A^\bullet(\mathbb{E}, \mathbb{F}') \xrightarrow{\circ} \mathrm{Hom}_A^\bullet(\mathbb{E}, \mathbb{G}'),$$

where $\mathbb{G} \to \mathbb{G}'$ and $\mathbb{F} \to \mathbb{F}'$ are K-injective resolutions, and

$$\mathbb{Q} \to \operatorname{Hom}^\bullet_A(\mathbb{F}', \mathbb{G}')$$

is a quasi-isomorphism, where $\mathbb{Q}$ is the inductive limit of an inductive system indexed by $\mathbb{N}$, of $A\#U$-projective complexes bounded above. Taking the cohomology, we have the Yoneda product of hyper-Ext groups

$$\operatorname{Ext}^i_A(\mathbb{F}, \mathbb{G}) \otimes_A \operatorname{Ext}^j_A(\mathbb{E}, \mathbb{F}) \to \operatorname{Ext}^{i+j}(\mathbb{E}, \mathbb{G}),$$

which is $A\#U$-linear. Concretely, it is given as the composite map

$$\operatorname{Ext}^i_A(\mathbb{F}, \mathbb{G}) \otimes_A \operatorname{Ext}^j_A(\mathbb{E}, \mathbb{F}) = H^i(\mathbb{Q}) \otimes_A H^j(\operatorname{Hom}^\bullet_A(\mathbb{E}, \mathbb{F}'))$$
$$\to H^{i+j}(\mathbb{Q} \otimes^\bullet_A \operatorname{Hom}^\bullet_A(\mathbb{E}, \mathbb{F}')) \xrightarrow{H^{i+j}(y)} H^{i+j}(\operatorname{Hom}^\bullet_A(\mathbb{E}, \mathbb{G}')) = \operatorname{Ext}^{i+j}(\mathbb{E}, \mathbb{G}).$$

Clearly, this is nothing but the usual Yoneda product as an A-linear map, and hence it is associative. In particular, for an $A\#U$-module M, the Yoneda algebra $\bigoplus_{i\geq 0} \operatorname{Ext}^i_A(M, M)$ admits an $A\#U$-action.

5.3 (G, A)-modules

(5.3.1) Let R be a commutative ring, $G = \operatorname{Spec} H$ an affine flat R-group scheme, and A a commutative G-algebra (or an H-comodule algebra). Set $X := \operatorname{Spec} A$. If $f : A \to B$ is an H-comodule algebra map, then we say that f is a *G-algebra map*, or sometimes that B is a *(G, A)-algebra*.

We call an (H, A)-Hopf module a *(G, A)-module*, and instead of using ${}_A\mathbb{M}^H$, we use ${}_{G,A}\mathbb{M}$. Note that $\mathbb{M}^H = {}_{G,R}\mathbb{M}$ is nothing but ${}_G\mathbb{M}$. Instead of writing $\operatorname{Ext}^i_{{}_G\mathbb{M}}$, $H^i(\mathbb{M}^H, ?)$, or $\operatorname{Ext}^i_{{}_{G,A}\mathbb{M}}$, we use the notation Ext^i_G, $H^i(G, ?)$, and $\operatorname{Ext}^i_{G,A}$, respectively. The functor $H^0(G, ?)$ is also written $(?)^G$, and is called the *G-invariance functor.* A (G, A)-submodule of the (G, A)-module A is called a *G-ideal.* If I is a G-ideal of A, then A/I is a (G, A)-algebra in a natural way. The following is checked easily.

Lemma 5.3.2 *Let R, A and G be as above. Then ${}_{G,A}\mathbb{M}$ is an abelian category which satisfies the* (AB5) *condition. The forgetful functors ${}_{G,A}\mathbb{M} \to {}_G\mathbb{M}$ and ${}_{G,A}\mathbb{M} \to {}_A\mathbb{M}$ are faithful exact and preserve inductive limits.*

(5.3.3) Let R, A and G be as above. Let $M, N \in {}_{G,A}\mathbb{M}$ and $V, W \in {}_G\mathbb{M}$. Then $M \otimes W$ and $V \otimes N$ are G-modules. They are A-modules with the actions given by $a(m \otimes w) := am \otimes w$ and $a(v \otimes n) := v \otimes an$, respectively, and are easily checked to be (G, A)-modules. Then we have two different (G, A)-module structures of $M \otimes N$. As in the case of $A\#U$-modules, we give priority to the one in which A acts on M, unless otherwise specified. The A-module G-module $M \otimes_A N$ is a (G, A)-module, which is a quotient object of $M \otimes N$ via the natural projection. If B and C are commutative (G, A)-algebras, then the tensor product $B \otimes_A C$ is both a (G, A)-module and a commutative A-algebra, which is easily checked to be a (G, A)-algebra. This tensor product corresponds to the fiber product $\operatorname{Spec} B \times_X \operatorname{Spec} C$ with the diagonal G-action, which is a fiber product in the category of G-schemes, where $X = \operatorname{Spec} A$.

(5.3.4) Let R, G, A, M, N, V, and W be as in (5.3.3). If M (resp. V) is of finite presentation as an R-module, then $\operatorname{Hom}(M, W)$ (resp. $\operatorname{Hom}(V, N)$) is a G-module A-module, which is easily checked to be a (G, A)-module.

If M is of finite presentation as an A-module, then as we have

$$\operatorname{Hom}_A(M, N) \otimes H \cong \operatorname{Hom}_A(M, N \otimes H)$$

by Lemma 2.1.7, $\operatorname{Hom}_A(M, N)$ is a G-module A-module, which is checked to be a (G, A)-module. In this case, we have

$$\operatorname{Hom}_{G,A}(M, N) = \operatorname{Hom}_G(M, N) \cap \operatorname{Hom}_A(M, N) = H^0(G, \operatorname{Hom}_A(M, N)).$$

(5.3.5) Let R, G, and A be as in (5.3.3), and B a commutative G-algebra. Let $\varphi : A \to B$ be a G-algebra map.

If M is a (G, B)-module and V is a (G, A)-module, then $M \otimes_A V$ and $V \otimes_A M$ are (G, B)-modules in a natural way. If, moreover, V (resp. M) is of finite presentation as an A-module, then $\operatorname{Hom}_A(V, M)$ (resp. $\operatorname{Hom}_A(M, V)$) is a (G, B)-module.

For (G, B)-modules M, N and (G, A)-modules V, W, the maps in (5.1.5)–(5.1.19) are (G, B)-homomorphisms natural with respect to M, N, V and W, provided the first argument(s) of Hom_A (resp. Hom_B) are of finite presentation as A-module(s) (resp. B-module(s)). For example, in the map (5.1.11), we require that V is of finite presentation as an A-module, and M is of finite presentation as a B-module.

However, as for (5.1.8–5.1.13), when we only assume that the first argument of Hom inside another Hom is of finite presentation, the map in question is an isomorphism of B-modules, and the set of G-linear maps is mapped bijectively onto the set of G-linear maps.

In particular, we have isomorphisms

(5.3.6) $\Psi : \operatorname{Hom}_{G,B}(V \otimes_A M, N) \cong \operatorname{Hom}_{G,A}(V, \operatorname{Hom}_B(M, N))$

(5.3.7) $\Psi : \operatorname{Hom}_{G,B}(M \otimes_A V, N) \cong \operatorname{Hom}_{G,B}(M, \operatorname{Hom}_A(V, N))$

natural with respect to V, M and N. Here we require that M is of finite presentation as a B-module in (5.3.6), and V is of finite presentation as an A-module in (5.3.7). In particular, we have

Lemma 5.3.8 *Let R, G, and $\varphi : A \to B$ be as in* (5.3.5). *Let M and N be (G, B)-modules, and V a (G, A)-module. Then the following hold:*

1 *If M is a (G, B)-module which is A-flat and of finite presentation as a B-module, and N is an injective (G, B)-module, then $\operatorname{Hom}_B(M, N)$ is an injective (G, A)-module.*

2 *Let V be an A-finite projective (G, A)-module, and N an injective (G, B)-module. Then $\operatorname{Hom}_A(V, N)$ is an injective (G, B)-module.*

Corollary 5.3.9 *If B is A-flat, then any (G, B)-injective module is (G, A)-injective.*

Lemma 5.3.10 *For any $M \in {}_{G,A}\mathbb{M}$ and any A-finite A-submodule V of M, there exists some A-finite A-submodule N of M and a (G, A)-submodule L of M such that $V \subset L \subset N$. If, moreover, R or A is noetherian, then we can take L to be A-finite.*

Proof. Let $x_1, \ldots, x_n$ be a set of generators of V as an A-module. By Lemma 3.7.1, there exists some R-finite R-submodule N_0 of M and some G-submodule L_0 of M such that $\{x_1, \ldots, x_n\} \subset L_0 \subset N_0$. We set $L_1 := A \otimes L_0$, and $N_1 := A \otimes N_0$. When we define L and N to be the image of the canonical maps $L_1 \to M$ and $N_1 \to M$, respectively, then the required conditions are satisfied. □

Corollary 5.3.11 *The category ${}_{G,A}\mathbb{M}$ is Grothendieck. In particular, ${}_{G,A}\mathbb{M}$ has injective hulls. If A is noetherian, then ${}_{G,A}\mathbb{M}$ is locally noetherian, and hence ${}_{G,A}\mathbb{M}$ satisfies the (AB3*) condition. In this case, ${}_{G,A}\mathbb{M}_f$ consists of A-finite (G, A)-modules.*

Proof. This is similar to Corollary 3.7.3, and we omit the proof. □

Corollary 5.3.12 *If either of R and A is noetherian, then any (G, A)-module is the filtered inductive limit of its A-finite (G, A)-submodules.*

Corollary 5.3.13 *Assume that A is noetherian. Let $M \in {}_{G,A}\mathbb{M}_f$ and let (N_λ) be a filtered inductive system in ${}_{G,A}\mathbb{M}$. Then the canonical map*

$$\varinjlim \operatorname{Ext}^i_{G,A}(M, N_\lambda) \to \operatorname{Ext}^i_{G,A}(M, \varinjlim N_\lambda)$$

is an isomorphism for $i \geq 0$.

Remark 5.3.14 Even if $R = k$ is a field and $A = k$, the forgetful functor ${}_G\mathbb{M} = {}_{G,k}\mathbb{M} \to {}_k\mathbb{M}$ does not preserve direct products in general.

Notes and References. All the results stated here are straightforward, and there is no reference.

Chapter II

Homological Algebra of Equivariant Modules and Matijevic-Roberts Type Theorem

1 Homological aspects of (G, A)-modules

1.1 Construction of Ext_A

(1.1.1) Let R be a commutative ring, G an affine R-flat group scheme, A and B be G-algebras, and $\varphi : A \to B$ a G-algebra map.

If M is a (G, B)-module, then M is a (G, A)-module. So we have a restriction functor $\varphi^{\#} : {}_{G,B}\mathbb{M} \to {}_{G,A}\mathbb{M}$. If V is a (G, A)-module, then $V \otimes_A B$ is a (G, B)-module, which we denote by $\varphi_{\#}V$. Note that $\varphi_{\#} : {}_{G,A}\mathbb{M} \to {}_{G,B}\mathbb{M}$ is a left adjoint functor of $\varphi^{\#}$. When we consider the case $G = \{e\}$ (or equivalently, when we forget the G-action), we have an adjoint pair $\varphi^{\#} : {}_B\mathbb{M} \to {}_A\mathbb{M}$ and $\varphi_{\#} : {}_A\mathbb{M} \to {}_B\mathbb{M}$ for an R-algebra map $\varphi : A \to B$.

(1.1.2) Let R, G, and A be as in (1.1.1). We define $\alpha : A \to A \otimes H$ by $\alpha(a) := a \otimes 1$, $\beta : A \to A \otimes H$ by $\beta := \omega_A$, and $\gamma : A \otimes H \to A \otimes H$ by $\gamma(a \otimes h) := \sum_{(a)} a_0 \otimes a_1 h$. Note that α, β, and γ are R-algebra maps, and we have $\gamma \circ \alpha = \beta$. Note also that γ is an automorphism with the inverse

$$\gamma^{-1} : A \otimes H \to A \otimes H \qquad (a \otimes h \mapsto \sum_{(a)} a_0 \otimes (Sa_1)h).$$

For an R-algebra C, we consider the trivial G-action on C, and C is a G-algebra. However, if C has another G-algebra structure, we denote the trivial G-algebra C by C', to avoid confusion. For a C-module M, when we consider the trivial G-action, then M is a (G, C')-module. We denote this by M', to avoid confusion. Note that $\alpha : A \to A \otimes H$ and $\gamma : A \otimes H \to A' \otimes H$ are G-algebra maps, whence so is $\beta : A \to A' \otimes H$. Note also that $\alpha : A' \to A' \otimes H$ is a G-algebra map, which we denote by α', to avoid confusion.

Lemma 1.1.3 *Let C and D be G-algebras, and V a (G, C)-module. Let $\alpha : C \to C \otimes D$ be the G-algebra map given by $\alpha(c) = c \otimes 1$. Then $\alpha_{\#}V$ is isomorphic to the $(G, C \otimes D)$-module $V \otimes D$, where the $C \otimes D$-action is given by $(c \otimes d)(v \otimes d') := cv \otimes dd'$.*

Proof. When we define

$$\psi : \alpha_{\#}V = V \otimes_C (C \otimes D) \to V \otimes D$$

by $v \otimes (c \otimes d) \mapsto cv \otimes d$, then it is a G-isomorphism. It is $C \otimes D$-linear as well, by the definition of $C \otimes D$-action on $V \otimes D$. □

By the lemma, $\alpha_{\#}V = V \otimes H$.

Lemma 1.1.4 *Let V be a (G, A)-module. Then $\beta_{\#}V$ is isomorphic to the $(G, A' \otimes H)$-module $V \otimes H$ with the $A' \otimes H$-action $(a \otimes h)(v \otimes h') := \sum_{(a)} a_0 v \otimes (Sa_1)hh'$.*

Proof. We have $\beta_{\#} = \gamma_{\#} \circ \alpha_{\#} \cong (\gamma^{-1})^{\#} \circ \alpha_{\#}$. Hence, $\beta_{\#}V$ is the $(G, A' \otimes H)$-module

$$(\gamma^{-1})^{\#}(V \otimes_A (A \otimes H)) \xrightarrow{\cong} (\gamma^{-1})^{\#}(V \otimes H)$$

by Lemma 1.1.3. Now the assertion is trivial. □

Lemma 1.1.5 *Let A be a G-algebra, and M a G-module A-module. Then M is a (G, A)-module if and only if $\omega_M : M \to M' \otimes H$ is a G-linear A-linear map from M to $M \otimes H = \beta^{\#}\alpha'_{\#}M'$.*

Proof. As M is a G-module, ω_M is G-linear. For $a \in A$ and $m \in M$, we have $\omega_M(am) = \sum_{(am)} (am)_0 \otimes (am)_1$, and $a\omega_M(m) = \sum_{(a),(m)} a_0 m_0 \otimes a_1 m_1$, and we have ω_M is A-linear if and only if M is a (G, A)-module. □

Lemma 1.1.6 *We have the following.*

1 *Let M be a (G, A)-module. When we define $\square_M : M \otimes H \to M' \otimes H$ by $\square_M(m \otimes h) := \sum_{(m)} m_0 \otimes m_1 h$, then we have that $\square : \beta_{\#} \to \alpha_{\#}$ is a natural isomorphism between functors from ${}_{G,A}\mathbb{M}$ to ${}_{A\otimes H}\mathbb{M}$. Moreover, the composite map*

$$[(\beta \otimes 1_H) \circ \beta]_{\#}M = [(1_A \otimes \Delta_H) \circ \beta]_{\#}M \xrightarrow{(1_A \otimes \Delta_H)_{\#}\square_M} [(1_A \otimes \Delta_H) \circ \alpha]_{\#}M$$

agrees with the composite map

$$[(\beta \otimes 1_H) \circ \beta]_\# M \xrightarrow{(\beta\otimes 1_H)_\# \square_M} [(\beta \otimes 1_H) \circ \alpha]_\# M$$
$$= [\alpha_{12} \circ \beta]_\# M \xrightarrow{(\alpha_{12})_\# \square_M} [\alpha_{12} \circ \alpha]_\# M = [(1_A \otimes \Delta_H) \circ \alpha]_\# M,$$

where $\alpha_{12} : A \otimes H \to A \otimes H \otimes H$ *is given by* $\alpha_{12}(a \otimes h) := a \otimes h \otimes 1$.

2 *Conversely, if an A-module M and an isomorphism* $\square_M : \beta_\# M \to \alpha_\# M$ *such that the commutativity in* **1** *is satisfied are given, then defining* $\omega_M : M \to M \otimes H$ *to be the composite map*

$$M \to \beta^\# \beta_\# M \xrightarrow{\beta^\# \square_M} \beta^\# \alpha_\# M = M \otimes H,$$

M is a G-module, and hence is a (G, A)*-module.*

Proof. The map $\square_M$ in **1** is nothing but the $A \otimes H$-linear map which corresponds to $\omega_M : M \to \beta^\# \alpha_\# M$ by the adjunction between $\beta_\#$ and $\beta^\#$. This map is an isomorphism, as the inverse is given by $\square_M^{-1}(m \otimes h) = \sum_{(m)} m_0 \otimes S(m_1)h$. The naturality is obvious. The map ω_M in **2** corresponds to $\square_M$ (given in **2**) by the adjunction.

To say that the diagram in question is commutative is the same as to say that the corresponding maps $M \to M \otimes H \otimes H$ to the two composite maps (by adjunction) $(1_M \otimes \Delta_H) \circ \omega_M$ and $(\omega_M \otimes 1_H) \circ \omega_M$ agree. Hence, this is equivalent to the coassociativity of ω_M. Hence, **1** is obvious.

We show **2**. The base change of $\square_M$ by $A \otimes H \xrightarrow{1_A \otimes \varepsilon} A \otimes R = A$ is an A-isomorphism, and it agrees with the composite map

$$M \xrightarrow{\omega_M} M \otimes H \xrightarrow{1_M \otimes \varepsilon} M \otimes R \cong M.$$

When we denote it by ρ, we have $\rho \circ \rho = \rho$ by the coassociativity. As ρ is an isomorphism, we have $\rho = 1_M$, and the counit law follows. Now we know that M is a G-module. As ω_M is A-linear, we have that M is a (G, A)-module by Lemma 1.1.5. □

(1.1.7) Let R, G, and A be as in (1.1.1). As $\alpha' : A' \to A' \otimes H$ and $\beta : A \to A' \otimes H$ are G-algebra maps, we have $\beta^\#(\alpha'_\#(V'))$ is a (G, A)-module for any A-module V. More explicitly, $\beta^\#(\alpha'_\#(V'))$ is the G-module $V' \otimes H$, equipped with the A-module structure by $a(v \otimes h) := \sum_{(a)} a_0 v \otimes a_1 h$. If M is a (G, A)-module, then $\omega_M : M \to M' \otimes H = \beta^\#(\alpha'_\#(M'))$ is A-linear (Lemma 1.1.5) and G-linear (Example I.3.6.6), and hence it is (G, A)-linear.

Lemma 1.1.8 *Let* $M \in {}_{G,A}\mathbb{M}$, *and* $V \in {}_A\mathbb{M}$. *Then the map*

$$\Phi : \mathrm{Hom}_A(M, V) \to \mathrm{Hom}_{G,A}(M, \beta^\#(\alpha'_\#(V')))$$

defined by $\Phi(f)(m) := \sum_{(m)} f m_0 \otimes m_1$ *is an isomorphism which is natural with respect to M and V. In particular, the forgetful functor* ${}_{G,A}\mathbb{M} \to {}_A\mathbb{M}$ *has* $\Xi_A := \beta^\# \alpha'_\#(?')$ *as its right adjoint.*

Proof. As we have $\Phi(f) = \beta^{\#}(\alpha'_{\#}f') \circ \omega$, $\Phi(f)$ is certainly a (G, A)-linear map by the remark above. We define $\Psi : \mathrm{Hom}_{G,A}(M, \beta^{\#}(\alpha'_{\#}(V'))) \to \mathrm{Hom}_A(M, V)$ by $\Psi(g)(m) := E(gm)$, where $E : V \otimes H \to V$ is given by $E(v \otimes h) := \varepsilon(h)v$. As can be checked easily, E is A-linear, and hence so is $\Psi(g)$. Thus, Ψ is well-defined. So it suffices to show that $\Phi \circ \Psi = \mathrm{id}$ and $\Psi \circ \Phi = \mathrm{id}$, which is straightforward, and is similar to the proof of Lemma I.3.5.6. □

Corollary 1.1.9 *If I is an injective A-module, then $\Xi(I) = \beta^{\#}\alpha'_{\#}(I')$ is an injective (G, A)-module. The set $\{\Xi(I) \,|\, I$ is an injective A-module$\}$ is an injective cogenerator of ${}_{G,A}\mathbb{M}$.*

Proof. The first part is obvious from the lemma. We show the latter part. Let M be a (G, A)-module. We take an injective A-linear map $i : M \hookrightarrow I$ with I an injective A-module. Then we have that the composite map

$$M \xrightarrow{\omega} \Xi(M) \xrightarrow{\Xi(i)} \Xi(I)$$

is (G, A)-linear and injective. □

Corollary 1.1.10 *Let $M \in {}_{G,A}\mathbb{M}$, $V \in {}_A\mathbb{M}$, and $i \geq 0$. Then there is an isomorphism*

$$\Phi : \mathrm{Ext}^i_A(M, V) \cong \mathrm{Ext}^i_{G,A}(M, \Xi(V))$$

natural with respect to M and V.

Proof. Let $\mathbb{I}$ be an A-injective resolution of V. Then as Ξ is exact and preserves injectives, $\Xi(\mathbb{I})$ is an injective resolution of $\Xi(V)$. Hence, we have a sequence of isomorphisms

$$\mathrm{Ext}^i_A(M, V) \cong H^i(\mathrm{Hom}_A(M, \mathbb{I})) \cong H^i(\mathrm{Hom}_{G,A}(M, \Xi(\mathbb{I}))) \cong \mathrm{Ext}^i_{G,A}(M, \Xi(V)),$$

which is natural with respect to M and V. □

Lemma 1.1.11 *Let A be noetherian. If J is an injective (G, A)-module, then $\mathrm{Hom}_A(?, J)$ is an exact functor on ${}_{G,A}\mathbb{M}$.*

Proof. First, we prove that $\mathrm{Hom}_A(?, J)$ is an exact functor on ${}_{G,A}\mathbb{M}_f$. By Corollary 1.1.9, we may assume that $J = \Xi(I)$, with I an A-injective module. Hence, we have isomorphisms of functors

$$\mathrm{Hom}_A(?, J) = \mathrm{Hom}_A(?, \beta^{\#}\alpha_{\#}I) \cong \mathrm{Hom}_A(\beta_{\#}?, \alpha_{\#}I) \cong \mathrm{Hom}_A(\alpha_{\#}?, \alpha_{\#}I),$$

where the last isomorphism is $\operatorname{Hom}_A(\square, \alpha_\# I)$, see Lemma 1.1.6. Note that the G-structures of modules are irrelevant here, because the question is the exactness of $\operatorname{Hom}_A(?, J)$, so the symbol $(?)'$ is omitted here. As ? in question is of finite presentation as an A-module, we have

$$\operatorname{Hom}_A(\alpha_\# ?, \alpha_\# I) \cong \alpha_\# \operatorname{Hom}_A(?, I)$$

by Lemma I.2.1.7. As $\alpha_\#$ and $\operatorname{Hom}_A(?, I)$ are exact, the exactness of $\operatorname{Hom}_A(?, J)$ as a functor over ${}_{G,A}\mathbb{M}_f$ is proved.

Hence, we have that $\operatorname{Hom}_{G,A}(?, \Xi(J))$ is an exact functor over ${}_{G,A}\mathbb{M}_f$ by Lemma 1.1.8. By Lemma I.9.4, $\Xi(J)$ is an injective object of ${}_{G,A}\mathbb{M}$, and hence $\operatorname{Hom}_A(?, J)$ is exact over ${}_{G,A}\mathbb{M}$, by Lemma 1.1.8 again. □

Now we can prove the first theorem in this chapter.

Theorem 1.1.12 *Let R be a commutative ring, G an affine flat R-group scheme, and A a noetherian commutative G-algebra. Then for any $M \in {}_{G,A}\mathbb{M}$ and an injective (G, A)-module J, we have $\operatorname{Ext}^i_A(M, J) = 0$ $(i > 0)$.*

Proof. We proceed by induction on $i \geq 1$. First, consider the case $M = V' \otimes A$, with $V \in {}_R\mathbb{M}$, where V' denotes the trivial G-module V. Then there exists an R-free module F and a surjective R-linear map $p : F \to V$, and we have an exact sequence

$$0 \to K \to F' \otimes A \to V' \otimes A \to 0$$

in ${}_{G,A}\mathbb{M}$, where $K := \operatorname{Ker}(p' \otimes 1)$. As we have $\operatorname{Ext}^1_A(F \otimes A, J) = 0$, it follows that $\operatorname{Ext}^1_A(V \otimes A, J) = 0$ by Lemma 1.1.11. If $i \geq 2$, then by induction assumption, we have $\operatorname{Ext}^i_A(V \otimes A, J) \cong \operatorname{Ext}^{i-1}(K, J) = 0$.

Now consider the general case. Consider the exact sequence of (G, A)-modules

$$0 \to L \to M \otimes A \to M \to 0.$$

As the A-module structures of $M \otimes A$ and $M' \otimes A$ are the same, we have $\operatorname{Ext}^1_A(M \otimes A, J) = 0$ by the above. Hence, by Lemma 1.1.11 again, we have $\operatorname{Ext}^1_A(M, J) = 0$. If $i \geq 2$, then $\operatorname{Ext}^i_A(M, J) \cong \operatorname{Ext}^{i-1}(L, J) = 0$ by induction assumption. □

(1.1.13) Let R, G and A be as in Theorem 1.1.12. Set $F : {}_{G,A}\mathbb{M} \to {}_A\mathbb{M}$ to be the forgetful functor. Then for $M \in {}_{G,A}\mathbb{M}_f$, the functor $F \circ R^i \operatorname{Hom}_A(M, ?) : {}_{G,A}\mathbb{M} \to {}_A\mathbb{M}$ is canonically isomorphic to $\operatorname{Ext}^i_A(M, ?)$ by Theorem 1.1.12, where $R^i \operatorname{Hom}_A(M, ?)$ is the derived functor of the left exact functor $\operatorname{Hom}_A(M, ?) : {}_{G,A}\mathbb{M} \to {}_{G,A}\mathbb{M}$. Thus, we denote the functor $R^i \operatorname{Hom}_A(M, ?) : {}_{G,A}\mathbb{M} \to {}_{G,A}\mathbb{M}$ simply by $\operatorname{Ext}^i_A(M, ?)$. That is, for $N \in {}_{G,A}\mathbb{M}$, the A-module $\operatorname{Ext}^i_A(M, N)$ is equipped with the (G, A)-module structure of $R^i \operatorname{Hom}_A(M, ?)(N)$.

Proposition 1.1.14 *Let R be a commutative ring, G an affine flat R-group scheme, $A \to B$ a G-algebra map between commutative noetherian G-algebras, $V \in {}_{G,A}\mathbb{M}$, and $M, N \in {}_{G,B}\mathbb{M}$. Then the following hold.*

1 *If M is A-flat and B-finite, then there is a spectral sequence of R-modules*

$$E_2^{p,q} = \operatorname{Ext}^p_{G,A}(V, \operatorname{Ext}^q_B(M,N)) \Rightarrow \operatorname{Ext}^{p+q}_{G,B}(V \otimes_A M, N).$$

If, moreover, V is A-finite, then there is a spectral sequence of (G, A)-modules

$$E_2^{p,q} = \operatorname{Ext}^p_A(V, \operatorname{Ext}^q_B(M,N)) \Rightarrow \operatorname{Ext}^{p+q}_B(V \otimes_A M, N).$$

2 *If V is A-finite projective, then there are isomorphisms*

$$\operatorname{Ext}^i_{G,B}(M \otimes V, N) \cong \operatorname{Ext}^i_{G,B}(M, \operatorname{Hom}_A(V,N)) \cong \operatorname{Ext}^i_{G,B}(M, V^\star \otimes N),$$

where $V^\star := \operatorname{Hom}_A(V, A)$.

Proof. This follows easily from (I.5.3.6), (I.5.3.7), the (G, A)-module version of (I.5.1.11), and Lemma I.5.3.8. □

If V is A-finite and we have either B is A-flat or there is a resolution $\mathbb{F} \to V$ of V in ${}_{G,A}\mathbb{M}$ such that each term of $\mathbb{F}$ is A-finite projective (e.g., the case G satisfies FPCP, see the next subsection), then the second spectral sequence in **1** can be taken to be that of (G, B)-modules. This is proved similarly to Proposition I.5.2.4.

(1.1.15) Let R, G and A be as in Theorem 1.1.12. Let I be a G-ideal of A. Then $\Gamma_I = \varinjlim \operatorname{Hom}_A(A/I^n, ?)$ is a left exact functor from ${}_{G,A}\mathbb{M}$ to itself. By Theorem 1.1.12, $F \circ (R^i\Gamma_I)$ agrees with the local cohomology functor H^i_I, where $F : {}_{G,A}\mathbb{M} \to {}_A\mathbb{M}$ is the forgetful functor. Hence, for $M \in {}_{G,A}\mathbb{M}$, the local cohomology $H^i_I(M)$ is equipped with a (G, A)-module structure. The canonical map $\varinjlim \operatorname{Ext}^i_A(A/I^n, M) \to H^i_I(M)$ is an isomorphism of (G, A)-modules.

1.2 Equivariant modules of a split torus

(1.2.1) Let R be a commutative ring, $n \geq 1$, and T the n-fold split torus $\mathbb{G}_m^n$ over R. A T-module and an $X(T)$-graded R-module (i.e., a $\mathbb{Z}^n$-graded R-module) are the same thing (I.4.3). For two T-modules V and W, the tensor product $V \otimes W$ is again a T-module, and it is graded. It is easy to see that the grading is given by

$$(V \otimes W)_\lambda = \bigoplus_{\mu,\nu \in X(T),\ \mu+\nu=\lambda} V_\mu \otimes W_\nu.$$

That is, the grading of the tensor product $V \otimes W$ is given by the total grading.

Let A be a T-module, and an R-algebra. To say that A is a T-algebra is the same as to say that $A \otimes A \to A$ preserves the grading. Hence, A is a T-algebra if and only if A is an $X(T)$-graded R-algebra. Similarly, a (T, A)-module is nothing but a graded A-module.

1.3 FPCP groups and IFP groups

(1.3.1) Let R be a noetherian commutative ring. Let $G = \operatorname{Spec} H$ be a flat affine R-group scheme, and A a commutative G-algebra.

Definition 1.3.2 We say that G satisfies PCP (resp. FPCP) if the coordinate Hopf algebra H satisfies PCP (resp. FPCP) (see Definition I.3.11.4), as an R-coalgebra.

By Lemma I.3.11.5, we have that if G satisfies FPCP, then G satisfies PCP, and the converse is true if $\operatorname{gl.dim} R \leq 2$.

Lemma 1.3.3 *If the R-group scheme G satisfies FPCP (resp. PCP), then for any A-finite (G, A)-module (resp. any (G, A)-module) M, there exists some R-projective $P \in {}_G\mathbb{M}_f$ (resp. ${}_G\mathbb{M}$) such that there is a surjective (G, A)-linear map $A \otimes P \to M$.*

Proof. Let M_0 be an R-finite G-submodule (resp. G-submodule) of M which generates M as an A-module. As G satisfies FPCP (resp. PCP), there exists some surjective G-linear map $P \to M_0$, with P a G-module which is R-finite projective (resp. R-projective). Now, the composite map

$$A \otimes P \to A \otimes M_0 \to M$$

is surjective. □

Proposition 1.3.4 *Let G be an affine flat R-group scheme which satisfies FPCP. Assume that R and A are noetherian, and A is R-flat. Then for a (G, A)-module I, the following are equivalent.*

1 *I is G-injective, and for any R-finite G-module V, any (G, A)-submodule N of $A \otimes V$, and any (G, A)-linear map $f : N \to I$, f is extended to a (G, A)-linear map $A \otimes V \to I$.*

2 *I is (G, A)-injective.*

Proof. **1⇒2** Let M be an A-finite (G, A)-module. Then there exists an exact sequence of (G, A)-modules

$$0 \to N \to A \otimes V \to M \to 0$$

such that V is an R-finite G-module. By Proposition 1.1.14, we have $\operatorname{Ext}^1_{G,A}(A \otimes V, I) \cong \operatorname{Ext}^1_G(V, I) = 0$. As the map $\operatorname{Hom}_{G,A}(A \otimes V, I) \to \operatorname{Hom}_{G,A}(N, I)$ is surjective by assumption, it follows that $\operatorname{Ext}^1_{G,A}(M, I) = 0$. By Lemma I.1.9.4 and Corollary I.5.3.11, we have that I is injective.

We show **2⇒1**. By Corollary I.5.3.9, we have that I is G-injective. When we set M to be the cokernel of $N \to A \otimes V$, then as we have $\operatorname{Ext}^1_{G,A}(M, I) = 0$, the map $\operatorname{Hom}_{G,A}(A \otimes V, I) \to \operatorname{Hom}_{G,A}(N, I)$ is surjective. □

(1.3.5) Let R be a noetherian commutative ring, and G an affine flat R-group scheme. If G satisfies PCP, then by Theorem I.1.4.9, a left derived functor

$$\underline{L}(? \otimes_A N) : D^-({}_{G,A}\mathbb{M}) \to D^-({}_{G,A}\mathbb{M})$$

exists. For $M, N \in {}_{G,A}\mathbb{M}$, we have $L_i(? \otimes_A N)(M) \cong L_i(M \otimes_A ?)(N)$, and they coincide with $\operatorname{Tor}_i^A(M, N)$, simply as A-modules. Hence, if G satisfies PCP, then $\operatorname{Tor}_i^A(M, N)$ is equipped with a (G, A)-module structure in a natural way.

If G satisfies FPCP, A is noetherian, and $N \in {}_{G,A}\mathbb{M}$, then there is a derived functor

$$\underline{R}\operatorname{Hom}_A(?, N) : D^-({}_{G,A}\mathbb{M}_f) \to D^+({}_{G,A}\mathbb{M}),$$

and for $M \in {}_{G,A}\mathbb{M}_f$, we have $R^i \operatorname{Hom}(?, N)(M) \cong \operatorname{Ext}^i_A(M, N)$, as (G, A)-modules.

Definition 1.3.6 Let R be a commutative ring, and G an affine R-group scheme. We say that G is *IFP* if the coordinate ring $R[G]$ of G is IFP as an R-coalgebra (Definition I.3.11.1).

By definition and Lemma I.3.11.6, an IFP R-group scheme satisfies FPCP.

Lemma 1.3.7 *Let R be a noetherian commutative ring, and $G = \operatorname{Spec} H$ an affine IFP R-group scheme of finite type. Then H is R-projective.*

Proof. Follows immediately from Lemma I.3.11.3. □

2 Matijevic–Roberts type theorem

2.1 Stability of various loci

(2.1.1) Throughout this subsection, let R be a noetherian commutative ring, G a flat R-group scheme *of finite type*, and X a locally noetherian right G-action.

Lemma 2.1.2 *Assume that G has integral geometric fibers. If Y is an irreducible (resp. reduced, integral) locally noetherian R-scheme, then so is $Y \times G$.*

Proof. As $Y \times G$ is flat over Y and any fiber is reduced by assumption, $Y \times G$ is reduced if Y is, by [110, Theorem 23.9, Corollary].

Now it suffices to consider only the irreducibility. If $Y = \operatorname{Spec} k$ with k a field, then the assertion is obvious, as $\operatorname{Spec} \bar{k} \times G \to \operatorname{Spec} k \times G$ is faithfully flat, where $\bar{k}$ denotes the algebraic closure of k.

Now consider the general case. Let Y be irreducible, with its generic point η. As the fiber G_η is irreducible by the previous paragraph, we have its generic point, say, ξ. It suffices to show that $\overline{\{\xi\}} = Y \times G$. Assume the contrary. Then there is an affine open set $U = \operatorname{Spec} B$ of $Y \times G$ which does not intersect G_η. We take an irreducible component of U, and let ζ be its generic point. As U is flat over Y, we have that ζ must be mapped to η by the going-down theorem, which contradicts $\zeta \notin G_\eta$. □

(2.1.3) Let the notation be as in (2.1.1). We denote the action of G on X (resp. the first projection) $X \times G \to X$ by $a = a_X$ (resp. $p = p_X = p_1$). We define $h_X : X \times G \to X \times G$ by $h_X(x, g) := (xg, g)$. As h_X is an isomorphism with $h_X^{-1}(x, g) = (xg^{-1}, g)$ and $p_X \circ h_X = a_X$, we have that a_X is also flat of finite type.

For a subscheme Y of X, we say that Y is *G-stable* if the action $Y \times G \to X$ $((y, g) \mapsto yg)$ factors through $Y \hookrightarrow X$, see (I.4.1.5). If Y is a G-stable subscheme of X, then we have a unique action $Y \times G \to Y$ such that $Y \hookrightarrow X$ is a G-morphism.

Lemma 2.1.4 *The union of a family of G-stable open subsets is G-stable. The intersection of finitely many G-stable subschemes is G-stable.*

Thus, X is a topological space, taking G-stable open sets as its open sets. However, we will not use this (usually too coarse) topology later.

For a subscheme Y of X, the image closure of $Y \times G \to X$ is denoted by Y^*.

Lemma 2.1.5 *If Y is a subscheme of X, then Y^* is the smallest G-stable closed subscheme of X which contains Y. In particular, Y is a G-stable closed subscheme of X if and only if $Y = Y^*$. If Y' is also a subscheme of X and $Y \subset Y'$, then we have $Y^* \subset (Y')^*$. If G has integral geometric fibers and Y is irreducible (resp. reduced, integral), then so is Y^*.*

Proof. By definition, Y^* is a closed subscheme of X. As the composition $Y = Y \times \{e\} \hookrightarrow Y \times G \to X$ agrees with the inclusion $Y \hookrightarrow X$, Y^* contains Y.

As G is R-flat, the image closure of $a \times 1 : Y \times G \times G \to X \times G$ is $Y^* \times G$. On the other hand, as $a \circ (a \times 1) = a \circ (1_Y \times \mu_G) : Y \times G \times G \to X$ factors through Y^*, we have that $a : Y^* \times G \to X$ also factors through Y^*, which shows that Y^* is G-stable.

If Z is a G-stable closed subscheme of X which contains Y, then, as the action $Y \times G \to X$ factors through Z, we have $Y^* \subset Z$. Hence, Y^* is the smallest among such Z. Assume $Y \subset Y'$. Then $Y \subset (Y')^*$, and hence $Y^* \subset (Y')^*$.

Finally, assume that G has integral geometric fibers. If Y is irreducible (resp. reduced), then so is $Y \times G$ by Lemma 2.1.2. By the definition of Y^*, we have that Y^* is irreducible (resp. reduced). □

(2.1.6) An argument similar to the proof above shows that if U is an open subscheme of X, then the image of $U \times G \to X$ is the smallest G-stable open subscheme of X which contains U (as a_X is flat of finite type, it is an open map).

Lemma 2.1.7 *X has a covering consisting of quasi-compact G-stable open subschemes.*

Proof. Let (U_i) be a covering consisting of quasi-compact open subschemes of X. Then $(a_X(U_i))$ is a covering of the desired type. □

If (V_i) is a covering of X consisting of quasi-compact G-stable open subschemes, then for any subscheme Y of X, we have that Y is G-stable if and only if $Y \cap V_i$ is a G-stable subscheme of V_i for any i.

(2.1.8) Let the notation be as in (2.1.1). Assume that G has integral geometric fibers. For $x \in X$, we denote the generic point of the G-stable integral closed subscheme $\overline{\{x\}}^*$ of X by x^*. If we have $x = x^*$, then we say that x is a *G-stable point* of X.

Lemma 2.1.9 *Let $X = X_0 \times G$ be a principal G-bundle, where X_0 is locally noetherian. Then any G-stable open subset of X is of the form $V \times G$, with V an open subset of X_0.*

Proof. Let U be a G-stable open subset of X. Then $V := p_{X_0}(U)$ is an open subset of X_0, as p_{X_0} is flat of finite type. It suffices to show that $U = p^{-1}(V) = V \times G$. Assume the contrary. Then there exists some geometric point $\xi = \operatorname{Spec} K \to \operatorname{Spec} R$ of $\operatorname{Spec} R$, and some $(x, g) \in V(\xi) \times G(\xi) \setminus U(\xi)$. As $U \to V$ is faithfully flat of finite type and K is algebraically closed, we have that $U(\xi) \to V(\xi)$ is surjective by Hilbert's theorem. Hence, there exists some $g' \in G(\xi)$ such that $(x, g') \in U(\xi)$. This shows $(x, g) = (x, g')((g')^{-1}g) \in U(\xi)$, which is a contradiction. □

Lemma 2.1.10 *Let X and X' be locally noetherian G-actions, and $\varphi : X \to X'$ a G-morphism locally of finite type. Let $\mathbb{P}$ be any of the following properties: flat, locally of finite flat dimension, locally Cohen–Macaulay, locally Gorenstein, local complete intersection, and smooth. Then the $\mathbb{P}$-locus $U(\mathbb{P}, \varphi)$ of φ is a G-stable open subset of X.*

Proof. Note that $U = U(\mathbb{P}, \varphi)$ is an open set of X by Corollary I.2.12.4. The right square of the diagram

$$\begin{array}{ccccc} X \times G & \xrightarrow{a_X} & X & \xleftarrow{p_X} & X \times G \\ \downarrow{\scriptstyle \varphi \times 1_G} & & \downarrow{\scriptstyle \varphi} & & \downarrow{\scriptstyle \varphi \times 1_G} \\ X' \times G & \xrightarrow{a_{X'}} & X' & \xleftarrow{p_{X'}} & X' \times G \end{array}$$

is a fiber square. As we have $(\varphi \times 1_G) \circ h_X = h_{X'} \circ (\varphi \times 1_G)$ and both h_X and $h_{X'}$ are isomorphisms, the left square is also a fiber square. Hence, by Theorem I.2.7.11, we have $a_X^{-1}(U) = U(\mathbb{P}, \varphi \times 1_G) = p_X^{-1}(U)$, and U is G-stable. □

Corollary 2.1.11 *Let G be an R-flat group scheme of finite type. Then G is locally Cohen–Macaulay over R.*

Proof. We may assume that R is a field. By the lemma, the Cohen–Macaulay locus U of G, which agrees with the Cohen–Macaulay locus of the canonical projection $G \to \operatorname{Spec} R$, is G-stable open. By Lemma 2.1.9, we have that $U = \emptyset$ or $U = G$. When we take the generic point γ of an irreducible component of G, then $\mathcal{O}_{G,\gamma}$ is artinian, and hence is Cohen–Macaulay. This shows that $\gamma \in U$ and hence $U \neq \emptyset$, and we have $U = G$. □

Remark 2.1.12 More is true. Under the same assumption as in the corollary, we have that G is l.c.i. over R. To verify this, we may assume that R is a field again. Then by Lemma I.4.2.8, G is a closed subgroup of some $P = GL_n(R)$. By [43, III.3.5.4], we have that the quotient P/G is a scheme.

By [43, III.3.2.5–6], the quotient map $\pi : P \to P/G$ is faithfully flat. By Theorem I.2.7.10, **2**′, we have that π is l.c.i. Hence, the fiber $\pi^{-1}(\pi(e)) = G$ is l.c.i., as desired.

Assume moreover that $R \supset \mathbb{Q}$. Then G is smooth. To verify this, we may assume that R is a field of characteristic zero. In this case, both $P = GL_n(R)$ and P/G are smooth varieties, and the field extension $k(P)/k(P/G)$ is separable. This shows that the smooth locus of π is non-empty. As π is a P-morphism and the only P-stable non-empty open subset of P is P, we have that π is smooth. Hence, the fiber G is geometrically regular.

Lemma 2.1.13 *Let G be an R-group scheme flat of finite type. Then G has reduced geometric fibers if and only if G is R-smooth.*

Proof. Assume that G has reduced geometric fibers. As G is flat of finite type over R, it suffices to show that G has regular geometric fibers. Hence, it suffices to show that G is regular, assuming that R is an algebraically closed field and G is reduced. The regular locus of G is G-stable open, and it is empty or G itself by Lemma 2.1.9. As G is reduced, G satisfies Serre's (R_0) condition, and hence the regular locus of G is non-empty. The opposite direction is trivial. □

Corollary 2.1.14 *Let G be an R-flat group scheme of finite type. Then G has integral geometric fibers if and only if G is R-smooth with connected geometric fibers.*

Lemma 2.1.15 *Let k be a field, and G a group scheme over k of finite type. Then G is a disjoint union of its irreducible components. In particular, if G is connected, then it is irreducible.*

Proof. Let K be the perfect field consisting of p^rth roots of elements of k for some r. As $(G \otimes_k K)_{\mathrm{red}}$ is a reduced group scheme over K, it is K-smooth by an argument similar to the proof of Lemma 2.1.13. Let I be the ideal of $H \otimes_k K$ consisting of its nilpotent elements. Then there exists some field k' such that $k \subset k' \subset K$, $[k' : k] < \infty$, and there exists some ideal J of $H \otimes_k k'$ such that $I = (H \otimes_k K)J$. Then J consists of nilpotent elements, and $(H \otimes_k k')/J(H \otimes_k k')$ is k'-smooth, and hence we have $(H \otimes_k k')_{\mathrm{red}} = (H \otimes_k k')/J(H \otimes_k k')$. In particular, we have that $G \otimes_k k'$ is the disjoint union of its irreducible components.

Now assume that there are two different irreducible components G_i and G_j $(i \neq j)$ of G which intersect at a point, say, g. As the canonical map $\pi : G_{k'} \to G$ is finite flat, there exist two different generic points of irreducible components of $G_{k'}$, say, ξ and η, such that $\pi(\xi) = \gamma_i$ and $\pi(\eta) = \gamma_j$, where γ_i and γ_j denote the generic points of G_i and G_j, respectively. By the going-up theorem, $\pi^{-1}(g)$ has points in two different irreducible components of $G_{k'}$. In particular, $\pi^{-1}(g)$ has at least two points. This contradicts the fact π is radical [65, IV.6.15.3.1]. □

Lemma 2.1.16 *If the Cohen–Macaulay locus $U = U(\mathrm{CM}, X)$ of X is an open subset of X, then U is G-stable. Similarly for the Gorenstein and l.c.i. properties. If, moreover, $G \to \operatorname{Spec} R$ has regular fibers, then we also have a similar result for the regular locus.*

Proof. If $f : X' \to X$ is a locally Cohen–Macaulay flat morphism, $x' \in X'$, and $x = f(x')$, then $\mathcal{O}_{X,x}$ is Cohen–Macaulay if and only if $\mathcal{O}_{X',x'}$ is Cohen–Macaulay. Hence, we have $f^{-1}(U) = U(\mathrm{CM}, X')$. Applying this to the Cohen–Macaulay morphisms $a_X : X \times G \to X$ and $p_X : X \times G \to X$, we have $a^{-1}(U) = U(\mathrm{CM}, X \times G) = p^{-1}(U)$. This shows that U is G-stable. Similarly for the other properties. □

Lemma 2.1.17 *Let Y be a reduced closed subscheme of X. If G has integral geometric fibers, then the following are equivalent.*

1 *Y is G-stable.*

2 *Any irreducible component of Y (i.e., maximal integral closed subscheme of Y) is G-stable.*

3 *The open set $X - Y$ is G-stable.*

Proof. **1⇒2** Let Z be an irreducible component of Y. By Lemma 2.1.5, Z^* is integral, and we have $Z \subset Z^* \subset Y^* = Y$. As Z is maximal among integral closed subschemes of Y, we have $Z = Z^*$.

2⇒3 Let (U_i) be a quasi-compact G-stable open covering of X. It suffices to show that $U_i \setminus Y$ is G-stable. As an irreducible component of $U_i \cap Y$ is nothing but a non-empty intersection of an irreducible component Z of Y with U_i, we may assume that X is quasi-compact, replacing X by U_i. As Y is noetherian this case, we can express Y as a finite union $Y_1 \cup \cdots \cup Y_r$ of its irreducible components. As we have $X - Y = \bigcap(X - Y_i)$, it suffices to show that each $X - Y_i$ is G-stable, and we may and shall assume that Y is integral.

We set $U := X - Y$. What we want to prove is $U' := UG = U$. Assume the contrary. Then there exists some geometric point $\eta : \operatorname{Spec} K \to \operatorname{Spec} R$ such that $U(\eta) \to U'(\eta)$ is not surjective. On the other hand, the action $U \times G \to U'$ is surjective of finite type. By Hilbert's theorem, we have that $U(\eta) \times G(\eta) \to U'(\eta)$ is surjective. Hence, there exist some $u \in U(\eta)$ and $g \in G(\eta)$ such that $ug \in U'(\eta) \setminus U(\eta)$. As we have $Y(\eta) = X(\eta) \setminus U(\eta)$, we have $ug \in Y(\eta)$. As Y is G-stable, this shows $u = (ug)g^{-1} \in Y(\eta)$, and we have $u \in Y(\eta) \cap U(\eta) = \emptyset$. This is a contradiction.

3⇒1 An argument similar to above shows that the image of $Y \times G \to X$ does not intersect $U := X \setminus Y$. Hence, Y^* is contained in Y set-theoretically. As Y^* is reduced, we have $Y = Y^*$. □

Lemma 2.1.18 *Assume that G has integral geometric fibers, and the Cohen–Macaulay (resp. Gorenstein, l.c.i., regular) locus U of X is an open subset of X. If any G-stable point of X is contained in U, then we have $U = X$.*

Proof. Assume the contrary. Consider $Y := X - U$ with the reduced structure. Then Y is a non-empty G-stable closed subscheme of X by Lemma 2.1.17. Let η be the generic point of an irreducible component of Y, which does exist. Then η is a G-stable point of Y by Lemma 2.1.17, but it is not contained U, which contradicts the assumption. □

The assumption that U is open in the lemma is not necessary, in fact. This will be proved in (2.4).

2.2 Universal density of hyperalgebras

(2.2.1) Let R be a noetherian commutative ring, and $G = \operatorname{Spec} H$ be an affine flat R-group scheme of finite type over R.

Definition 2.2.2 Let $I := \operatorname{Ker} \varepsilon_H$ be the kernel of the counit map. We define the *hyperalgebra* of G, denoted by $\operatorname{Hy} G$, to be $\varinjlim (H/I^n)^*$.

Note that $U = \operatorname{Hy} G$ is an R-subalgebra of the dual algebra H^* of H. In fact, we have $1_{H^*} = \varepsilon \in (H/I)^* \subset U$, and $(H/I^n)^*(H/I^m)^* \subset (H/I^{n+m-1})^*$.

Definition 2.2.3 We say that G is *infinitesimally flat* if H/I^n is R-finite R-projective for any $n \geq 1$.

This is equivalent to saying that I^n/I^{n+1} is R-finite R-projective for $n \geq 1$. As we are assuming that G is of finite type over R, this is also equivalent to saying that H is normally flat along I.

(2.2.4) Let R and $G = \operatorname{Spec} H$ be be as in (2.2.1). Set $I := \operatorname{Ker} \varepsilon_H$ and $U := \operatorname{Hy} G$. We assume that G is infinitesimally flat. Then as we have $(H/I^n)^*(I^n \cdot H + H \cdot I^n) = 0$ and $(H/I^n \otimes H/I^n)^* \cong (H/I^n)^* \otimes (H/I^n)^*$, we have an induced map $(H/I^n)^* \to (H/I^n)^* \otimes (H/I^n)^*$ for $n \geq 1$, and this gives an R-coalgebra structure to U. It is straightforward to see that U is an R-Hopf algebra in fact. For $n \geq 1$, we have that $(H/I^n)^*$ is an R-subcoalgebra of U, and hence U is IFP.

Lemma 2.2.5 *Let G be an affine flat, infinitesimally flat R-group scheme of finite type over R. Then U is R-projective.*

Proof. Follows immediately from Lemma I.3.11.3. □

(2.2.6) Let R, G, H, I and U be as in (2.2.4). As we have $U = \varinjlim (H/I^n)^*$ by definition, $U^* = \varprojlim H/I^n$. In other words, U^* is the I-adic completion of H.

Lemma 2.2.7 *Let the notation be as above. Assume that $R = k$ is a field, and G is connected. Then the canonical inclusion $U \to H^*$ is universally dense.*

Proof. It suffices to show that the I-adic completion $H \to U^*$ is injective, in other words, $\bigcap_{i \geq 0} I^i = 0$. By assumption and Lemma 2.1.15, G is irreducible. By Corollary 2.1.11, G is Cohen–Macaulay, and hence H has a unique associated prime by [110, Theorem 17.3]. This shows that any zero-divisor of H is nilpotent. If b is a non-zero element of $\bigcap_{i \geq 0} I^i$, then by Krull's intersection theorem [110, Theorem 8.9], there exists some $a \in H$ such that $a - 1 \in I$ and $ab = 0$. As a is a zero-divisor, a is nilpotent. As I is a prime ideal, $a \in I$. As $a - 1 \in I$, it follows that $1 = a - (a-1) \in I \neq H$, which is a contradiction. □

Theorem 2.2.8 *Let $G = \operatorname{Spec} H$ be an R-flat, infinitesimally flat affine R-group scheme of finite type with connected fibers. Then $U = \operatorname{Hy}(G)$ is a generalized hyperalgebra of H. In particular, $U \to H^*$ is universally dense. Moreover, the coordinate ring H is R-projective.*

Proof. As the fiber $G(\mathfrak{p})$ is connected, we have that the map

$$H(\mathfrak{p}) \to \operatorname{Hom}_{\kappa(\mathfrak{p})}(\operatorname{Hy}(G(\mathfrak{p})), \kappa(\mathfrak{p})) = \varprojlim (H/I^n)(\mathfrak{p})$$

is injective by Lemma 2.2.7. As this map factors through

$$(U^*)(\mathfrak{p}) = (\varprojlim H/I^n)(\mathfrak{p}),$$

we have that $U \to H^*$ is universally dense by Lemma I.3.8.4. Now it is easy to see that U is a generalized hyperalgebra of H.

Next, we show that H is R-projective. By assumption, H/I^n is R-projective for any n. Since the projective system (H/I^n) consists only of surjective maps, it satisfies the Mittag-Leffler condition. Hence, by Proposition I.2.2.15, $U^* = \varprojlim H/I^n$ is also Mittag-Leffler. As $U \to H^*$ is universally dense, the canonical map $H \to U^*$ is R-pure by Lemma I.3.8.5. Hence, by Corollary I.2.2.8, H is also a Mittag-Leffler R-module. As H is of finite type over R, it is countably generated as an R-module. By Proposition I.2.2.6, H is R-projective. □

Corollary 2.2.9 *If G is R-smooth with connected fibers, then $U \to H^*$ is universally dense, and H is R-projective.*

2.3 A generalization to equivariant sheaves

(2.3.1) Let R be a noetherian commutative ring, and $G = \operatorname{Spec} H$ an affine flat R-group scheme of finite type.

We denote the category of G-schemes (i.e., an R-scheme with a right G-action) and G-morphisms by G-sch. For $U \in G$-sch, we define $\operatorname{cov} U$ to be the set of families of G-morphisms $\{\varphi_i : U_i \to U\}$ with the same codomain U, such that each φ_i is flat locally of finite presentation, and that $\bigcup_i \varphi_i(U_i) = U$. We define the subcategory T_G of G-sch by $\operatorname{ob}(T_G) = \operatorname{ob}(G\text{-sch})$, and $\operatorname{Mor}(T_G)$ is defined to be the set of G-morphisms locally of finite presentation.

Lemma 2.3.2 G-sch *is a site with the Grothendieck topology above. Any base change of a morphism of* T_G *is again a morphism of* T_G*, and* $\{\varphi_i : U_i \to U\} \in \operatorname{cov} U$ *implies that* $\varphi_i \in T_G$ *for any* i*. If* $X \in G$-sch*, then* T_G/X *has a small topology.*

Proof. Everything is trivial, except for the last assertion. Let (V_i) be an affine open covering of X. Then as $\{V_i \times G \to X\} \in \operatorname{cov} X$, it suffices to prove the assertion for each $V_i \times G$. Hence, we may assume that $X = \operatorname{Spec} A$ is affine. In this case, the set of isomorphism classes of A-algebras of finite presentation is indexed by a small set, say Λ. Let $(B_\lambda)_{\lambda\in\Lambda}$ be a set of complete representatives of isomorphism classes of A-algebras of finite presentation. When we define T_0 to be the full subcategory of T_G/X with the object set $\{(\operatorname{Spec} B_\lambda)\times G \mid \lambda \in \Lambda\}$, T_0 is obviously small. Let $Y \in T_G/X$. When we take an affine open covering (U_i), then for each i, there exists some $\lambda(i) \in \Lambda$ such that $U_i \cong \operatorname{Spec} B_{\lambda(i)}$ [65, IV.1.4]. This shows that $\{\operatorname{Spec} B_{\lambda(i)} \times G \to Y\}$ is a covering of Y, and we have $\operatorname{Spec} B_{\lambda(i)} \times G \in T_0$. This shows that T_G/X has a small topology. □

(2.3.3) Let R and G be as in (2.3.1). Let $X \in G$-sch. We denote the full subcategory of T_G/X consisting of $f : Y \to X$ with Y affine by $(T_G/X)^a$, and the subcategory of T_G/X with objects consisting of affine Y and flat f and morphisms consisting of flat morphisms by (T_G/X)-aff. Note that $(T_G/X)^a$ and (T_G/X)-aff are closed under fiber products, and hence they are sites with the same Grothendieck topology as that of T_G/X.

Let $\mathcal{C}$ be a category with projective limits. A sheaf (resp. presheaf) on T_G/X with values in $\mathcal{C}$ is called a $\mathcal{C}$-valued G-*faisceau* (resp. G-presheaf) on X. By (I.1.8.5), any G-faisceau is identified with its restriction to $(T_G/X)^a$. A sheaf (resp. presheaf) on (T_G/X)-aff with values in $\mathcal{C}$ is called a $\mathcal{C}$-valued aff-G-faisceau (resp. aff-G-presheaf) on X.

As ${}_G\mathbb{M}$ is locally noetherian, it satisfies the (AB3*) condition. By Corollary I.1.8.10 and Lemma 2.3.2, we have that $\operatorname{sh}(T_G/X, {}_G\mathbb{M})$ is a Grothendieck category which satisfies the (AB3*) condition.

(2.3.4) Let R, G and X be as in (2.3.3). Let $\mathcal{A}$ be a G-faisceau of G-algebras over X. We say that $\mathcal{M}$ is a $(G, \mathcal{A})$-module (resp. $(G, \mathcal{A})$-module presheaf) if $\mathcal{M}$ is both a G-faisceau (resp. G-presheaf) of G-modules and a G-faisceau (resp. G-presheaf) of $\mathcal{A}$-modules ($\mathcal{A}$ is viewed as a G-faisceau of R-algebras), and for any $(f : Y \to X) \in T_G/X$, the G-module $\mathcal{A}(Y)$-module $\mathcal{M}(Y)$ is a $(G, \mathcal{A}(Y))$-module. By definition, if $\mathcal{M}$ is a $(G, \mathcal{A})$-module presheaf, then for any $Y' \to Y$, the canonical map $\mathcal{M}(Y) \to \mathcal{M}(Y')$ is $(G, \mathcal{A}(Y))$-linear.

The notion of aff-$(G, \mathcal{B})$-module (presheaf) is defined similarly, for an aff-G-faisceau of G-algebras $\mathcal{B}$, replacing T_G/X by (T_G/X)-aff.

(2.3.5) We define $\mathcal{O}_X$ by $\mathcal{O}_X(Y) = \Gamma(Y, \mathcal{O}_Y)$ for $Y \in T_G/X$. Note that $\mathcal{O}_X$ is a G-faisceau of G-algebras. By restriction, $\mathcal{O}_X$ is also an aff-G-faisceau of G-algebras. Using the same symbol $\mathcal{O}_X$ is an abuse of notation.

We say that $\mathcal{M}$ is a *quasi-coherent* $\mathcal{O}_X$-module if $\mathcal{M}$ is an $\mathcal{O}_X$-module (where $\mathcal{O}_X$ is viewed as a G-faisceau of commutative rings), and for any morphism $f : Y \to Z$ in $(T_G/X)^a$, the canonical map $\mathcal{O}_X(Y) \otimes_{\mathcal{O}_X(Z)} \mathcal{M}(Z) \to \mathcal{M}(Y)$ is an isomorphism. Assume that X is locally noetherian. $\mathcal{M}$ is a *coherent* $\mathcal{O}_X$-module if $\mathcal{M}$ is quasi-coherent, and if $Y \in (T_G/X)^a$ then $\mathcal{M}(Y)$ is a finitely generated $\mathcal{O}_X(Y)$-module. In the sequel, when we mention something on coherent $\mathcal{O}_X$-modules, we always assume that X is locally noetherian.

A $(G, \mathcal{O}_X)$-module is called quasi-coherent (resp. coherent), if it is quasi-coherent as an $\mathcal{O}_X$-module.

Similarly, quasi-coherent and coherent (for X locally noetherian) aff-G-$\mathcal{O}_X$-modules and aff-$(G, \mathcal{O}_X)$-modules are defined.

(2.3.6) Let R, G and X be as in (2.3.3), and let $\mathcal{A}$ be a G-faisceau of G-algebras on X.

Let $\mathcal{M}$ be a $(G, \mathcal{A})$-module presheaf. Then the sheafification $a(\mathcal{M})$ of $\mathcal{M}$ is a $(G, \mathcal{A})$-module (faisceau) in a natural way.

For $(G, \mathcal{A})$-modules $\mathcal{M}$ and $\mathcal{N}$, the tensor product $\mathcal{M} \otimes_{\mathcal{A}} \mathcal{N}$ is defined to be the sheafification of the $(G, \mathcal{O}_X)$-module presheaf given by $(\mathcal{M} \otimes_{\mathcal{A}} \mathcal{N})(Y) = \mathcal{M}(Y) \otimes_{\mathcal{A}(Y)} \mathcal{N}(Y)$.

For a G-morphism $f : Y \to X$ and a $(G, \mathcal{O}_Y)$-module $\mathcal{M}$, $f_*\mathcal{M}$ has a $(G, \mathcal{O}_X)$-module structure in a natural way. If $\mathcal{N}$ is a $(G, \mathcal{O}_X)$-module, then $f^*\mathcal{N} := \mathcal{O}_Y \otimes_{f^{-1}\mathcal{O}_X} f^{-1}\mathcal{N}$ is a $(G, \mathcal{O}_Y)$-module.

Similar definitions are also made for $\mathcal{O}_X$-modules, aff-$(G, \mathcal{O}_X)$-modules, and aff-G-$\mathcal{O}_X$-modules.

(2.3.7) Let R and G be as in (2.3.3). Let us denote by $\mu : G \times G \to G$ the product of G. For $X \in G$-sch, we denote by $a_X, p_X : X \times G \to X$ the action of G and the first projection, respectively. The following definition is due to Mumford [115].

Definition 2.3.8 A G-linearized $\mathcal{O}_X$-module is a pair $(\mathcal{M}, \phi)$, where $\mathcal{M}$ is an $\mathcal{O}_X$-module (with respect to the Zariski topology), and $\phi : a_X^* M \to p_1^* M$ is an $\mathcal{O}_{X\times G}$-isomorphism such that the following condition is satisfied: The isomorphism

$$[a_X \circ (a_X \times 1_G)]^* M = [a_X \circ (1_X \times \mu_G)]^* M \xrightarrow{(1_X\times\mu_G)^*\phi} [p_1 \circ (1_X \times \mu_G)]^* M$$

agrees with the composite

$$[a_X \circ (a_X \times 1_G)]^* M \xrightarrow{(a_X\times 1_G)^*\phi} [p_1 \circ (a_X \times 1_G)]^* M = [a_X \circ p_{12}]^* M \xrightarrow{p_{12}^*\phi} [p_1 \circ p_{12}]^* M = [p_1 \circ (1_X \times \mu_G)]^* M.$$

By Lemma 1.1.6, if $X = \operatorname{Spec} A$ is affine, then a quasi-coherent G-linearized $\mathcal{O}_X$-module is the same as a (G, A)-module. It is easy to see that the category of G-linearized $\mathcal{O}_X$-modules is abelian, and the forgetful functor to the category ${}_X\mathbb{M}$ is exact. Moreover, direct image functors and inverse image functors are enriched to the G-linearized versions, which are compatible with forgetful functors.

(2.3.9) Let R, G and X be as in (2.3.3). If $\mathcal{M} = (\mathcal{M}, \phi)$ is a quasi-coherent G-linearized $\mathcal{O}_X$-module and $(f : Y \to X) \in (T_G/X)^a$, then $f^*\mathcal{M} = (f^*\mathcal{M}, (f \times 1_G)^*\phi)$ is a quasi-coherent G-linearized $\mathcal{O}_Y$-module. Hence, letting $W(\mathcal{M})(f) := \Gamma(Y, f^*\mathcal{M})$ for $(f : Y \to X) \in T_G/X$, we have that $W(\mathcal{M})$ is a $(G, \mathcal{O}_X)$-module. By (I.2.3), it is easy to see that $W(\mathcal{M})$ is quasi-coherent. If X is locally noetherian, then W preserves the coherence.

(2.3.10) Let R, G and X be as in (2.3.3). A $(G, \mathcal{O}_X)$-module is an aff-$(G, \mathcal{O}_X)$-module by restriction. Note that this restriction preserves the quasi-coherence, and coherence.

(2.3.11) Let R, G and X be as in (2.3.3). Let us assume that $\mathcal{M}$ is a quasi-coherent aff-$(G, \mathcal{O}_X)$-module. We want to construct a G-linearized quasi-coherent $\mathcal{O}_X$-module $Q(\mathcal{M})$, so that Q is a quasi-inverse of the functor W in (2.3.9).

First, we construct $Q(\mathcal{M})$, assuming that X is quasi-compact.

Take a finite affine open covering $(V_1, \ldots, V_r)$ of X, and set $Y := (\coprod_i V_i) \times G$. Note that $Y \in (T_G/X)$-aff. We denote the canonical map $Y \to X$ by π. Note that π is faithfully flat, and $Y \times_X Y$ is also affine. We denote the ith projection $Y \times_X Y \to Y$ by π_i. Note that π_i is a morphism of (T_G/X)-aff for $i = 1, 2$. Let $\tau : Y \times_X Y \to Y \times_X Y$ be the isomorphism of (T_G/X)-aff given by $(y_1, y_2) \mapsto (y_2, y_1)$. We denote the composite map

$$(\pi_1)^*(\mathcal{M}(Y)) \cong \mathcal{M}(Y \times_X Y) \xrightarrow{\mathcal{M}(\tau)} \mathcal{M}(Y \times_X Y) \cong (\pi_2)^*(\mathcal{M}(Y))$$

by ψ. Then it is easy to verify that the cocycle condition $\pi_{31}^*\psi = \pi_{32}^*\psi \circ \pi_{21}^*\psi$ holds, where $\pi_{ji} : Y \times_X Y \times_X Y \to Y \times_X Y$ is given by $(y_1, y_2, y_3) \mapsto$

(y_i, y_j). By Proposition I.2.3.8, we have a quasi-coherent $\mathcal{O}_X$-module $Q(\mathcal{M})$ corresponding to the descent datum $(\mathcal{M}(Y), \psi)$, and we have $\mathcal{M}(Y) \cong \pi^*(Q(\mathcal{M}))$. (as Y is affine, we identify $\mathcal{M}(Y)$ with its associated Zariski sheaf). Now $Q(\mathcal{M})$, as an $\mathcal{O}_X$-module (in the usual Zariski topology), has been defined.

Next, we give a G-linearization to $Q(\mathcal{M})$. By definition, $Q(\mathcal{M})$ is $\mathrm{Ker}(h_1 - h_2)$, where h_1 and h_2 are as in (I.2.3.7). As ψ is G-linear, both h_1 and h_2 are compatible with G-linearizations. Hence, $Q(\mathcal{M})$ has a natural G-linearization $\phi_{\mathcal{M}}$ such that $\pi^*\phi_{\mathcal{M}} = \phi$, where ϕ is the G-linearization of $\mathcal{M}(Y)$.

Thus, we have a G-linearized quasi-coherent $\mathcal{O}_X$-module $(Q(\mathcal{M}), \phi_{\mathcal{M}})$ associated to the aff-$(G, \mathcal{O}_X)$-module $\mathcal{M}$, if X is quasi-compact. If $f : X' \to X$ is a morphism of T_G/X, both X and X' are quasi-compact, and f is flat, then $(f^*(Q(\mathcal{M})), (f \times 1_G)^*\phi_{\mathcal{M}})$ is identified with $(Q(f^*\mathcal{M}), \phi_{f^*\mathcal{M}})$ in a natural way.

Now consider the general case. X admits a covering consisting of quasi-compact G-stable open subschemes. We construct $Q(\mathcal{M})$ over each G-stable open subset, and we know that the patching goes well by the last paragraph, so we can construct $(Q(\mathcal{M}), \phi_{\mathcal{M}})$ in the case that X is not quasi-compact as well.

By (2.3.9), (2.3.10) and (2.3.11), we have the following.

Proposition 2.3.12 *Let $X \in G$-sch. Then the following categories are abelian, and equivalent to one another.*

1 *The category of quasi-coherent $(G, \mathcal{O}_X)$-modules.*

2 *The category of quasi-coherent* aff-*$(G, \mathcal{O}_X)$-modules.*

3 *The category of quasi-coherent G-linearized $\mathcal{O}_X$-modules.*

The equivalence **1**⇒**2** *is given by the restriction. The equivalence* **2**⇒**3** *is given by Q. The equivalence* **3**⇒**1** *is given by W. If X is locally noetherian, then the coherence is preserved by these equivalences.*

The argument above is still valid (and much easier) for quasi-coherent $\mathcal{O}_X$-modules without G-action. Hence,

Proposition 2.3.13 *Let $X \in G$-sch. Then the following categories are abelian, and equivalent to one another.*

1 *The category of quasi-coherent $\mathcal{O}_X$-module G-faisceaux.*

2 *The category of quasi-coherent $\mathcal{O}_X$-module* aff-*G-faisceaux.*

3 *The category of quasi-coherent $\mathcal{O}_X$-modules in the Zariski topology.*

The equivalence **1**⇒**2** *is given by the restriction. The equivalence* **2**⇒**3** *is given by Q. The equivalence* **3**⇒**1** *is given by W. If X is locally noetherian, then the coherence is preserved by these equivalences.*

(2.3.14) If A is a commutative noetherian ring, B an A-flat commutative algebra, M an A-finite module, and N an A-module, then the canonical map

$$\mathrm{Ext}^i_A(M,N)\otimes_A B \to \mathrm{Ext}^i_B(M\otimes B, N\otimes B)$$

is an isomorphism.

Let R, G and X be as in (2.3.3). Assume that X is locally noetherian. If $\mathcal{M}$ is a coherent $(G,\mathcal{O}_X)$-module, and $\mathcal{N}$ a quasi-coherent $(G,\mathcal{O}_X)$-module, then we define a quasi-coherent aff-$(G,\mathcal{O}_X)$-module $\underline{\mathrm{Ext}}^i_{\mathcal{O}_X}(\mathcal{M},\mathcal{N})$ by

$$\underline{\mathrm{Ext}}^i_{\mathcal{O}_X}(\mathcal{M},\mathcal{N})(Y) := \mathrm{Ext}^i_{\mathcal{O}_X(Y)}(\mathcal{M}(Y),\mathcal{N}(Y))$$

for $Y \in (T_G/X)$-aff. Hence, $\underline{\mathrm{Ext}}^i_{\mathcal{O}_X}(\mathcal{M},\mathcal{N})$ has a quasi-coherent $(G,\mathcal{O}_X)$-module structure.

If G satisfies PCP and $\mathcal{M}$ and $\mathcal{N}$ are quasi-coherent $(G,\mathcal{O}_X)$-modules, then we can also define a quasi-coherent $(G,\mathcal{O}_X)$-module structure of the sheaf $\underline{\mathrm{Tor}}_i^{\mathcal{O}_X}(\mathcal{M},\mathcal{N})$ in a similar way.

Proposition 2.3.15 *Let R be a noetherian ring, G an affine flat R-group scheme of finite type, X a locally noetherian right G-action, $\mathcal{M}$ a coherent $(G,\mathcal{O}_X)$-module, and $r \geq 0$. Then the set $U_r(\mathcal{M})$ defined by*

$$U_r(\mathcal{M}) := \{x \in X \mid \mathcal{M}_x \cong \mathcal{O}_{X,x}^r \quad (\text{as } \mathcal{O}_{X,x}\text{-modules})\}$$

is a G-stable open subset of X. If X is Cohen–Macaulay, then the maximal Cohen–Macaulay locus of $\mathcal{M}$ is also G-stable.

Proof. It is well-known that $U_r(\mathcal{M})$ is an open set of X. Let $A \to B$ be a flat local homomorphism of noetherian local rings, and M a finite A-module. Then $M \cong A^r$ if and only if $M \otimes_A B \cong B^r$ (easy). Hence, as we have a G-linearization $a_X^*(\mathcal{M}) \cong p_X^*(\mathcal{M})$, we have that

$$a_X^{-1}(U_r(\mathcal{M})) = U_r(a_X^*(\mathcal{M})) = U_r(p_X^*(\mathcal{M})) = p_X^{-1}(U_r(\mathcal{M})),$$

and $U_r(\mathcal{M})$ is G-stable.

The last assertion is proved similarly, as we know that G is Cohen–Macaulay over R. □

By the proposition, we have that the complement of the support of a coherent $(G,\mathcal{O}_X)$-module is G-stable open. More precisely, we have:

Lemma 2.3.16 *Let R, G, X and $\mathcal{M}$ be as in* Proposition 2.3.15. *Then the annihilator* $\mathrm{ann}\,\mathcal{M}$ *of $\mathcal{M}$ is a G-ideal sheaf of X, namely, a $(G,\mathcal{O}_X)$-submodule of $\mathcal{O}_X$.*

Proof. As $\mathrm{ann}\,\mathcal{M}$ is nothing but the kernel of the $(G,\mathcal{O}_X)$-linear map

$$\mathcal{O}_X \to \underline{\mathrm{Hom}}_{\mathcal{O}_X}(\mathcal{M},\mathcal{M}),$$

the assertion is obvious. □

(2.3.17) Let R be noetherian, G be R-smooth with connected geometric fibers, and A be a commutative noetherian G-algebra. We say that a G-ideal I is a *G-maximal ideal* if it is maximal among G-ideals with respect to the incidence relation. It is easy to see that a G-maximal ideal is a prime ideal. The intersection of all G-maximal ideals of A is called the *G-radical* of A. Note that the G-radical of A is a G-ideal by Corollary 2.2.9 and Exercise I.3.10.6.

Lemma 2.3.18 *Let G and A be as above, M an A-finite (G, A)-module, and J an ideal of A contained in the G-radical of A. If $M = JM$, then we have $M = 0$.*

Proof. We may assume that J is the G-radical of A. Then any G-maximal ideal is not contained in $\operatorname{supp} M$ by Nakayama's lemma. If $M \neq 0$, then $\operatorname{supp} M$ is a non-empty G-stable closed subset of $\operatorname{Spec} A$, and by Lemma 2.1.17 there is some G-prime ideal which supports M. So there is a maximal one among such, and it is a G-maximal ideal, which is a contradiction. □

(2.3.19) Let X be a locally noetherian G-scheme. We denote the category of aff-$(G, \mathcal{O}_X)$-modules by ${}_{G,X}\mathbb{M}$, and the full subcategory of quasi-coherent $(G, \mathcal{O}_X)$-modules by $\mathrm{Qco}(G, X)$. Note that $\mathrm{Qco}(G, X)$ is an abelian category in a natural way, so that $\Gamma(Y, ?)$ is exact for each $Y \in (T_G/X)$-aff. If $Y \in (T_G/X)$-aff is faithfully flat over X, then $\Gamma(Y, ?)$ is also faithful.

The category $\mathrm{Qco}(G, X)$ satisfies (AB5). For each $Y \in (T_G/X)$-aff, $\Gamma(Y, ?)$ preserves arbitrary inductive limits. Hence, for each quasi-compact $Y \in T_G/X$, the functor $\Gamma(Y, ?)$ preserves filtered inductive limits.

Obviously, we have a forgetful functor $F_X^G : \mathrm{Qco}(G, X) \to \mathrm{Qco}(X)$, which is faithful exact and preserves inductive limits.

Let X' denote the R-scheme X with the trivial G-action, and $\mathcal{M} \in \mathrm{Qco}(X)$. As $\mathcal{M}$ together with the trivial G-linearization lies in $\mathrm{Qco}(G, X')$, and $p_X : X' \times G \to X'$ is a G-morphism, we have that $p_X^* \mathcal{M}$ is a quasi-coherent $(G, \mathcal{O}_{X' \times G})$-module.

Hence, $(a_X)_* p_X^* \mathcal{M} \in \mathrm{Qco}(G, X)$, as $a_X : X' \times G \to X$ is an affine G-morphism. Taking

$$\omega_{\mathcal{M}} : \mathcal{M} \to (a_X)_* a_X^* \mathcal{M} \xrightarrow{(a_X)_* \phi \mathcal{M}} (a_X)_* p_X^* \mathcal{M}$$

as the unit of adjunction and

$$(a_X)_* p_X^* \mathcal{N} \to (a_X)_* e_* e^* p_X^* \mathcal{N} = \mathcal{N}$$

as the counit of adjunction (where $e : \operatorname{Spec} R \to G$ is the unit element of G), we have that $(a_X)_* p_X^*$ is right adjoint to the forgetful functor F_X^G. It follows that $(a_X)_* p_X^*$ preserves injectives. As $\omega_{\mathcal{M}}$ is a monomorphism, it follows that the set of all objects of the form $(a_X)_* p_X^* \mathcal{I}$ with $\mathcal{I}$ an $\mathcal{O}_X$-injective quasi-coherent $\mathcal{O}_X$-module is an injective cogenerator of $\mathrm{Qco}(G, X)$, see [69, Theorem II.7.18]. Hence, we have

Lemma 2.3.20 *The category* $\mathrm{Qco}(G, X)$ *has enough injectives.*

We also have

Lemma 2.3.21 *Let* $\mathcal{J}$ *be an injective object in* $\mathrm{Qco}(G, X)$. *If* U *is a* G-*stable open subset of* X, *then* $\mathcal{I}|_U$ *is injective in* $\mathrm{Qco}(G, U)$.

Proof. We may assume that $\mathcal{J} = (a_X)_* p_X^* \mathcal{I}$, with $\mathcal{I}$ an $\mathcal{O}_X$-injective quasi-coherent $\mathcal{O}_X$-module. As $\mathcal{I}|_U$ is injective in ${}_U\mathbb{M}$, the assertion follows easily. □

(2.3.22) Let $f : Y \to X$ be a quasi-compact G-morphism of locally noetherian G-schemes. By (2.3.6), we have canonical functors

$$f_*^G : \mathrm{Qco}(G, Y) \to \mathrm{Qco}(G, X)$$

and

$$f_G^* : \mathrm{Qco}(G, X) \to \mathrm{Qco}(G, Y),$$

and these functors are obviously compatible with the forgetful functors: $F_X^G \circ f_*^G = f_* \circ F_Y^G$ and $F_Y^G \circ f_G^* = f^* \circ F_X^G$.

Proposition 2.3.23 *Let* $f : Y \to X$ *be a quasi-compact* G-*morphism of locally noetherian* G-*schemes. If* $\mathcal{J}$ *is an injective object of* $\mathrm{Qco}(G, Y)$, *then we have* $R^i f_* \mathcal{J} = 0$ *for* $i > 0$.

Proof. As the question is local on X and X is covered by quasi-compact G-stable open subschemes, we may assume that both X and Y are quasi-compact. We may and shall assume that $\mathcal{J} = (a_Y)_* p_Y^* \mathcal{I}$, with $\mathcal{I}$ an $\mathcal{O}_Y$-injective quasi-coherent $\mathcal{O}_Y$-module. Then we have

$$R^i f_* (a_Y)_* p_Y^* \mathcal{I} \cong (a_X)_* R^i (f \times 1_G)_* p_Y^* \mathcal{I} \cong (a_X)_* p_X^* R^i f_* \mathcal{I} = 0.$$

The first isomorphism is by the Leray spectral sequence (note that a_Y and a_X are affine and $f \circ a_Y = a_X \circ (f \times 1_G)$). The second one comes from the flat base change of higher direct images, see [64, (1.4.15)]. Thus, we have $R^i f_* \mathcal{J} = 0$ for $i > 0$, as desired. □

By the proposition, we have

$$F_X^G R^i f_*^G \mathcal{M} \cong R^i (F_X^G \circ f_*^G) \mathcal{M} \cong R^i (f_* \circ F_Y^G) \mathcal{M} \cong R^i f_* (F_Y^G \mathcal{M})$$

for any $\mathcal{M} \in \mathrm{Qco}(G, Y)$. Hence, $R^i f_* \mathcal{M}$ is endowed with a $(G, \mathcal{O}_X)$-module structure of $R^i f_*^G \mathcal{M}$ in a natural way.

Corollary 2.3.24 *Let* f *be as in the proposition. Then* $R^i f_*^G$ *preserves filtered inductive limits.*

2.4 Matijevic–Roberts type theorem

(2.4.1) Throughout this subsection, let R be a noetherian commutative ring, and G an affine smooth R-group scheme with connected geometric fibers. Let X be a locally noetherian G-action. In this subsection, we prove the following, which is the main theorem in this chapter.

Theorem 2.4.2 *Let $x \in X$. Then the following hold:*

1 *If $\mathcal{O}_{X,x^*}$ is a regular local ring (resp. complete intersection), then so is $\mathcal{O}_{X,x}$. In particular, if X is regular (resp. l.c.i.) at its all stable points, then X is regular (resp. l.c.i.).*

2 *Let $\mathcal{M}$ be a coherent $(G, \mathcal{O}_X)$-module. If $\mathcal{M}_{x^*}$ is Gorenstein (resp. Cohen–Macaulay, free), then so is $\mathcal{M}_x$. In particular, if for any stable point x, $\mathcal{M}_x$ is Gorenstein (resp. Cohen–Macaulay, free), then $\mathcal{M}$ is Gorenstein (resp. Cohen–Macaulay, locally free).*

Let us consider the case that $X = \operatorname{Spec} A$ in the theorem is affine. For an ideal I of A, the sum of all G-ideals contained in I is again a G-ideal, and is the largest G-ideal contained in I, which we denote by I^*. Clearly, we have $\operatorname{Spec} A/I^* = (\operatorname{Spec} A/I)^*$. Hence, if I is a prime ideal, then so is I^*. If I corresponds to $x \in X$, then I^* corresponds to x^*.

As a split torus $G = \mathbb{G}_m^n$ satisfies the assumption of the theorem (one can consider that $R = \mathbb{Z}$), we have the following by (1.2).

Corollary 2.4.3 *Let A be a $\mathbb{Z}^n$-graded noetherian commutative ring, and $P \in \operatorname{Spec} A$. Then the following hold:*

1 *If A_{P^*} is regular (resp. l.c.i.), then so is A_P. In particular, if the localization of A at any graded prime ideal is regular (resp. l.c.i.), then A is regular (resp. l.c.i.).*

2 *Let M be an A-finite graded A-module. If M_{P^*} is Gorenstein (resp. Cohen–Macaulay, free), then so is M. In particular, if the localization of M at any graded prime is Gorenstein (resp. Cohen–Macaulay, free), then M is Gorenstein (resp. Cohen–Macaulay, projective).*

Note that Corollary 2.4.3 was originally conjectured by Nagata [119] (the case $n = 1$ for the Cohen–Macaulay property), and proved in [85, 109, 7, 108, 60, 34, 16], except for the l.c.i. property in **1** for the case $n \geq 2$. So we could say that this has been well-known. The theorem generalizes Corollary 2.4.3 to more general G.

Now we start proving the theorem.

Lemma 2.4.4 *Let Y be a subvariety of X, and η its generic point. Then the following hold:*

1 *$\mathcal{O}_{Y^*,\eta}$ is a regular local ring.*

2 *Let $\mathcal{M}$ be a coherent $(G, \mathcal{O}_{Y^*})$-module. Then $\mathcal{M}_\eta$ is $\mathcal{O}_{Y^*,\eta}$-free. If $\mathcal{N}$ is a coherent $(G, \mathcal{O}_X)$-module, then $\operatorname{Ext}^i_{\mathcal{O}_{X,\eta}}(\mathcal{M}_\eta, \mathcal{N}_\eta)$ is also $\mathcal{O}_{Y^*,\eta}$-free for $i \geq 0$.*

Proof. We show **1**. Let $\varphi : Y \times G \to Y^*$ be the action $(y, g) \mapsto yg$. Then we have $\operatorname{Flat}(\varphi)$ is a G-stable open subset of $Y \times G$ by Lemma 2.1.10. Hence, by Lemma 2.1.9, there exists some open subset F of Y such that $\operatorname{Flat}(\varphi) = F \times G$. As Y^* is integral and φ is dominating by the definition of Y^*, we have $\operatorname{Flat}(\varphi) \neq \emptyset$, and hence we have $F \neq \emptyset$ and $\eta \in F$. It follows that the composite morphism

$$\psi : \operatorname{Spec} \kappa(\eta) \times G \to Y \times G \xrightarrow{\varphi} Y^*$$

is flat, because the first morphism is flat, and the second morphism is flat on the image of the first. As G is R-smooth, we have that $\operatorname{Spec} \kappa(\eta) \times G$ is $\kappa(\eta)$-smooth, and in particular, it is a regular scheme. As the unit element $\operatorname{Spec} \kappa(\eta) \times \{e\}$ is mapped to η by ψ, we have that $\mathcal{O}_{Y^*,\eta}$ is also regular local.

We show **2**. As Y^* is integral, the free locus U of $\mathcal{M}$ is non-empty, and is a G-stable open subset of Y^* (Proposition 2.3.15). Assume that $U \cap Y = \emptyset$. Then $Y^* \setminus U$ with the reduced structure is a G-stable closed subscheme of X containing Y (Lemma 2.1.17). As $U \neq \emptyset$, this contradicts the minimality of Y^*. Hence, $U \cap Y \neq \emptyset$. As $\eta \in U \cap Y$, $\mathcal{M}_\eta$ is $\mathcal{O}_{Y^*,\eta}$-free. We prove the last statement. As $\underline{\operatorname{Ext}}^i_{\mathcal{O}_X}(\mathcal{M}, \mathcal{N})$ is a coherent $(G, \mathcal{O}_{Y^*})$-module for $i \geq 0$,

$$\operatorname{Ext}^i_{\mathcal{O}_{X,\eta}}(\mathcal{M}_\eta, \mathcal{N}_\eta) \cong \underline{\operatorname{Ext}}^i_{\mathcal{O}_X}(\mathcal{M}, \mathcal{N})_\eta$$

is $\mathcal{O}_{Y^*,\eta}$-free, by the first part. □

Let X, x and $\mathcal{M}$ be as in the theorem, and set $Y := \overline{\{x\}}$, $B := \mathcal{O}_{X,x}$, $B/P = \mathcal{O}_{Y^*,x}$, and $M := \mathcal{M}_x$.

Lemma 2.4.5 *The following hold:*

1 *B/P is a regular local ring.*

2 *B is normally flat along the prime ideal P.*

3 *$\operatorname{Ext}^i_B(B/P, B/P)$ is B/P-free for $i \geq 0$.*

4 *B_P is regular local (resp. a complete intersection), by assumption of the theorem.*

5 *M is normally flat along P.*

6 *$\operatorname{Ext}^i_B(B/P, M)$ is B/P-free for $i \geq 0$.*

7 *M_P is Cohen–Macaulay (resp. Gorenstein, free) by assumption of the theorem.*

Proof. Note that **4,7** are assumptions of the theorem. Note that $M_P = \mathcal{M}_{x^*}$ and $B_P = \mathcal{O}_{X,x^*}$. The assertion **1** is obvious by Lemma 2.4.4. We denote the defining ideal sheaf of Y^* by $\mathcal{P}$. Note that $\mathcal{P}$ is a $(G, \mathcal{O}_X)$-submodule of $\mathcal{O}_X$, and $\mathcal{P}^n\mathcal{M}/\mathcal{P}^{n+1}\mathcal{M}$ is a coherent $(G, \mathcal{O}_{Y^*})$-module for $n \geq 0$. Hence, $(\mathcal{P}^n\mathcal{M}/\mathcal{P}^{n+1}\mathcal{M})_x = P^n M/P^{n+1}M$ is B/P-free, by Lemma 2.4.4. That is, **5** holds. In particular, considering the case $\mathcal{M} = \mathcal{O}_X$, we have **2**. Similarly, **3,6** follow from Lemma 2.4.4 easily. □

Now statement **2** of the theorem is obvious by Lemma I.2.13.5. On the other hand, statement **1** in the theorem is obvious by Lemma I.2.13.9. So the proof of the theorem is now complete. □

Chapter III

Highest Weight Theory over an Arbitrary Base

So-called highest weight theory plays an important role not only in the theory of algebraic groups, but also in various areas of representation theory. This chapter is devoted to reviewing and studying the theory of good filtrations, Schur algebras and the theory of quasi-hereditary algebras over an arbitrary base ring, from the viewpoint of comodules. As the central purpose is to reconstruct S. Donkin's Schur algebra over an arbitrary base, we will not go into the theory of highest weight category originated by Cline–Parshall–Scott [37], but give a more concrete treatment using coalgebras. For highest weight theory over an arbitrary base from a different approach, see [48, 51, 145].

1 Highest weight theory over a field

This section is devoted to reviewing the highest weight theory over a field. In view of the later characteristic-free treatment, simple comodules, which depend on characteristic, do not appear in the first definition.

1.1 Weak split highest weight coalgebras

(1.1.1) For an ordered set P and $x, y \in P$, we use the interval notation such as $[x, y] := \{z \in P \mid x \leq z \leq y\}$ and $(-\infty, x) := \{z \in P \mid z < x\}$.

A subset Q of an ordered set P is called a *poset ideal* of P if $q \in Q$, $p \in P$ and $p \leq q$ together imply $p \in Q$. For $p \in P$, the intervals $(-\infty, p]$ and $(-\infty, p)$ are poset ideals of P.

For a finite totally ordered set C, we define $\operatorname{rank} C := \#C - 1$. In general, we define

$$\operatorname{rank} P := \sup_{C \subset P} \operatorname{rank} C$$

for any ordered set P, where C runs through all finite totally ordered subsets of P. Thus, $\operatorname{rank} P$ is an integer greater than or equal to -1, or ∞. For $p \in P$, we define

$$\operatorname{ht} p := \operatorname{rank}(-\infty, p], \qquad \text{and} \qquad \operatorname{coht} p := \operatorname{rank}[p, \infty).$$

Lemma 1.1.2 *Let X^+ be an ordered set. Then the following are equivalent.*

1 *There exists an injective order-preserving map $f : X^+ \to \mathbb{N}$.*

2 *There exists an injective order-preserving map $g : X^+ \to \mathbb{N}$ such that the image $g(X^+)$ is a poset ideal of $\mathbb{N}$.*

2′ *Either $\#X^+ < \infty$ or there exists some order-preserving bijective map $g : X^+ \to \mathbb{N}$.*

3 *X^+ is countable, and $\#(-\infty, \lambda] < \infty$ for any $\lambda \in X^+$.*

Proof. (By Y. Yoshinobu) **1**⇒**2** Define $g(x) := \#(f(X^+) \cap [1, f(x)])$. **2** ⇒ **2′** is obvious, as any infinite poset ideal of $\mathbb{N}$ is $\mathbb{N}$ itself. **2′** ⇒ **3** We may assume that X^+ is infinite. In this case, X^+ is obviously countable. As we have $(-\infty, \lambda] \subset g^{-1}([1, g(\lambda)])$, it follows that $\#(-\infty, \lambda] < \infty$.

We show **3**⇒**1**. We set $X_n^+ := \{x \in X^+ \mid \operatorname{ht} x = n\}$ $(n \in \mathbb{N}_0)$, and $Y(n) := \bigcup_{i=0}^n X_i^+$. As each X_n^+ is countable, we can label

$$X_n^+ = \{x_{n1}, x_{n2}, x_{n3}, \ldots\}.$$

As we have $\bigcup_{n \geq 0} Y(n) = X^+$ by assumption, it suffices to construct an injective order-preserving map $g_n : Y(n) \to \mathbb{N}$ such that the restriction of g_n to $Y(i)$ $(i < n)$ is g_i. We define such a g_n inductively on n, so that the image of g_n does not contain a multiple of 2^{n+1}. So assume that g_i $(i < n)$ are already defined so that the image of g_i does not contain any multiple of 2^{i+1}. We define $(g_n)|_{Y(n-1)} = g_{n-1}$. We define $g_n(x_{nj})$ $(j = 1, 2, \ldots, \#X_n^+)$ inductively on j. Assuming that $g_n(x_{nl})$ is defined for $l < j$, we define $g_n(x_{nj})$ to be the minimum natural number which is larger than $g_n(x_{nl})$ $(l < j)$ and $g_{n-1}(y)$ $(y < x_{nj})$ and is divisible by 2^n but not divisible by 2^{n+1}.

Now g_n is defined, and it is easy to see that g_n is an extension of g_i $(i < n)$, injective and order-preserving, and the image of g_n does not contain any multiple of 2^{n+1}, as desired. □

(1.1.3) Let k be a field.

Definition 1.1.4 We say that $(C, X^+, \Delta, \nabla) = (C, X_C^+, \Delta_C, \nabla_C)$ is a *split highest weight coalgebra* (resp. *weak split highest weight coalgebra*) over k if

0 C is a k-coalgebra.

1 X^+ is an ordered set which satisfies the conditions in Lemma 1.1.2.

2 $\Delta = (\Delta_C(\lambda))_{\lambda \in X^+}$ is a family of finite dimensional C-comodules indexed by X^+.

2* $\nabla = (\nabla_C(\lambda))_{\lambda \in X^+}$ is a family of finite dimensional C-comodules indexed by X^+.

3 For any finite dimensional C-comodule V and any $\lambda \in X^+$, if

$$\mathrm{Hom}_C(\Delta_C(\mu), V) = 0$$

for any $\mu > \lambda$, then

$$\mathrm{Ext}^i_C(\Delta_C(\lambda), V) = 0$$

for $i = 1, 2$ (resp. $i = 1$).

3* For any finite dimensional C-comodule V and any $\lambda \in X^+$, if

$$\mathrm{Hom}_C(V, \nabla_C(\mu)) = 0$$

for any $\mu > \lambda$, then

$$\mathrm{Ext}^i_C(V, \nabla_C(\lambda)) = 0$$

for $i = 1, 2$ (resp. $i = 1$).

4 If V is a finite dimensional C-comodule and $\mathrm{Hom}_C(\Delta_C(\lambda), V) = 0$ for any $\lambda \in X^+$, then $V = 0$.

4* If V is a finite dimensional C-comodule and $\mathrm{Hom}_C(V, \nabla_C(\lambda)) = 0$ for any $\lambda \in X^+$, then $V = 0$.

5 For $\lambda, \mu \in X^+$, $\mathrm{Hom}_C(\Delta_C(\lambda), \nabla_C(\mu))$ equals 0 if $\lambda \neq \mu$, and is one-dimensional as a k-space if $\lambda = \mu$.

An element of X^+ is called a *dominant weight.* The C-comodule $\Delta_C(\lambda)$ is called the *Weyl module* of highest weight λ, and $\nabla_C(\lambda)$ is called the *induced module* of highest weight λ.

(1.1.5) Let k be a field, and (C, X^+, Δ, ∇) a weak split highest weight coalgebra over k.

Lemma 1.1.6 *We have*

$$\mathrm{Ext}^1_C(\Delta_C(\lambda), \nabla_C(\mu)) = 0$$

for $\lambda, \mu \in X^+$. *If, moreover,* (C, X^+, Δ, ∇) *is a split highest weight coalgebra, then we have* $\mathrm{Ext}^2_C(\Delta_C(\lambda), \nabla_C(\mu)) = 0$.

Proof. We have either $\mu \not< \lambda$ or $\mu \not> \lambda$. Assume that $\mu \not< \lambda$. Then as we have $\nu \neq \lambda$ for any $\nu > \mu$, we have the desired formula by **5** and **3*** in Definition 1.1.4. If $\mu \not> \lambda$, then by **5** and **3**, we have the result. □

For a poset ideal π of X^+, we denote by $\mathbb{M}^C(\pi)$ the full subcategory of $\mathbb{M}^C$ consisting of C-comodules V such that $\mathrm{Hom}_C(\Delta_C(\mu), V) = 0$ for any $\mu \in X^+$ with $\mu \notin \pi$.

Lemma 1.1.7 *Let π and π' be poset ideals of X^+. Then the following hold:*

1 $\mathbb{M}^C(\emptyset) = 0$.

2 *If $\pi \subset \pi'$, then $\mathbb{M}^C(\pi) \subset \mathbb{M}^C(\pi')$.*

3 $\mathbb{M}^C(\pi)$ *is very thick (i.e., closed under subquotients and extensions) and closed under inductive limits in $\mathbb{M}^C$.*

Proof. **1** and **2** are obvious. We prove **3**. It is clear that $\mathbb{M}^C(\pi)$ is closed under extensions and subcomodules. Let $V_1 \subset V$, with $V \in \mathbb{M}^C(\pi)$. Then $V_1 \in \mathbb{M}^C(\pi)$. As for any $\lambda \notin \pi$, $\mu > \lambda$ implies $\mu \notin \pi$, we have that $\mathrm{Ext}^1_C(\Delta_C(\lambda), V_1) = 0$ by Definition 1.1.4, **3**. By the obvious long exact sequence argument on Ext, we have $\mathrm{Hom}_C(\Delta_C(\lambda), V/V_1) = 0$, and hence $V/V_1 \in \mathbb{M}^C(\pi)$. This shows that $\mathbb{M}^C(\pi)$ is closed under quotients.

Next, we show that $\mathbb{M}^C(\pi)$ is closed under inductive limits. Let (V_i) be an inductive system in $\mathbb{M}^C(\pi)$. Then for $\lambda \notin \pi$, we have

$$\mathrm{Hom}_C(\Delta_C(\lambda), \bigoplus_{i \in I} V_i) \cong \bigoplus_{i \in I} \mathrm{Hom}_C(\Delta_C(\lambda), V_i) = 0$$

by Lemma I.1.9.3. Hence, $\bigoplus_i V_i \in \mathbb{M}^C(\pi)$. As $\varinjlim V_i$ is a quotient of $\bigoplus_i V_i$, we have $\varinjlim V_i \in \mathbb{M}^C(\pi)$, and $\mathbb{M}^C(\pi)$ is closed under inductive limits. □

Lemma 1.1.8 *Let $\lambda \in X^+$. Then* $\mathrm{top}(\Delta_C(\lambda)) \cong \mathrm{soc}(\nabla_C(\lambda))$. *Let us denote these isomorphic modules by $L_C(\lambda)$. Then $L_C(\lambda)$ is simple, and* $\mathrm{End}_C(L_C(\lambda)) \cong k$. *The set $\{L_C(\lambda) \mid \lambda \in X^+\}$ is a complete set of representatives of the isomorphism classes of simple objects of $\mathbb{M}^C$.*

Proof. As $\mathrm{Hom}_C(\Delta_C(\lambda), \nabla_C(\lambda)) \cong k \neq 0$, we have $\Delta_C(\lambda) \neq 0$ and $\nabla_C(\lambda) \neq 0$. Hence, $\mathrm{top}(\Delta_C(\lambda)) \neq 0$ and $\mathrm{soc}(\nabla_C(\lambda)) \neq 0$, because both of them are finite dimensional.

Let L be a simple object which is a direct summand of the semisimple object $\mathrm{top}(\Delta_C(\lambda))$. Then we have

$$\mathrm{Hom}_C(L, \nabla_C(\mu)) \subset \mathrm{Hom}_C(\Delta_C(\lambda), \nabla_C(\mu)) = 0$$

for $\mu \neq \lambda$. As L is simple (in particular, non-zero), we have

$$\mathrm{Hom}_C(L, \mathrm{soc}(\nabla_C(\lambda))) \cong \mathrm{Hom}_C(L, \nabla_C(\lambda)) \neq 0$$

by Definition 1.1.4, **4***. Hence, when we set $\mathrm{top}(\Delta_C(\lambda)) = L_1 \oplus L_2 \oplus \cdots \oplus L_r$, with each L_i simple, then we have

$$r \leq \dim_k \mathrm{Hom}_C(\mathrm{top}(\Delta_C(\lambda)), \mathrm{soc}(\nabla_C(\lambda))) \leq \dim_k \mathrm{Hom}_C(\Delta_C(\lambda), \nabla_C(\lambda)) = 1.$$

As $r > 0$, this shows $r = 1$, and hence $\mathrm{top}(\Delta_C(\lambda))$ is simple.

Similarly, utilizing condition **4** in Definition 1.1.4, we have that $\mathrm{soc}(\nabla_C(\lambda))$ is also simple.

As $\mathrm{Hom}_C(\mathrm{top}(\Delta_C(\lambda)), \mathrm{soc}(\nabla_C(\lambda))) \cong k$ as above, we must have

$$L_C(\lambda) := \mathrm{top}(\Delta_C(\lambda)) \cong \mathrm{soc}(\nabla_C(\lambda))$$

and $\mathrm{End}_C\, L_C(\lambda) \cong k$.

Assume $\lambda, \mu \in X^+$ and $\lambda \neq \mu$. Then we have

$$\mathrm{Hom}_C(L_C(\lambda), L_C(\mu)) \subset \mathrm{Hom}_C(\Delta_C(\lambda), \nabla_C(\mu)) = 0.$$

This shows that a different λ yields a non-isomorphic $L_C(\lambda)$.

Lastly, assume that L is a simple object of $\mathbb{M}^C$. As $L \neq 0$, there exists some $\lambda \in X^+$ such that $\mathrm{Hom}_C(\Delta_C(\lambda), L) \neq 0$. Then clearly L is a homomorphic image of $L_C(\lambda) = \mathrm{top}(\Delta_C(\lambda))$, and $L \cong L_C(\lambda)$. □

Lemma 1.1.9 *Let π be a poset ideal of X^+, and V a finite dimensional C-comodule. Then $V \in \mathbb{M}^C(\pi)$ if and only if $\mathrm{Hom}_C(V, \nabla_C(\mu)) = 0$ for any $\mu \in X^+$ with $\mu \notin \pi$.*

Proof. As in (I.1.9.2), we denote the full subcategory of $\mathbb{M}^C(\pi)$ consisting of finite dimensional C-comodules by $\mathbb{M}^C(\pi)_f$. For $\mu \in X^+$, we denote the full subcategory of $\mathbb{M}^C$ consisting of finite dimensional C-comodules V such that $\mathrm{Hom}_C(V, \nabla_C(\mu)) = 0$ for $\mu \notin \pi$ by $\mathcal{C}$. By Lemma 1.1.7, we have that $\mathbb{M}^C(\pi)_f$ is closed under subquotients and extensions in $\mathbb{M}^C$.

On the other hand, $\mathcal{C}$ is also closed under subquotients and extensions in $\mathbb{M}^C$. Indeed, $\mathcal{C}$ is obviously closed under extensions and quotients. Let $V_1 \subset V$, with $V, V/V_1 \in \mathcal{C}$. Then by **3*** of Definition 1.1.4, we have $\mathrm{Ext}^1_C(V/V_1, \nabla_C(\mu)) = 0$ for $\mu \notin \pi$, and hence $\mathrm{Hom}_C(V_1, \nabla_C(\mu)) = 0$.

Thus, in order to prove the lemma, we may and shall assume that $V = L_C(\lambda)$ is a simple C-comodule. If $\lambda \notin \pi$, then we have

$$\mathrm{Hom}_C(L_C(\lambda), \nabla_C(\lambda)) \supset \mathrm{End}_C\, L_C(\lambda) \neq 0, \qquad \mathrm{Hom}_C(\Delta_C(\lambda), L_C(\lambda)) \neq 0,$$

and hence $L_C(\lambda)$ belongs to neither $\mathbb{M}^C(\pi)_f$ nor $\mathcal{C}$.

On the other hand, if $\lambda \in \pi$, then

$$\mathrm{Hom}_C(L_C(\lambda), \nabla_C(\mu)) \subset \mathrm{Hom}_C(\Delta_C(\lambda), \nabla_C(\mu)) = 0 \supset \mathrm{Hom}_C(\Delta_C(\mu), L_C(\lambda))$$

for $\mu \notin \pi$, and hence $L_C(\lambda)$ belongs to both $\mathbb{M}^C(\pi)_f$ and $\mathcal{C}$. □

Lemma 1.1.10 *Let $\lambda, \mu \in X^+$. If*

$$\mathrm{Ext}^1_C(L_C(\lambda), \nabla_C(\mu)) \neq 0$$

or $\mathrm{Ext}^1_C(\Delta_C(\mu), L_C(\lambda)) \neq 0$, *then we have $\mu < \lambda$. For $\lambda \in X^+$,* $\mathrm{rad}\, \Delta_C(\lambda)$ *and $\nabla_C(\lambda)/\mathrm{soc}(\nabla_C(\lambda))$ belong to $\mathbb{M}^C((-\infty, \lambda))$. Moreover, $\Delta_C(\lambda)$ and $\nabla_C(\lambda)$ belong to $\mathbb{M}^C((-\infty, \lambda])$.*

Proof. Assume $\mu \notin (-\infty, \lambda)$, or equivalently, $\mu \not< \lambda$. Then for any $\nu > \mu$, we have $\nu \neq \lambda$ and hence $\mathrm{Ext}^1_C(L_C(\lambda), \nabla_C(\mu)) = 0$. As the sequence

$$0 = \mathrm{Hom}_C(\Delta_C(\lambda), \nabla_C(\mu)) \to \mathrm{Hom}(\mathrm{rad}(\Delta_C(\lambda)), \nabla_C(\mu)) \to \mathrm{Ext}^1_C(L_C(\lambda), \nabla_C(\mu)) = 0$$

is exact, $\mathrm{rad}(\Delta_C(\lambda)) \in \mathbb{M}^C((-\infty, \lambda))$ by Lemma 1.1.9. Similarly,

$$\mathrm{Ext}^1_C(\Delta_C(\mu), L_C(\lambda)) = 0$$

and $\nabla_C(\lambda)/\mathrm{soc}(\nabla_C(\lambda)) \in \mathbb{M}^C((-\infty, \lambda))$ are proved using the dual argument.

Next, $\nabla_C(\lambda) \in \mathbb{M}^C((-\infty, \lambda])$ is obvious. As we have $L_C(\lambda) \subset \nabla_C(\lambda)$, it follows that $L_C(\lambda) \in \mathbb{M}^C((-\infty, \lambda])$. As $\mathbb{M}^C((-\infty, \lambda])$ is closed under extensions, we have $\Delta_C(\lambda) \in \mathbb{M}^C((-\infty, \lambda])$. □

Lemma 1.1.11 *Let π be a poset ideal of X^+, and $V \in \mathbb{M}^C$. Then $V \in \mathbb{M}^C(\pi)$ if and only if any simple subquotient of V is isomorphic to $L_C(\lambda)$ with $\lambda \in \pi$.*

Proof. Easy.

1.2 Weak highest weight theory

(1.2.1) Let k be a field.

Proposition 1.2.2 *Let C be a k-coalgebra, and assume $(C, X^+, \Delta, \nabla, L)$ satisfies the following.*

a *X^+ satisfies the equivalent conditions in* Lemma 1.1.2.

b *$\Delta = (\Delta_C(\lambda))$, $\nabla = (\nabla_C(\lambda))$ and $L = (L_C(\lambda))$ are families of finite dimensional C-comodules indexed by X^+.*

c *For each $\lambda \in X^+$, $L_C(\lambda)$ is simple, and any simple C-comodule is isomorphic to some $L_C(\lambda)$.*

d $\mathrm{top}(\Delta_C(\lambda)) \cong L_C(\lambda) \cong \mathrm{soc}(\nabla_C(\lambda))$.

e *Any simple subquotient of* $\operatorname{rad}\Delta_C(\lambda)$ *is isomorphic to* $L_C(\mu)$ *for some* $\mu < \lambda$.

e* *Any simple subquotient of* $\nabla_C(\lambda)/\operatorname{soc}(\nabla_C(\lambda))$ *is isomorphic to* $L_C(\mu)$ *for some* $\mu < \lambda$.

f $\lambda, \mu \in X^+$ *and* $\lambda \neq \mu$ *imply* $\operatorname{Hom}_C(\Delta_C(\lambda), \nabla_C(\mu)) = 0$.

g $\lambda, \mu \in X^+$ *and* $i = 1$ *(resp.* $i = 1, 2$*) imply* $\operatorname{Ext}^i_C(\Delta_C(\lambda), \nabla_C(\mu)) = 0$.

h *If* $\lambda \in X^+$*, then* $\operatorname{End}_C L_C(\lambda) \cong k$.

Then we have that (C, X^+, Δ, ∇) *is a weak split highest weight coalgebra (resp. split highest weight coalgebra), and* $L_C(\lambda) = \operatorname{top}(\Delta_C(\lambda))$ *for any* $\lambda \in X^+$*. Conversely, if* (C, X^+, Δ, ∇) *is a weak split highest weight coalgebra (resp. split highest weight coalgebra), then defining* $L_C(\lambda) := \operatorname{top}(\Delta_C(\lambda))$*, conditions* **a–h** *are satisfied by* $(C, X^+, \Delta, \nabla, L)$.

The converse was proved in the last subsection. We prove the first assertion. In the rest of this subsection, C is a k-coalgebra, and $(C, X^+, \Delta, \nabla, L)$ is assumed to satisfy conditions **a–f** and condition **g** for $i = 1$ in the proposition.

The axioms **1,2,2*** in Definition 1.1.4 are obvious. The proof is divided into several steps.

Lemma 1.2.3 *If* $\lambda, \mu \in X^+$ *and* $\lambda \neq \mu$*, then* $L_C(\lambda) \not\cong L_C(\mu)$.

Proof. $\operatorname{Hom}_C(L_C(\lambda), L_C(\mu)) \subset \operatorname{Hom}_C(\Delta_C(\lambda), \nabla(\mu)) = 0$. □

Lemma 1.2.4 *If* $\lambda, \mu \in X^+$ *and* $\mu \not> \lambda$*, then* $\operatorname{Ext}^i_C(\Delta_C(\lambda), L_C(\mu)) = 0$ *for* $i = 1$*. If, moreover, we assume condition* **g** *in the proposition for* $i = 1, 2$*, then we have* $\operatorname{Ext}^i_C(\Delta_C(\lambda), L_C(\mu)) = 0$ *for* $i = 1, 2$.

Proof. If $\nu < \mu$, then $\nu \neq \lambda$. Hence, we have

$$\operatorname{Hom}_C(\Delta_C(\lambda), L_C(\nu)) \subset \operatorname{Hom}_C(\Delta_C(\lambda), \nabla_C(\nu)) = 0.$$

This shows

$$\operatorname{Hom}_C(\Delta_C(\lambda), \nabla_C(\mu)/\operatorname{soc}(\nabla_C(\mu))) = 0.$$

As $\operatorname{Ext}^1_C(\Delta_C(\lambda), \nabla_C(\mu)) = 0$, it follows that $\operatorname{Ext}^1_C(\Delta_C(\lambda), L_C(\mu)) = 0$ by an easy long exact sequence argument. If $\nu < \mu$, then $\nu \not> \lambda$, so we have $\operatorname{Ext}^1_C(\Delta_C(\lambda), L_C(\nu)) = 0$. Hence,

$$\operatorname{Ext}^1_C(\Delta_C(\lambda), \nabla_C(\mu)/\operatorname{soc}(\nabla_C(\mu))) = 0.$$

Hence, if we also assume $\operatorname{Ext}^2_C(\Delta_C(\lambda), \nabla_C(\mu)) = 0$, then

$$\operatorname{Ext}^2_C(\Delta_C(\lambda), L_C(\mu)) = 0$$

by an easy long exact sequence argument. □

Lemma 1.2.5 *If $\lambda, \mu \in X^+$ and $\mu \not> \lambda$, then* $\mathrm{Ext}^i_C(L_C(\mu), \nabla_C(\lambda)) = 0$ *for $i = 1$. If we also assume that condition* **g** *for $i = 1, 2$ in the proposition is satisfied, then we have* $\mathrm{Ext}^i_C(L_C(\mu), \nabla_C(\lambda)) = 0$ *for $i = 1, 2$.*

Proof. Similar to Lemma 1.2.4. □

Lemma 1.2.6 *Let V be a finite dimensional C-comodule, and π a poset ideal of X^+. We have* $\mathrm{Hom}_C(\Delta_C(\lambda), V) = 0$ *for any $\lambda \notin \pi$ if and only if any composition factor of V is isomorphic to $L_C(\mu)$ for some $\mu \in \pi$.*

Proof. The 'if' part is obvious, as $\lambda \notin \pi$ and $\mu \in \pi$ imply

$$\mathrm{Hom}_C(\Delta_C(\lambda), L_C(\mu)) \subset \mathrm{Hom}_C(\Delta_C(\lambda), \nabla_C(\mu)) = 0.$$

We prove the 'only if' part. We use induction on $\dim V$. As the case $V = 0$ is obvious, we may assume that $V \neq 0$. If M is a simple subcomodule of V, then we have $M \cong L_C(\mu)$, with $\mu \in \pi$. As we have $\mu \not> \lambda$ for any $\lambda \notin \pi$, we have $\mathrm{Ext}^1_C(\Delta_C(\lambda), M) = 0$ by Lemma 1.2.4. Hence, we have $\mathrm{Hom}_C(\Delta_C(\lambda), V/M) = 0$. By induction assumption, we are done. □

Lemma 1.2.7 *The equivalent conditions in* Lemma 1.2.6 *are also equivalent to the condition:* $\mathrm{Hom}_C(V, \nabla_C(\lambda)) = 0$ *for $\lambda \notin \pi$.*

Proof. Proved similarly to Lemma 1.2.6, using Lemma 1.2.5. □

Now we continue the proof of the proposition. Axiom **3** in Definition 1.1.4 is obvious by Lemma 1.2.4 and Lemma 1.2.6 (applied to $\pi := X^+ \setminus (\lambda, \infty)$). **3*** is obvious by Lemma 1.2.5 and Lemma 1.2.7. Axiom **4** (resp. axiom **4***) is nothing but the case $\pi = \emptyset$ in Lemma 1.2.6 (resp. Lemma 1.2.7). In axiom **5**, **f** is assumed in the proposition. On the other hand, we have $\mathrm{Hom}_C(\mathrm{rad}\,\Delta_C(\lambda), L_C(\lambda)) = 0$ by **e** and Lemma 1.2.3. As we have

$$\mathrm{Hom}_C(\Delta_C(\lambda), \nabla_C(\lambda)/\operatorname{soc}(\nabla_C(\lambda))) = 0$$

applying Lemma 1.2.6 to $\pi = X^+ \setminus [\lambda, \infty)$, we have the two canonical maps

$$\mathrm{End}_C\, L_C(\lambda) \to \mathrm{Hom}_C(\Delta_C(\lambda), L_C(\lambda)) \to \mathrm{Hom}_C(\Delta_C(\lambda), \nabla_C(\lambda))$$

are isomorphisms. After we invoke assumption **h**, which has not been used so far, axiom **5** is also proved. Hence, (C, X^+, Δ, ∇) is a weak split highest weight coalgebra (resp. split highest weight coalgebra). The fact $L_C(\lambda) = \mathrm{top}(\Delta_C(\lambda))$ is assumed in **e**. This completes the proof of the proposition.

□

Definition 1.2.8 When conditions **a**–**f** and **g** for $i = 1$ (resp. $i = 1, 2$) in the proposition are satisfied, we say that $(C, X^+, \Delta, \nabla, L)$ is a *weak highest weight coalgebra* (resp. *highest weight coalgebra*). If, moreover, **h** in the proposition is also satisfied, then we say that $(C, X^+, \Delta, \nabla, L)$ is *split* (by the proposition, this definition is justified).

Definition 1.2.9 We say that $(A, X^+, \Delta, \nabla, L)$ is a *weak quasi-hereditary algebra* (resp. *quasi-hereditary algebra*) over k if A is a finite dimensional k-algebra, and $(A^*, X^+, \Delta, \nabla, L)$ is a weak highest weight coalgebra (resp. highest weight coalgebra) over k.

(1.2.10) Let k be a field, and $(C, X^+, \Delta, \nabla, L)$ a weak highest weight coalgebra over k.

For a poset ideal π of X^+, we denote by $\mathbb{M}^C(\pi)$ the full subcategory of $\mathbb{M}^C$ consisting of C-comodules V such that $\operatorname{Hom}_C(\Delta_C(\lambda), V) = 0$ if $\lambda \notin \pi$.

Remark 1.2.11 Lemmas 1.1.7, 1.1.9, 1.1.10, and 1.1.11 in the last subsection are still valid in our situation. This is obvious from the argument so far.

Lemma 1.2.12 *Let* $V \in \mathbb{M}^C$, *and* π *be a finite poset ideal of* X^+, *and* $n \geq 1$. *Assume the following:*

a *For any* $\mu \in \pi$, *we have* $\operatorname{Ext}^i_C(\Delta_C(\mu), V) = 0$ $(1 \leq i \leq n)$.

b *For* $\mu \in \pi$, *we have* $\operatorname{Hom}_C(\Delta_C(\mu), V) = 0$ *unless* μ *is a maximal element of* π.

Then we have for any $\lambda \in \pi$:

1 $\operatorname{Ext}^i_C(L_C(\lambda), V) = 0$ *for* $1 \leq i \leq n$.

2 *The canonical maps*

$$\operatorname{Hom}_C(\nabla_C(\lambda), V) \to \operatorname{Hom}_C(L_C(\lambda), V) \to \operatorname{Hom}_C(\Delta_C(\lambda), V)$$

are all isomorphisms.

3 $\operatorname{Hom}_C(L_C(\lambda), V) = 0$ *unless* λ *is a maximal element of* π.

Proof. We use induction on $\operatorname{ht}(\lambda)$. If $\mu < \lambda$, then $(-\infty, \mu]$ and μ (instead of π and λ) also satisfy the assumptions of the lemma, and by induction assumption, we have $\operatorname{Ext}^i_C(L_C(\mu), V) = 0$ $(1 \leq i \leq n)$ and $\operatorname{Hom}_C(L_C(\mu), V) \cong \operatorname{Hom}_C(\Delta_C(\mu), V)$. As μ is not a maximal element of π, we have $\operatorname{Ext}^i_C(L_C(\mu), V) = 0$ $(0 \leq i \leq n)$. Hence, we have $\operatorname{Ext}^i_C(W, V) = 0$ $(0 \leq i \leq n)$ for any $W \in \mathbb{M}^C((-\infty, \lambda))$. In particular, we have

$$\operatorname{Ext}^i_C(\operatorname{rad}(\Delta_C(\lambda)), V) = 0 \qquad (0 \leq i < n).$$

Combining this with $\operatorname{Ext}^i_C(\Delta_C(\lambda), V) = 0$ $(1 \le i \le n)$, assertion **1** follows.

As $n \ge 1$, we have $\operatorname{Ext}^1_C(W, V) = 0$ and $\operatorname{Hom}_C(W, V) = 0$ for $W \in \mathbb{M}^C((-\infty, \lambda))$ with $\dim W < \infty$. Considering the cases $W = \operatorname{rad}(\Delta_C(\lambda))$ and $W = \nabla_C(\lambda)/\operatorname{soc}(\nabla_C(\lambda))$, assertion **2** follows easily. **3** is obvious from **2** and assumption **b**. □

(1.2.13) Let k and $(C, X^+, \Delta, \nabla, L)$ be as in (1.2.10). Let π be a poset ideal of X^+, and $V \in \mathbb{M}^C$. We denote the sum of all C-subcomodules of V which belong to $\mathbb{M}^C(\pi)$ by $V(\pi)$. Note that $V(\pi)$ is the largest C-subcomodule of V which belongs to $\mathbb{M}^C(\pi)$. Note also that $V \in \mathbb{M}^C(\pi)$ if and only if $V = V(\pi)$. Note also that $?(\pi)$ is a functor from $\mathbb{M}^C$ to $\mathbb{M}^C(\pi)$, which is right adjoint to the canonical embedding $\mathbb{M}^C(\pi) \hookrightarrow \mathbb{M}^C$, see (I.1.10).

Lemma 1.2.14 *If (V_i) is a filtered inductive system of C-comodules, then the canonical map*

$$\varinjlim V_i(\pi) \to (\varinjlim V_i)(\pi)$$

is an isomorphism.

Proof. Follows immediately from Lemma I.1.10.3. □

For a finite dimensional C-comodule V, $V^* = \operatorname{Hom}(V, k)$ is a *left* C-comodule (hence is a right C^{op}-comodule), see (I.3.5). The duality $\operatorname{Hom}(?, k)$ gives a contravariant equivalence between the category of finite dimensional C-comodules $\mathbb{M}^C_f$ and the category of finite dimensional left C-comodules ${}^C\mathbb{M}_f$.

Lemma 1.2.15 *The quintuple*

$$(C^{\mathrm{op}}, X^+, \nabla^*, \Delta^*, L^*)$$

is a weak highest weight coalgebra over k. If $(C, X^+, \Delta, \nabla, L)$ is split, then so is $(C^{\mathrm{op}}, X^+, \nabla^*, \Delta^*, L^*)$, where

$$\Delta^* := (\Delta_C(\lambda)^*)_{\lambda \in X^+}, \quad \nabla^* := (\nabla_C(\lambda)^*)_{\lambda \in X^+}, \quad \textit{and} \quad L^* := (L_C(\lambda)^*)_{\lambda \in X^+}.$$

Proof. If V and W are finite dimensional C-comodules, then we have

$$\operatorname{Ext}^n_C(V, W) \cong \operatorname{Ext}^n_{\mathbb{M}^C_f}(V, W) \cong \operatorname{Ext}^n_{\mathbb{M}^{C^{\mathrm{op}}}_f}(W^*, V^*) \cong \operatorname{Ext}^n_{\mathbb{M}^{C^{\mathrm{op}}}}(W^*, V^*)$$

for $n = 0, 1$ by Lemma I.1.5.7. The lemma follows from this. □

Lemma 1.2.16 *Let k' be an extension field of k, C a k-coalgebra, X^+ an ordered set, and Δ and ∇ families of C-comodules indexed by X^+. Then (C, X^+, Δ, ∇) is a weak split highest weight coalgebra over k if and only if $(C', X^+, \Delta', \nabla')$ is a weak split highest weight coalgebra over k', where $C' := C \otimes_k k'$, $\Delta' := (\Delta_C(\lambda) \otimes_k k')_{\lambda \in X^+}$, and $\nabla' := (\nabla_C(\lambda) \otimes_k k')_{\lambda \in X^+}$.*

Proof. For a finite dimensional C-comodule V and a C-comodule W, we have

$$\operatorname{Ext}^n_C(V,W)\otimes_k k' \cong \operatorname{Ext}^n_{C'}(V\otimes_k k', W\otimes_k k') \qquad (n\geq 0)$$

by Proposition I.3.6.20, **3**. Hence, the 'if' part is obvious.

We prove the 'only if' part. We check the conditions in Definition 1.1.4. The conditions **0,1,2** and **2*** are obvious. Condition **5** is also obvious from the isomorphism above. If V is a finite dimensional C-comodule and W is a C'-comodule, then we have

$$\operatorname{Ext}^n_C(V,W)\cong \operatorname{Ext}^n_{C\otimes_k k'}(V\otimes_k k', W)$$

by Proposition I.3.6.20, **4**. Hence, conditions **3** (for $i=1$) and **4** are obvious. By Lemma 1.2.15, $(C, X^+, \nabla^*, \Delta^*)$ is a weak split highest weight coalgebra over k. Combining this with **3,4**, we have **3*,4***. □

Note that a base change of a non-split highest weight coalgebra is not necessarily a highest weight coalgebra again, as $L_C(\lambda)\otimes_k k'$ may not be simple.

(1.2.17) Let k and $(C, X^+, \Delta, \nabla, L)$ be as in (1.2.10). As $\mathbb{M}^C$ is locally finite, any object admits an injective hull by Theorem I.1.7.6. We denote the injective hull of $L_C(\lambda)$ by $Q_C(\lambda)$. As $L_C(\lambda)\hookrightarrow \nabla_C(\lambda)$ is an essential mono, we have that $Q_C(\lambda)$ is also an injective hull of $\nabla_C(\lambda)$.

The following two lemmas show that Δ and ∇ are determined only by C, X^+ and L. Hence, we may say that (X^+, L), or an ordering X^+ of the set L of simples of C, is a (weak) *highest weight theory* on C, and Δ is its set of Weyl modules, and ∇ is its set of induced modules.

Lemma 1.2.18 *Let M be a finite dimensional C-comodule, and $\lambda\in X^+$. Then the following are equivalent.*

1 $M\cong \Delta_C(\lambda)$.

2 top $M \cong L_C(\lambda)$, *any simple subquotient of* rad M *is of the form* $L_C(\mu)$ *with* $\mu<\lambda$, *and* $\operatorname{Ext}^1_C(M, L_C(\nu))=0$ *for any* $\nu\in X^+$ *with* $\nu\not>\lambda$.

3 top $M \cong L_C(\lambda)$, *any simple subquotient of M is of the form* $L_C(\mu)$ *with* $\mu\not>\lambda$, *and* $\operatorname{Ext}^1_C(M, L_C(\nu))=0$ *for* $\nu\in X^+$ *with* $\nu<\lambda$.

Proof. **1⇒2⇒3** is obvious. We show **3⇒1**. By assumption, there exists an exact sequence

$$0\to M'\to M\to L_C(\lambda)\to 0.$$

As any simple subquotient of M' is of the form $L_C(\mu)$ with $\mu\not>\lambda$, we have $\operatorname{Ext}^1_C(\Delta_C(\lambda), M')=0$ by Remark 1.2.11. Hence, the canonical map $\Delta_C(\lambda)\to L_C(\lambda)$ lifts to a map, say, $\varphi:\Delta_C(\lambda)\to M$. As $\dim M<\infty$, we have $\mathrm{top}(\mathrm{Coker}\,\varphi) = \mathrm{Coker}(\mathrm{top}\,\varphi) = 0$, and hence $\mathrm{Coker}\,\varphi = 0$ and φ is surjective. On the other hand, as $\mathrm{Ext}^1_C(M, \mathrm{rad}(\Delta_C(\lambda))) = 0$, the canonical map $M \to L_C(\lambda) \cong \mathrm{top}(\Delta_C(\lambda))$ lifts to a surjective map $\psi : M \to \Delta_C(\lambda)$. This shows $\dim_k M = \dim_k \Delta_C(\lambda)$, and hence φ is an isomorphism. □

Similarly, the following is also proved easily.

Lemma 1.2.19 *Let N be a C-comodule, and $\lambda \in X^+$. Then the following are equivalent.*

1 $N \cong \nabla_C(\lambda)$.

2 $\dim N < \infty$, $\mathrm{soc}\, N \cong L_C(\lambda)$, *any composition factor of $N/\mathrm{soc}\, N$ is of the form $L_C(\mu)$ with $\mu < \lambda$, and* $\mathrm{Ext}^1_C(L_C(\nu), N) = 0$ *if $\nu \in X^+$ with $\nu \not> \lambda$.*

3 $\mathrm{soc}\, N \cong L_C(\lambda)$, *any simple subquotient of N is of the form $L_C(\mu)$ with $\mu \not> \lambda$, and* $\mathrm{Ext}^1_C(L_C(\nu), N) = 0$ *if $\nu \in X^+$ with $\nu < \lambda$.* □

Lemma 1.2.20 *For any $\lambda \in X^+$, we have that*

$$\mathrm{End}_C(L_C(\lambda)), \quad \textit{and} \quad \mathrm{End}_C(\Delta_C(\lambda)), \quad \textit{and} \quad \mathrm{End}_C(\nabla_C(\lambda))$$

are division rings, and are isomorphic to one another as k-algebras. In particular, $\Delta_C(\lambda)$ and $\nabla_C(\lambda)$ are indecomposable.

Proof. As $L_C(\lambda)$ is an simple object, $\mathrm{End}_C(L_C(\lambda))$ is a division ring, by Schur's lemma. Next, as soc and top are k-endofunctors of $\mathbb{M}^C_f$, we have k-algebra maps

$$\mathrm{soc} : \mathrm{End}_C(\nabla_C(\lambda)) \to \mathrm{End}_C(L_C(\lambda))$$

and

$$\mathrm{top} : \mathrm{End}_C(\Delta_C(\lambda)) \to \mathrm{End}_C(L_C(\lambda)).$$

Note that soc is given by a restriction, and top is induced by an induced map.

Let $\varphi \in \mathrm{Ker}\,\mathrm{soc}$. Then as $\mathrm{Im}\,\varphi$ is a quotient of $\nabla_C(\lambda)/\mathrm{soc}(\nabla_C(\lambda))$, it does not have $L_C(\lambda)$ as its subquotient. On the other hand, as $\mathrm{soc}(\mathrm{Im}\,\varphi) \subset \mathrm{soc}(\nabla_C(\lambda)) = L_C(\lambda)$, it follows that $\mathrm{soc}(\mathrm{Im}\,\varphi) = 0$. Hence, $\varphi = 0$. This shows that soc is an injective k-algebra map. On the other hand, as

$$\mathrm{Ext}^1_C(\nabla_C(\lambda)/\mathrm{soc}(\nabla_C(\lambda)), \nabla_C(\lambda)) = 0,$$

the composite map

$$L_C(\lambda) \xrightarrow{f} L_C(\lambda) \hookrightarrow \nabla_C(\lambda)$$

is extended to a map defined on $\nabla_C(\lambda)$ for any $f \in \mathrm{End}_C(L_C(\lambda))$. This shows that soc is surjective. Similarly, top is also proved to be bijective. □

1.3 Highest weight coalgebras and good comodules

(1.3.1) Let $(C, X^+, \Delta, \nabla, L)$ be a highest weight coalgebra.

Proposition 1.3.2 *For $V \in \mathbb{M}^C$, the following are equivalent.*

1 *For any $\lambda \in X^+$ and $i > 0$, we have $\operatorname{Ext}^i_C(\Delta_C(\lambda), V) = 0$.*

2 *For any $\lambda \in X^+$, we have $\operatorname{Ext}^1_C(\Delta_C(\lambda), V) = 0$.*

3 *For any finite poset ideal π of X^+ and any maximal element λ of π, the canonical maps*

$$\begin{aligned}\operatorname{Hom}_C(\Delta_C(\lambda), V) &\to \operatorname{Hom}_C(\Delta_C(\lambda), V/V(\pi')) \\ &\leftarrow \operatorname{Hom}_C(L_C(\lambda), V/V(\pi')) \leftarrow \operatorname{Hom}_C(\nabla_C(\lambda), V/V(\pi'))\end{aligned}$$

are all isomorphisms, and the canonical pairing

$$\operatorname{Hom}_C(\nabla_C(\lambda), V/V(\pi')) \otimes_E \nabla_C(\lambda) \to V/V(\pi')$$

is injective, and its image agrees with $V(\pi)/V(\pi')$, where $\pi' := \pi \setminus \{\lambda\}$, and $E := \operatorname{End}_C(\nabla_C(\lambda))$. Moreover, $\operatorname{Hom}_C(\nabla_C(\lambda), V/V(\pi'))$ is E-free, and hence $V(\pi)/V(\pi')$ is isomorphic to a direct sum of copies of $\nabla_C(\lambda)$, as a C-comodule.

4 *For any injective order-preserving map $f : X^+ \to \mathbb{N}$ with $f(X^+)$ a poset ideal of $\mathbb{N}$, there is a unique filtration $0 = V_0 \subset V_1 \subset \cdots \subset V$ which satisfies the following conditions:*

a $V = \varinjlim V_i$

b *For $i \geq 1$, V_i/V_{i-1} is isomorphic to a direct sum of copies of $\nabla_C(f^{-1}(i))$.*

5 *The set of finite dimensional C-subcomodules of V and the set of C-subcomodules belonging to $\mathcal{F}(\nabla)$ (see (I.1.11)) are cofinal.*

6 *V is a filtered inductive limit of objects of $\mathcal{F}(\nabla)$.*

Proof. **1⇒2** is obvious.

We prove **2⇒3** by induction on $\#\pi$. As $V(\emptyset) = 0$, it is easy to see that $V(\pi') \in \mathcal{F}(\nabla)$, by induction assumption. Hence, by Lemma 1.1.6, $\operatorname{Ext}^i_C(\Delta_C(\mu), V(\pi')) = 0$ for $i = 1, 2$ and $\mu \in X^+$. As we are assuming **2**, we have $\operatorname{Ext}^1_C(\Delta_C(\mu), V/V(\pi')) = 0$ for any $\mu \in X^+$.

On the other hand,

$$\operatorname{Hom}_C(\Delta_C(\mu), V(\pi')) \to \operatorname{Hom}_C(\Delta_C(\mu), V)$$

is an isomorphism for any $\mu \in \pi'$, by the definition of $?(\pi')$ and the fact $\Delta_C(\mu) \in \mathbb{M}^C(\pi')$. Combining this observation with $\operatorname{Ext}^1_C(\Delta_C(\mu), V(\pi')) = 0$, we have $\operatorname{Hom}_C(\Delta_C(\mu), V/V(\pi')) = 0$ for $\mu \in \pi'$.

Hence, by Lemma 1.2.12,

$$\mathrm{Hom}_C(\Delta_C(\lambda), V/V(\pi')) \leftarrow \mathrm{Hom}_C(L_C(\lambda), V/V(\pi')) \\ \leftarrow \mathrm{Hom}_C(\nabla_C(\lambda), V/V(\pi'))$$

are both isomorphisms.

By definition, we have $V(\pi') \in \mathbb{M}^C(\pi')$ and hence

$$\mathrm{Hom}_C(\Delta_C(\lambda), V(\pi')) = 0.$$

As $\mathrm{Ext}^1_C(\Delta_C(\lambda), V(\pi')) = 0$, it follows that

$$\mathrm{Hom}_C(\Delta_C(\lambda), V) \to \mathrm{Hom}_C(\Delta_C(\lambda), V/V(\pi'))$$

is also an isomorphism.

Hence, the first part of **3** is proved.

Next, we remark that by Lemma 1.2.20 and its proof, $L_C(\lambda)$ is an E-module in a natural way. Also $\mathrm{Hom}_C(L_C(\lambda), V/V(\pi'))$ is E-free, and the pairing

$$\rho : \mathrm{Hom}_C(L_C(\lambda), V/V(\pi')) \otimes_E L_C(\lambda) \to V/V(\pi') \qquad (f \otimes v \mapsto fv)$$

is injective. In fact, when we set $\mathrm{Im}\,\rho = \bigoplus_{i\in I} L_i$ with $L_i \cong L_C(\lambda)$, then we have

$$\mathrm{Hom}_C(L_C(\lambda), V/V(\pi')) \cong \mathrm{Hom}_C(L_C(\lambda), \mathrm{Im}\,\rho) \cong \bigoplus_{i\in I} \mathrm{End}_C\, L_C(\lambda) \cong \bigoplus_{i\in I} E,$$

and we have the desired assertion. In particular, $\mathrm{Hom}_C(\nabla_C(\lambda), V/V(\pi'))$ is also E-free.

Next, we show that

$$\rho' : \mathrm{Hom}_C(\nabla_C(\lambda), V/V(\pi')) \otimes_E \nabla_C(\lambda) \to V/V(\pi')$$

is injective. Assume the contrary. Then $\mathrm{Ker}\,\rho' \neq 0$, and this implies

$$\mathrm{soc}(\mathrm{Ker}\,\rho') = \mathrm{soc}(\mathrm{Hom}_C(\nabla_C(\lambda), V/V(\pi')) \otimes_E \nabla_C(\lambda)) \cap \mathrm{Ker}\,\rho' \neq 0$$

by Lemma I.1.10.7. On the other hand,

$$\mathrm{soc}(\mathrm{Hom}_C(\nabla_C(\lambda), V/V(\pi')) \otimes_E \nabla_C(\lambda)) = \\ \mathrm{Hom}_C(\nabla_C(\lambda), V/V(\pi')) \otimes_E \mathrm{soc}(\nabla_C(\lambda)) \cong \mathrm{Hom}_C(L_C(\lambda), V/V(\pi')) \otimes_E L_C(\lambda)$$

by Lemma I.1.10.7 again, and hence ρ is not injective, which is a contradiction. Hence, ρ' is also injective.

We set $I := \mathrm{Im}\,\rho'$, and we define J to be the pull-back of I by the projection $V \to V/V(\pi')$. As I is a direct sum of copies of $\nabla_C(\lambda)$, we have $I \in \mathbb{M}^C(\pi)$. As $V(\pi') \in \mathbb{M}^C(\pi')$, it follows that $J \in \mathbb{M}^C(\pi)$. Hence, $J \subset V(\pi)$. As I is a direct sum of copies of $\nabla_C(\lambda)$, $\operatorname{Ext}^i_C(\Delta_C(\mu), I) = 0$ for $i = 1, 2$ and $\mu \in X^+$. As $I, V(\pi') \in \mathcal{F}(\nabla)$, we have $J \in \mathcal{F}(\nabla)$, and hence $\operatorname{Ext}^i_C(\Delta_C(\mu), J) = 0$ for $i = 1, 2$ and $\mu \in X^+$. It follows that $\operatorname{Ext}^1_C(\Delta_C(\mu), V/J) = 0$ for any $\mu \in X^+$.

Now assume that $V(\pi)/J \neq 0$. As $V(\pi)/J \in \mathbb{M}^C(\pi)$, there exists some $\mu \in \pi$ such that

$$\operatorname{Hom}_C(\Delta_C(\mu), V(\pi)/J) \neq 0.$$

Since $V(\pi)/J \subset V/J$, we have $\operatorname{Hom}_C(\Delta_C(\mu), V/J) \neq 0$, too. Let $\mu \in \pi$ be a minimal element among such, then $\operatorname{Hom}_C(\nabla_C(\mu), V/J) \neq 0$ by Lemma 1.2.12.

On the other hand, as

$$0 \to I \to V/V(\pi') \to V/J \to 0$$

is exact and I is a direct sum of copies of $\nabla_C(\lambda)$ and hence $\operatorname{Ext}^1_C(\nabla_C(\mu), I) = 0$ (by Lemma 1.2.12), we conclude that

$$\operatorname{Hom}_C(\nabla_C(\mu), I) \to \operatorname{Hom}_C(\nabla_C(\mu), V/V(\pi'))$$

is not surjective. By definition of I, we have $\mu \neq \lambda$, and $\mu \in \pi'$. We have $\operatorname{Hom}_C(\Delta_C(\nu), I) = 0$ for $\nu < \mu$. On the other hand, by minimality of μ, we have $\operatorname{Hom}_C(\Delta_C(\nu), V/J) = 0$. These facts show $\operatorname{Hom}_C(\Delta_C(\nu), V/V(\pi')) = 0$ for $\nu < \mu$. Hence, again by Lemma 1.2.12, we have

$$\operatorname{Hom}_C(\nabla_C(\mu), V/V(\pi')) \cong \operatorname{Hom}_C(\Delta_C(\mu), V/V(\pi')) \neq 0.$$

On the other hand, as $\mu \in \pi'$, we have that

$$\operatorname{Hom}_C(\Delta_C(\mu), V(\pi')) \to \operatorname{Hom}_C(\Delta_C(\mu), V)$$

is an isomorphism. Combining this with $\operatorname{Ext}^1_C(\Delta_C(\mu), V(\pi')) = 0$, we have $\operatorname{Hom}_C(\Delta_C(\mu), V/V(\pi')) = 0$, which is a contradiction. Hence, $V(\pi)/J = 0$ is proved, and we have $V(\pi) = J$. It follows that $V(\pi)/V(\pi') = I = \operatorname{Im} \rho'$, and the desired assertion follows.

We prove **3⇒4**. First, we show the existence of a required filtration. To prove this, we set $V_i := V(f^{-1}([1, i]))$. By assumption, condition **b** is obvious. Hence, we have $V_i \in \mathcal{F}(\nabla)$ for each $i \geq 0$. This shows $\operatorname{Ext}^1_C(\Delta_C(\lambda), V_i) = 0$ for any $\lambda \in X^+$. And hence we have

$$\operatorname{Ext}^1_C(\Delta_C(\lambda), \varinjlim V_i) = 0.$$

On the other hand, for any $\lambda \in X^+$, any C-comodule map $\Delta_C(\lambda) \to V$ factors through $V_{f(\lambda)}$, hence it also factors through $\varinjlim V_i$. This shows

$$\operatorname{Hom}_C(\Delta_C(\lambda), \varinjlim V_i) \to \operatorname{Hom}_C(\Delta_C(\lambda), V)$$

is surjective. Hence, $\mathrm{Hom}_C(\Delta_C(\lambda), V/\varinjlim V_i) = 0$ for any $\lambda \in X^+$, which shows $V = \varinjlim V_i$. The proof of the existence is complete.

Next, we show the uniqueness, namely, $V_i = V(f^{-1}([1,i]))$ for such a filtration (V_i). As V_j/V_{j-1} is a direct sum of copies of $\nabla_C(f^{-1}(j))$, we have $V_i \in \mathbb{M}^C(f^{-1}([1,i]))$, and hence $V_i \subset V(f^{-1}[1,i])$.

As $\mathrm{Hom}_C(\Delta_C(f(l)), V_j/V_{j-1}) = 0$ for $j \neq l$, we have

$$\mathrm{Hom}_C(\Delta_C(f(l)), V_j/V_i) = 0 \qquad (j \geq i \geq l).$$

Hence,

$$\mathrm{Hom}_C(\Delta_C(f(l)), V(f^{-1}([1,i]))/V_i) \subset \mathrm{Hom}_C(\Delta_C(f(l)), V/V_i) = 0$$

for $i \geq l$. On the other hand, as $V(f^{-1}([1,i]))/V_i$ belongs to $\mathbb{M}^C(f^{-1}([1,i]))$, we have $\mathrm{Hom}_C(\Delta_C(f(l)), V(f^{-1}([1,i]))/V_i) = 0$ also for $l > i$. Thus, we have

$$\mathrm{Hom}_C(\Delta_C(\lambda), V(f^{-1}([1,i]))/V_i) = 0$$

for any $\lambda \in X^+$, and this shows $V_i = V(f^{-1}([1,i]))$.

4⇒5 is obvious, because each V_i belongs to $\mathcal{F}(\nabla)$, and we are assuming that there is an injective order-preserving map $f : X^+ \to \mathbb{N}$ with $f(X^+)$ a poset ideal of $\mathbb{N}$, see Lemma 1.1.2.

5⇒6 This is clear, as V is the filtered inductive limit of its finite dimensional C-subcomodules.

We prove **6⇒1** using induction on i. As the assertion is obvious for $i = 1, 2$, we may assume that $i \geq 3$. Let I be the injective hull of V. Then for any $\lambda \in X^+$ and $j \geq 1$, we have $\mathrm{Ext}^j_C(\Delta_C(\lambda), I/V) \cong \mathrm{Ext}^{j+1}_C(\Delta_C(\lambda), V)$. Hence, in particular, we have $\mathrm{Ext}^1_C(\Delta_C(\lambda), I/V) = 0$ for any $\lambda \in X^+$. As we have already proved **2⇒6**, we know that I/V is a filtered inductive limit of objects of $\mathcal{F}(\nabla)$. By induction assumption, we have $\mathrm{Ext}^{i-1}_C(\Delta_C(\lambda), I/V) = 0$ for any $\lambda \in X^+$, which shows $\mathrm{Ext}^i_C(\Delta_C(\lambda), V) = 0$. □

We say that a C-comodule V is *good* if the equivalent conditions in Proposition 1.3.2 are satisfied. By the proposition, it is not so difficult to show that $\mathbb{M}^C$ is a highest weight category [37] with X^+ its set of weights, ∇ its set of induced modules, and $L = (L_C(\lambda))_{\lambda \in X^+}$ its set of simple modules.

Corollary 1.3.3 *Let $\lambda, \mu \in X^+$ and $i \geq 0$. Then we have*

$$\mathrm{Ext}^i_C(\Delta_C(\lambda), \nabla_C(\mu)) = 0$$

unless $\lambda = \mu$ and $i = 0$.

Proof. The case $i = 0$ is contained in the definition of a highest weight coalgebra. If $i > 0$, then as we have $\nabla_C(\mu) \in \mathcal{F}(\nabla)$, the required vanishing holds by the proposition. □

Lemma 1.3.4 *The full subcategory of good C-comodules is closed under extensions, monocokernels, direct summands and filtered inductive limits in* $\mathbb{M}^C$.

Proof. Obvious from condition **1** in Proposition 1.3.2. □

Lemma 1.3.5 *Let* V *be a good* C*-comodule. Then* $V(\pi)$ *is also good for any poset ideal* π *of* X^+.

Proof. The case $\#\pi < \infty$ is proved easily by induction on $\#\pi$. The general case follows from this, as we have $V(\pi) = \varinjlim V(\rho)$, where ρ runs through all finite poset ideals of π. □

Lemma 1.3.6 *There exists a unique family of* k*-subcoalgebras* $(C_\pi)_\pi$ *of* C *indexed by the set of finite poset ideals of* X^+*, satisfying the following conditions:*

a $C_\emptyset = 0$.

b *If* $\pi \subset \pi'$*, then* $C_\pi \subset C_{\pi'}$.

c *If* π *is a finite poset ideal of* X^+ *and* λ *is its maximal element, then we have an isomorphism of* (C, C)*-bicomodules*

$$C_\pi / C_{\pi'} \cong \Delta_C(\lambda)^* \otimes_E \nabla_C(\lambda),$$

where $\pi' := \pi \setminus \{\lambda\}$ *and* $E := \operatorname{End}_C(\nabla_C(\lambda))$*. Here we regard* $\Delta_C(\lambda)$ *as an* E*-module through the canonical isomorphisms*

$$E \xrightarrow{\text{soc}} \operatorname{End}_C(L_C(\lambda)) \xrightarrow{\text{top}^{-1}} \operatorname{End}_C(\Delta_C(\lambda)).$$

In fact, $C_\pi := C(\pi)$ *satisfies the conditions, and we must define* C_π *thus. We also have*

d $\varinjlim C_\pi = C$.

Proof. We prove conditions **a,b,c,d** after setting $C_\pi := C(\pi)$. Note that $C(\pi)$ is a k-subcoalgebra of C by Lemma I.3.6.10 (applied to $\mathcal{B} = \mathbb{M}^C(\pi)$ and $j = ?(\pi)$). Conditions **a,b,d** are obvious.

We prove that condition **c** is satisfied by $C(\pi)$. As condition **1** in Proposition 1.3.2 is satisfied by the injective C-comodule C, we have that C is good. We may utilize condition **3** in Proposition 1.3.2, and hence the composite

$$\begin{aligned}\Delta_C(\lambda)^* \otimes_E \nabla_C(\lambda) &\cong \operatorname{Hom}_C(\Delta_C(\lambda), C) \otimes_E \nabla_C(\lambda)\\ &\cong \operatorname{Hom}_C(\Delta_C(\lambda), C/C(\pi')) \otimes_E \nabla_C(\lambda) \cong \operatorname{Hom}_C(\nabla_C(\lambda), C/C(\pi')) \otimes_E \nabla_C(\lambda)\\ &\hookrightarrow C/C(\pi')\end{aligned}$$

is an injective (C, C)-bicomodule map. As the image of this map agrees with $C(\pi)/C(\pi')$, **c** is also satisfied.

We prove the uniqueness, i.e., $C_\pi = C(\pi)$ for all π, assuming **a,b,c**, using induction on $\#\pi$. If $\pi = \emptyset$, then the assertion is obvious. If $\pi \neq \emptyset$, then we take a maximal element λ of π, and set $\pi' := \pi \setminus \{\lambda\}$. Then we have $C_{\pi'} = C(\pi')$ by induction assumption, and $C_\pi \subset C(\pi)$, as $C_\pi/C_{\pi'}$ is a direct sum of copies of $\nabla_C(\lambda)$ as a C-comodule. By dimension counting, we have $C_\pi = C(\pi)$, as desired. □

Assume that π is a poset ideal of X^+, not necessarily finite. Then as we have $C(\pi) = \varinjlim C(\rho)$, where ρ runs through all finite poset ideals of π, we have that $C(\pi)$ is a good subcoalgebra of C. We call $C(\pi)$ the *Donkin subcoalgebra* of C with respect to π.

If π is finite, then $C(\pi)$ is finite dimensional. In this case, we denote the dual algebra $C(\pi)^*$ of $C(\pi)$ by $S_C(\pi)$, and call it the *Schur algebra* of C with respect to π.

Lemma 1.3.7 *Let V be a C-comodule. Then $V \in \mathbb{M}^C(\pi)$ if and only if $V \in \mathbb{M}^{C(\pi)}$.*

Proof. Follows immediately from Lemma I.3.6.10 (applied to $\mathcal{B} = \mathbb{M}^C(\pi)$ and $j =?(\pi)$). □

Lemma 1.3.8 *Let $\lambda, \mu \in X^+$, and assume $\mu \not> \lambda$. Then we have*

$$\operatorname{Ext}^i_C(L_C(\mu), \nabla_C(\lambda)) = 0, \qquad \operatorname{Ext}^i_C(\Delta_C(\lambda), L_C(\mu)) = 0,$$
$$\operatorname{Ext}^i_C(\nabla_C(\mu), \nabla_C(\lambda)) = 0, \quad \textit{and} \quad \operatorname{Ext}^i_C(\Delta_C(\lambda), \Delta_C(\mu)) = 0$$

for $i > 0$.

Proof. We set $\pi = X^+ \setminus (\lambda, \infty)$ and $n = i$, and apply Lemma 1.2.12. Then we have $\operatorname{Ext}^i_C(L_C(\mu), \nabla_C(\lambda)) = 0$ by Corollary 1.3.3. The vanishing $\operatorname{Ext}^i_C(\Delta_C(\lambda), L_C(\mu)) = 0$ is the dual assertion, and is proved similarly.

The vanishings $\operatorname{Ext}^i_C(\nabla_C(\mu), \nabla_C(\lambda)) = 0$ and $\operatorname{Ext}^i_C(\Delta_C(\lambda), \Delta_C(\mu)) = 0$ for $i > 0$ are obvious, as we have $\Delta_C(\mu), \nabla_C(\mu) \in \mathbb{M}^C(X^+ \setminus (\lambda, \infty))$. □

1.4 Weak highest weight coalgebras and good filtrations

(1.4.1) Let k be a field, and $(C, X^+, \Delta, \nabla, L)$ a weak highest weight coalgebra over k. For a countable set Y, the symbol $[Y]$ stands for the interval $[1, \#Y]$ if $\#Y < \infty$, and $[Y] = \mathbb{N}$ if Y is countably infinite.

emma 1.4.2 *Let $V \in \mathbb{M}^C$. Then the following are equivalent.*

1 *$V \in \mathcal{F}(\nabla)$.*

2 *For any order-preserving bijective map $f : X^+ \to [X^+]$, there exist some $n \in [X^+]$ and some filtration*

$$0 = V_0 \subset V_1 \subset V_2 \subset \cdots \subset V_n = V$$

such that for any i, V_i/V_{i-1} is a finite direct sum of copies of $\nabla_C(f^{-1}(i))$.

3 *V is finite dimensional, and for any finite poset ideal π of X^+ and any maximal element λ of π, we have that $V(\pi)/V(\pi \setminus \{\lambda\})$ is a finite direct sum of copies of $\nabla_C(\lambda)$.*

Proof. **1⇒2** Using induction on i, we construct a filtration

$$0 = V_0 \subset V_1 \subset \cdots \subset V_i \subset V,$$

such that for any j with $0 < j \leq i$, V_j/V_{j-1} is a finite direct sum of copies of $\nabla_C(f^{-1}(j))$, and $V/V_i \in \mathcal{F}(\{\nabla_C(f^{-1}(j)) \mid i < j \leq \#X\})$.

This is enough to prove **1⇒2**. In fact, as V is finite dimensional, there are only finitely many $\mu \in X^+$ such that $L_C(\mu)$ appears in the composition factor of V. If we take such a μ with $f(\mu)$ maximal, and we set $n := f(\mu)$, then we have $V = V_n$, as we have $V/V_n \in \mathcal{F}(\{\nabla_C(f^{-1}(j)) \mid j > f(\mu)\})$.

If $i = 0$, the required conditions are satisfied by letting $V_0 = 0$. Assuming $i > 0$, and the construction is already done up to V_{i-1}, we construct V_i. By induction assumption, there exists some filtration

$$0 = W_0 \subset W_1 \subset \cdots \subset W_r = V/V_{i-1}$$

such that for each $l = 1, 2, \ldots, r$, there exists some j_l such that $i \leq j_l \leq \#X$ and $W_l/W_{l-1} \cong \nabla_C(f^{-1}(j_l))$. Assume that the number of l such that $j_l = i$ is s. Then as we have $\operatorname{Ext}^1_C(\nabla_C(f^{-1}i), \nabla_C(f^{-1}j)) = 0$ for $j \geq i$ by Lemma 1.2.19, there is an exact sequence

$$0 \to \bigoplus_{t=1}^{s} \nabla_C(f^{-1}i) \to V/V_{i-1} \to W \to 0$$

such that $W \in \mathcal{F}(\{\nabla_C(f^{-1}(j)) \mid i < j \leq \#X\})$. Defining V_i to be the kernel of the composite $V \to V/V_{i-1} \to W$, the required conditions are satisfied.

2⇒3 Obviously, V is finite dimensional. If π is a finite poset ideal of X^+ with $\#\pi = m$ and λ is its maximal element, then applying Lemma 1.1.2 to the ordered sets $\pi \setminus \{\lambda\}$ and $X^+ \setminus \pi$, we have that there exists an order-preserving bijective map $f : X^+ \to [1, \#X^+]$ such that $f(\pi) = [1, m]$ and $f(\lambda) = m$. By the assumption of **2**, there exists some $n \geq 0$ and a filtration

$$0 = V_0 \subset V_1 \subset V_2 \subset \cdots \subset V_n = V$$

such that V_i/V_{i-1} is a finite direct sum of copies of $\nabla_C(f^{-1}(i))$. Letting $V_i := V$ for $i > n$ if necessary, we may assume that $m \leq n$. Then we have $V_i = V(f^{-1}[1,i])$ for $i = 0, 1, \ldots, n$. In fact, each $\nabla_C(f^{-1}(j))$ $(j \leq i)$ belongs to $\mathbb{M}^C(f^{-1}[1,i])$ and as $\mathbb{M}^C(f^{-1}[1,i])$ is closed under extensions, we have $V_i \in \mathbb{M}^C(f^{-1}[1,i])$, and hence $V_i \subset V(f^{-1}[1,i])$. On the other hand, we have

$$\mathrm{Hom}_C(L_C(f^{-1}(j)), \nabla_C(f^{-1}(l))) = \mathrm{Hom}_C(L_C(f^{-1}(j)), L_C(f^{-1}(l))) = 0$$

for $j \leq i < l$. Hence, we have $\mathrm{Hom}_C(L_C(f^{-1}(j)), V/V_i) = 0$ for $j \leq i$. This shows $\mathrm{soc}((V/V_i)(f^{-1}[1,i])) = 0$, and hence $(V/V_i)(f^{-1}[1,i]) = 0$. As we have $V(f^{-1}[1,i])/V_i \subset (V/V_i)(f^{-1}[1,i])$, it follows that $V(f^{-1}[1,i]) = V_i$. Hence, $V(\pi)/V(\pi \setminus \{\lambda\}) = V_m/V_{m-1}$ is a finite direct sum of copies of $\nabla_C(\lambda)$.

As **3**$\Rightarrow$**2**$\Rightarrow$**1** is obvious, the proof of the lemma is complete. □

Corollary 1.4.3 *Let $W \in \mathbb{M}^C$. Then the following are equivalent.*

1 *For any finite poset ideal π of X^+, $V = W(\pi)$ satisfies the equivalent conditions of the lemma.*

2 *For any finite poset ideal π of X^+ and its maximal element λ, $W(\pi)/W(\pi \setminus \{\lambda\})$ is a finite direct sum of copies of $\nabla_C(\lambda)$.*

3 *For any order-preserving bijective map $f : X^+ \to [X^+]$, there exists some filtration $0 \subset W_0 \subset W_1 \subset W_2 \subset \cdots \subset W$ such that $\varinjlim W_i = W$ and each W_i/W_{i-1} is a finite direct sum of copies of $\nabla_C(f^{-1}(i))$.*

4 *There exists a filtration*

$$0 = W_0 \subset W_1 \subset W_2 \subset \cdots \subset W \tag{1.4.4}$$

of W such that $\varinjlim W_i = W$, for $i \geq 1$, W_i/W_{i-1} is isomorphic to $\nabla_C(\lambda)$ for some $\lambda \in X^+$, and for any $\lambda \in X^+$, there are only finitely many i such that $W_i/W_{i-1} \cong \nabla_C(\lambda)$.

We say that $W \in \mathbb{M}^C$ has a *good filtration*, if W satisfies the equivalent conditions in the corollary. We call a filtration (1.4.4) of W as in **4** of the corollary a good filtration of W. Assume that $W \in \mathbb{M}^C$ has a good filtration. Then for any good filtration (1.4.4) of W and any $\lambda \in X^+$, the number of i such that $W_i/W_{i-1} \cong \nabla_C(\lambda)$ agrees with

$$\dim \mathrm{Hom}_C(\Delta_C(\lambda), W)/\dim(\mathrm{End}_C(L_C(\lambda))),$$

which is independent of the choice of good filtration. We denote this number by $[W : \nabla_C(\lambda)]$.

By definition, if V has a good filtration, then $V(\pi)$ and $V/V(\pi)$ also have good filtrations for any finite poset ideal π of X^+. If $(C, X^+, \Delta, \nabla, L)$ is a highest weight coalgebra and W has a good filtration, then W is good.

Lemma 1.4.5 *The class of comodules with good filtrations is closed under finite direct sums and direct summands.*

Proof. As the functor $?(\pi)/?(\pi')$ is additive, it suffices to show that any direct summand of a finite direct sum of copies of $\nabla_C(\lambda)$ is a finite direct sum of copies of $\nabla_C(\lambda)$ for any $\lambda \in X^+$. This is immediate from the indecomposability of $\nabla_C(\lambda)$ (Lemma 1.2.20) and the Krull–Schmidt theorem (Lemma I.1.12.6). □

For a finite dimensional C-comodule V, we denote the number of appearances of $L_C(\lambda)$ in a composition series of V by $(V : L_C(\lambda))$. Obviously, we have

$$(V : L_C(\lambda)) = \dim \mathrm{Hom}_C(V, Q_C(\lambda))/\dim \mathrm{End}_C(L_C(\lambda)),$$

where $Q_C(\lambda)$ denotes the injective hull of $L_C(\lambda)$.

Lemma 1.4.6 *Assume that the C-comodule C has a good filtration. Then for any $\lambda \in X^+$, $Q_C(\lambda)$ has a good filtration, and*

$$[Q_C(\lambda) : \nabla_C(\mu)] = (\Delta_C(\mu) : L_C(\lambda))$$

for any $\mu \in X^+$. In particular, we have $[Q_C(\lambda) : \nabla_C(\lambda)] = 1$, and $[Q_C(\lambda) : \nabla_C(\mu)] = 0$ for $\mu \not> \lambda$.

Proof. As $\omega : L_C(\lambda) \to L_C(\lambda) \otimes C$ is injective and $L_C(\lambda) \otimes C$ is C-injective, we have that $Q_C(\lambda)$ is a direct summand of a finite direct sum of copies of C. By Lemma 1.4.5, $Q_C(\lambda)$ has a good filtration. The equality

$$\begin{aligned}[Q_C(\lambda) : \nabla_C(\mu)] &= \dim \mathrm{Hom}_C(\Delta_C(\mu), Q_C(\lambda))/\dim \mathrm{End}_C(L_C(\lambda)) \\ &= (\Delta_C(\mu) : L_C(\lambda))\end{aligned}$$

is obvious. The last assertion follows from this. □

Theorem 1.4.7 *Let $(C, X^+, \Delta, \nabla, L)$ be a weak highest weight coalgebra over a field k. Then the following are equivalent.*

1 $\mathrm{Ext}^i_C(\Delta, \nabla) = 0$ $(i \geq 1)$.

2 *$(C, X^+, \Delta, \nabla, L)$ is a highest weight coalgebra.*

3 *The C-comodule C has a good filtration.*

4 *$(C^{\mathrm{op}}, X^+, \nabla^*, \Delta^*, L^*)$ is a highest weight coalgebra.*

Proof. **1⇒2⇒3** is obvious. We prove **3⇒1**, using induction on $i \geq 1$. If $i = 1$, then the assertion is obvious by the definition of weak highest weight coalgebra, so we may assume $i \geq 2$. By induction assumption, if V has a good filtration and $1 \leq j < i$, then $\operatorname{Ext}^j_C(\Delta, V) = 0$. Let $\lambda \in X^+$. When we set $\pi := (-\infty, \lambda]$, then we have a short exact sequence of C-comodules with good filtrations

$$0 \to Q_C(\lambda)(\pi) \to Q_C(\lambda) \to Q_C(\lambda)/Q_C(\lambda)(\pi) \to 0.$$

By Lemma 1.4.6, we have $Q_C(\lambda)(\pi) \cong \nabla_C(\lambda)$. We have

$$\operatorname{Ext}^{i-1}_C(\Delta_C(\mu), Q_C(\lambda)/Q_C(\lambda)(\pi)) = 0$$

by induction assumption, and obviously we have $\operatorname{Ext}^i_C(\Delta_C(\mu), Q_C(\lambda)) = 0$ for any $\mu \in X^+$. This shows $\operatorname{Ext}^i_C(\Delta_C(\mu), \nabla_C(\lambda)) = 0$.

Next, we show **2⇒4**. We already know that $(C^{\mathrm{op}}, X^+, \nabla^*, \Delta^*, L^*)$ is a weak highest weight coalgebra by Lemma 1.2.15. By Lemma 1.3.6, the C^{op}-comodule C has a good filtration. By **3⇒2**, which is already proved, we have that $(C^{\mathrm{op}}, X^+, \nabla^*, \Delta^*, L^*)$ is a highest weight coalgebra over k.

We prove **4⇒2**. By **2⇒4**, which is just proved, we have

$$(C^{\mathrm{opop}}, X^+, \Delta^{**}, \nabla^{**}, L^{**}) = (C, X^+, \Delta, \nabla, L)$$

is a highest weight coalgebra. □

Corollary 1.4.8 *Let k' be an extension field of k, C a k-coalgebra, X^+ an ordered set, and Δ and ∇ families of C-comodules indexed by X^+. Then (C, X^+, Δ, ∇) is a split highest weight coalgebra over k if and only if $(C', X^+, \Delta', \nabla')$ is a split highest weight coalgebra over k', where $C' := C \otimes_k k'$, $\Delta' := (\Delta_C(\lambda)')_{\lambda \in X^+}$, and $\nabla' := (\nabla_C(\lambda)')_{\lambda \in X^+}$. Moreover, we have $L_C(\lambda) \otimes_k k' \cong L_{C'}(\lambda)$ for any $\lambda \in X^+$ in this case.*

Proof. Obvious by Lemma 1.2.16 and the theorem. □

2 Donkin systems

2.1 U-acyclicity of flat complexes

(2.1.1) Let R be a commutative ring.

Lemma 2.1.2 (Universal coefficient theorem) *Let $\mathbb{F}$ be an R-flat complex, and M an R-module. Assume that $\mathbb{F}$ is bounded above or* $\operatorname{flat.dim}_R M < \infty$. *Then there is a spectral sequence*

$$E_2^{p,q} = \operatorname{Tor}^R_{-p}(H^q(\mathbb{F}), M) \Rightarrow H^{p+q}(\mathbb{F} \otimes M).$$

In particular, if flat.dim$_R M \leq 1$, *then we have* $E_2^{p,q} = E_\infty^{p,q}$, *and there is a short exact sequence*

$$0 \to H^n(\mathbb{F}) \otimes M \to H^n(\mathbb{F} \otimes M) \to \mathrm{Tor}_1^R(H^{n+1}(\mathbb{F}), M) \to 0.$$

Proof. Let $\mathbb{G}$ be a flat resolution of M. If flat.dim$_R M < \infty$, then we take $\mathbb{G}$ to be bounded. It is easy to see that $\mathbb{F} \otimes_R^\bullet \mathbb{G} \to \mathbb{F} \otimes M$ is a quasi-isomorphism. Now consider the spectral sequence associated with the double complex $\mathbb{F} \otimes_R^\bullet \mathbb{G}$, then the assertion follows. □

Definition 2.1.3 An R-complex

$$\mathbb{F} : 0 \to F^0 \to F^1 \to F^2 \to \cdots$$

is said to be *u-acyclic* (universally acyclic) if for any R-module M, we have $H^i(\mathbb{F} \otimes M) = 0$ $(i > 0)$ and the canonical map

$$\rho_M : H^0(\mathbb{F}) \otimes M \to H^0(\mathbb{F} \otimes M)$$

is an isomorphism.

We list some consequences of the definition.

Lemma 2.1.4 *Let* $\mathbb{F}$ *be a u-acyclic R-complex. Then* $R' \otimes \mathbb{F}$ *is a u-acyclic* R'*-complex for any base change* $R \to R'$. *Moreover, for any R-module M,* $\mathbb{F} \otimes M$ *is u-acyclic.*

Proof. Easy. □

Lemma 2.1.5 *Let*

$$0 \to \mathbb{F} \xrightarrow{f} \mathbb{G} \to \mathbb{H} \to 0$$

be an exact sequence of R-complexes. Assume that $\mathbb{F}$ *is u-acyclic and* $f^i : F^i \to G^i$ *is R-pure for* $i \geq 0$. *Then* $\mathbb{G}$ *is u-acyclic if and only if* $\mathbb{H}$ *is u-acyclic.*

Proof. Easy. □

Lemma 2.1.6 *Let* $(\mathbb{F}_\lambda)_{\lambda \in \Lambda}$ *be a filtered inductive system of u-acyclic R-complexes. Then* $\varinjlim \mathbb{F}_\lambda$ *is u-acyclic.*

Proof. Obvious.

(2.1.7) Let R be a noetherian commutative ring.

Lemma 2.1.8 *Let $(R, \mathfrak{m})$ be a local ring, and F a flat R-module. Then there is an exact sequence*

$$0 \to P \to F \to G \to 0 \tag{2.1.9}$$

of R-modules such that P is R-free, G is R-flat, and $G/\mathfrak{m}G = 0$.

Proof. Let B be an $R/\mathfrak{m}$-basis of $F/\mathfrak{m}F$, and consider the R-free module P freely generated by B over R. Then the canonical map $P \to P/\mathfrak{m}P \cong F/\mathfrak{m}F$ is lifted to some R-linear map $\varphi : P \to F$. We set $G := \operatorname{Coker} \varphi$. Then $G/\mathfrak{m}G = 0$ is obvious. By Lemma I.2.1.4, φ is injective, and G is R-flat. □

As is our general convention in these notes, we denote $M \otimes \kappa(\mathfrak{p})$ by $M(\mathfrak{p})$ for short, for $\mathfrak{p} \in \operatorname{Spec} R$.

Corollary 2.1.10 *Let $(R, \mathfrak{m})$ be a local ring, F an R-flat module, and c a non-negative integer. If $\dim_{\kappa(\mathfrak{p})} F(\mathfrak{p}) = c$ for any $\mathfrak{p} \in \operatorname{Spec} R$, then we have $F \cong R^c$.*

Proof. We take an exact sequence (2.1.9) as in the lemma. Then as $\dim_{\kappa(\mathfrak{m})} P(\mathfrak{m}) = c$ and P is R-free, we have $P \cong R^c$. For any $\mathfrak{p} \in \operatorname{Spec} R$,

$$0 \to P(\mathfrak{p}) \to F(\mathfrak{p}) \to G(\mathfrak{p}) \to 0$$

is exact by flatness of G, and we have $G(\mathfrak{p}) = 0$ for any $\mathfrak{p} \in \operatorname{Spec} R$ by dimension counting. By Corollary I.2.1.6, we have $G = 0$. Hence, we have $F \cong P \cong R^c$. □

Lemma 2.1.11 *Let*

$$\mathbb{F} : F^0 \xrightarrow{d^0} F^1 \xrightarrow{d^1} F^2$$

be an R-flat complex. Then we have the following:

1 *If $H^1(\mathbb{F} \otimes R/\mathfrak{p})$ is R-finite (resp. 0) for any $\mathfrak{p} \in \operatorname{Spec} R$, then $H^1(\mathbb{F} \otimes M)$ is R-finite (resp. 0) for any R-finite module M (resp. any R-module M).*

2 *If $H^1(\mathbb{F} \otimes R/\mathfrak{p}) = 0$ for any $\mathfrak{p} \in \operatorname{Spec} R$, then*

$$F^0 \to F^1 \to F^2 \to \operatorname{Coker} d^1 \to 0$$

is a u-acyclic R-flat complex, and $\operatorname{Ker} d^0$ is an R-pure submodule of F^0, hence it is R-flat.

Proof. **1** If M is R-finite, then M admits a filtration whose successive subquotients are of the form $R/\mathfrak{p}$ with $\mathfrak{p} \in \operatorname{Spec} R$. As $H^1(\mathbb{F}\otimes ?)$ is half-exact, $H^1(\mathbb{F} \otimes M)$ is also R-finite (resp. 0), as ${}_R\mathbb{M}_f$ and $\{0\}$ are very thick in ${}_R\mathbb{M}$. As $H^1(\mathbb{F}\otimes ?)$ preserves filtered inductive limits and $\{0\}$ is also closed under filtered inductive limits, $H^1(\mathbb{F} \otimes M) = 0$ also for general M.

2 We set $C := \operatorname{Coker} d^1$. As we have $H^1(\mathbb{F}) = 0$ by **1**, we can take an R-flat resolution of C of the form

$$\cdots \to F^{-2} \to F^{-1} \to F^0 \to F^1 \to F^2 \to C \to 0.$$

By **1**, we have $\operatorname{Tor}_1^R(C, M) = 0$ for any R-module M, and hence C is R-flat. As

$$0 \to \operatorname{Ker} d^0 \to F^0 \to F^1 \to F^2 \to C \to 0$$

is exact and R-flat modules are closed under epikernels, $\operatorname{Ker} d^0$ is also R-flat. As this sequence is a flat resolution of 0, the rest of the assertion follows. □

(2.1.12) Let R be a noetherian commutative ring, and M an R-module. We say that M is *R-metafinite* if there exist some noetherian commutative R-algebra A and an A-finite module structure of M which induces the original R-module structure of M via restriction.

Lemma 2.1.13 *Let M be a metafinite R-module. Then for any R-finite module N, the tensor product $N \otimes M$ is R-metafinite. If $R \to R'$ is a homomorphism of noetherian commutative rings essentially of finite type, then $R' \otimes M$ is R'-metafinite. If $M(\mathfrak{p}) = 0$ for any $\mathfrak{p} \in \operatorname{Spec} R$, then we have $M = 0$.*

Proof. There is a commutative noetherian R-algebra A such that M is a finite A-module. Then we have $N \otimes M \cong (N \otimes A) \otimes_A M$, which is A-finite. Hence, $N \otimes M$ is also R-metafinite. As $R' \otimes A$ is noetherian and $R' \otimes M$ is $R' \otimes A$-finite, we have that $R' \otimes M$ is R'-metafinite. For the last assertion, we refer the reader to [110, Theorem 4.9]. □

Proposition 2.1.14 *Let R be a noetherian commutative ring, and*

$$\mathbb{F} : 0 \to F^0 \xrightarrow{d^0} F^1 \xrightarrow{d^1} F^2 \to \cdots$$

an R-flat complex. Consider the following conditions:

1 *For any $\mathfrak{p} \in \operatorname{Spec} R$, $H^i(\mathbb{F}\otimes R/\mathfrak{p}) = 0$ $(i > 0)$ and $H^0(\mathbb{F}\otimes R/\mathfrak{p})$ is R-finite.*

2 *$H^0(\mathbb{F})$ is R-finite projective, and $\mathbb{F}$ is u-acyclic.*

3 *$H^0(\mathbb{F})$ is R-finite, and $H^i(\mathbb{F}) = 0$ $(i > 0)$.*

4 *For any* $\mathfrak{p} \in \operatorname{Spec} R$, *we have* $H^i(\mathbb{F}(\mathfrak{p})) = 0$ *for* $i > 0$, *and* $h^0_{\mathbb{F}}(\mathfrak{p}) := \dim_{\kappa(\mathfrak{p})} H^0(\mathbb{F}(\mathfrak{p}))$ *is finite. The function* $h^0_{\mathbb{F}}$ *is locally constant on* $\operatorname{Spec} R$.

5 $H^i(\mathbb{F}(\mathfrak{m})) = 0$ $(i > 0)$ *for any* $\mathfrak{m} \in \operatorname{Max} R$.

Then we have the following.

(2.1.15) **5⇐4⇐1⇔2⇒3**.

(2.1.16) *If* R *is a regular ring, or* $\mathbb{F}$ *is bounded, then* **3⇒1**.

(2.1.17) *If* F^0 *is* R*-projective, then* **4⇒1**.

(2.1.18) *Let* n *be a non-negative integer. Assume that* $H^i(\mathbb{F}(\mathfrak{p})) = 0$ $(i \geq n)$, *and* $H^n(\mathbb{F} \otimes R/\mathfrak{p})$ *is* R*-metafinite for any* $\mathfrak{p} \in \operatorname{Spec} R$. *Then* $H^i(\mathbb{F} \otimes M) = 0$ $(i \geq n)$ *for any* R*-module* M. *In particular, if* $H^i(\mathbb{F}(\mathfrak{p})) = 0$ $(i \geq 1)$ *and* $H^1(\mathbb{F} \otimes R/\mathfrak{p})$ *is* R*-metafinite for any* $\mathfrak{p} \in \operatorname{Spec} R$, *then* $\mathbb{F}$ *is u-acyclic.*

(2.1.19) *Let* n *be a non-negative integer. Assume that* $H^i(\mathbb{F}(\mathfrak{m})) = 0$ $(i \geq n)$, *and* $H^n(\mathbb{F} \otimes R/\mathfrak{p})$ *is* R*-finite for any* $\mathfrak{p} \in \operatorname{Spec} R$. *Then* $H^i(\mathbb{F} \otimes M) = 0$ $(i \geq n)$ *for any* R*-module* M. *In particular, if* $H^1(\mathbb{F} \otimes R/\mathfrak{p})$ *is* R*-finite for any* $\mathfrak{p} \in \operatorname{Spec} R$ *and* $H^0(\mathbb{F})$ *is* R*-finite, then we have* **5⇒1**.

Proof. The implication **1⇒2** is obvious by Lemma 2.1.11. It is trivial that **2** implies **1**,**3** and **4**. The implication **4⇒5** is also trivial. Hence, (2.1.15) follows.

The assertion (2.1.16) follows easily from Lemma 2.1.2.

We prove (2.1.17). For $\mathfrak{p} \in \operatorname{Spec} R$, as $H^0(\mathbb{F} \otimes R/\mathfrak{p})$ is an $R/\mathfrak{p}$-submodule of the projective module $F^0 \otimes R/\mathfrak{p}$, we have $H^0(\mathbb{F} \otimes R/\mathfrak{p})$ is $R/\mathfrak{p}$-finite, or equivalently, R-finite by Corollary I.3.11.8. So it suffices to show that $H^i(\mathbb{F} \otimes R/\mathfrak{p}) = 0$ $(i > 0)$ for $\mathfrak{p} \in \operatorname{Spec} R$. To prove this, we may localize at maximal ideals of R, and we may assume that $(R, \mathfrak{m})$ is local. So we shall assume that $(R, \mathfrak{m})$ is a d-dimensional local ring, and we proceed by induction on d. Next, we may replace R by $R/\mathfrak{p}$ with $\dim R/\mathfrak{p} = d$, and we may and shall assume that R is a domain. If $d = 0$, then R is a field and there is nothing to be proved. Hence, we consider the case $d > 0$.

By induction assumption, for any non-zero ideal I of R, $\mathbb{F} \otimes R/I$ is u-acyclic. Hence, it suffices to show $H^i(\mathbb{F}) = 0$ for $i > 0$. By induction assumption, any proper localization of $\mathbb{F}$ is also u-acyclic. This shows that $\operatorname{supp} H^i(\mathbb{F}) \subset \{\mathfrak{m}\}$ for $i > 0$. We take an element $0 \neq x \in \mathfrak{m}$. As R is a domain, we have $\operatorname{proj.dim}_R R/Rx = 1$.

By Lemma 2.1.2, we have an exact sequence

$$0 \to H^i(\mathbb{F}) \otimes R/Rx \to H^i(\mathbb{F} \otimes R/Rx) \to \operatorname{Tor}_1^R(H^{i+1}(\mathbb{F}), R/Rx) \to 0.$$

Hence, for $i \geq 2$, we have

$$\operatorname{soc} H^i(\mathbb{F}) = \operatorname{Hom}_R(R/\mathfrak{m}, H^i(\mathbb{F})) \subset \operatorname{Hom}_R(R/xR, H^i(\mathbb{F})) \\ \cong \operatorname{Tor}_1^R(R/xR, H^i(\mathbb{F})) = 0.$$

As $\operatorname{supp} H^i(\mathbb{F}) \subset \{\mathfrak{m}\}$, and the category of R-modules whose support is contained in $\{\mathfrak{m}\}$ is locally finite, we have $H^i(\mathbb{F}) = 0$ for $i \geq 2$. Hence, it suffices to show that $H^1(\mathbb{F}) = 0$.

By Lemma 2.1.11, we have that $\operatorname{Ker} d^1$ is an R-pure submodule of F^1, and is R-flat. Hence, replacing $\mathbb{F}$ by the R-flat complex

$$0 \to F^0 \to \operatorname{Ker} d^1 \to 0,$$

we may assume that $F^i = 0$ $(i \geq 2)$ without loss of generality.

As F^0 is R-projective and R is local, F^0 is R-free by Kaplansky's theorem [92]. We take a basis B of F^0. As $\dim_{\kappa(\mathfrak{m})} H^0(\mathbb{F}(\mathfrak{m})) < \infty$, there exists some finite subset B_0 of B such that $H^0(\mathbb{F}(\mathfrak{m}))$ is contained in the $\kappa(\mathfrak{m})$-span of B_0 in $F^0(\mathfrak{m}) = \kappa(\mathfrak{m}) \cdot B$. Now we set $G^0 := R \cdot B_0$ and $Q := R \cdot (B \setminus B_0)$. When we denote the composite map

$$Q \hookrightarrow F^0 \xrightarrow{d^0} F^1$$

by φ, we have that φ is injective and $G^1 := \operatorname{Coker} \varphi$ is R-flat by Lemma I.2.1.4.

The composite map

$$G^0 \hookrightarrow F^0 \xrightarrow{d^0} F^1 \to G^1$$

gives an R-flat complex $\mathbb{G}$ of length one, and we have a short exact sequence of R-flat complexes

$$0 \to (\mathrm{id}_Q : Q \xrightarrow{1_Q} Q) \to \mathbb{F} \xrightarrow{\pi} \mathbb{G} \to 0.$$

As π and $\pi(\mathfrak{p})$ $(\mathfrak{p} \in \operatorname{Spec} R)$ are quasi-isomorphisms we may assume, replacing $\mathbb{F}$ by $\mathbb{G}$, that F^0 is R-finite free without loss of generality.

As the sequence

$$0 \to H^0(\mathbb{F}(\mathfrak{p})) \to F^0(\mathfrak{p}) \to F^1(\mathfrak{p}) \to 0 = H^1(\mathbb{F}(\mathfrak{p}))$$

is exact, $\dim_{\kappa(\mathfrak{p})} F^1(\mathfrak{p})$ is finite and constant on $\operatorname{Spec} R$ by assumption. By Corollary 2.1.10, we have that F^1 is R-finite, and hence so is $H^1(\mathbb{F})$. As $H^1(\mathbb{F}) \otimes R/Rx \subset H^1(\mathbb{F} \otimes R/Rx) = 0$, we have $H^1(\mathbb{F}) = 0$ by Nakayama's lemma, and this completes the proof of (2.1.17).

We prove (2.1.18). We prove the first assertion. It suffices to show $H^i(\mathbb{F} \otimes R/\mathfrak{p}) = 0$ for any $\mathfrak{p} \in \operatorname{Spec} R$. By Lemma 2.1.13, we may assume that $(R, \mathfrak{m})$ is local, and proceed by induction on $d = \dim R$. We may also assume that R is a local domain and $d > 0$. By induction assumption, it suffices to show $H^i(\mathbb{F}) = 0$ for $i \geq n$. Note that $\operatorname{supp} H^i(\mathbb{F}) \subset \{\mathfrak{m}\}$. By the same argument as in the proof of (2.1.17), $H^i(\mathbb{F}) = 0$ for $i > n$ is proved easily, using Lemma 2.1.2.

We prove $H^n(\mathbb{F}) = 0$. As $H := H^n(\mathbb{F})$ is R-metafinite by assumption, there is a noetherian commutative R-algebra A such that H is A-finite. We have $H/xH \subset H^n(\mathbb{F} \otimes R/Rx) = 0$ for any $x \in R \setminus \{0\}$. This shows that $x : H \to H$ is a surjective A-linear map, which must be bijective. This shows that H is a $\kappa(0)$-module. As we have $H = H(0) \cong H^n(\mathbb{F}(0)) = 0$, we are done.

Now we prove the second assertion of (2.1.18). As we have $H^i(\mathbb{F} \otimes R/\mathfrak{p}) = 0$ for $i > 0$, $\mathbb{F}$ is u-acyclic by Lemma 2.1.11.

The assertion (2.1.19) is easier, and we omit the proof. □

2.2 The definition and the existence of a Donkin system

(2.2.1) Let R be a commutative ring, and C an R-flat coalgebra. For $\mathfrak{p} \in \operatorname{Spec} R$ and an R-module M, $M(\mathfrak{p})$ stands for $M \otimes \kappa(\mathfrak{p})$.

Definition 2.2.2 Let V and W be C-comodules. We say that W is *u-acyclic* with respect to V if for any R-module M, the conditions

1 $\operatorname{Ext}^i_C(V, W \otimes M) = 0$ $(i > 0)$, and

2 The canonical map $\rho : \operatorname{Hom}_C(V, W) \otimes M \to \operatorname{Hom}_C(V, W \otimes M)$ is an isomorphism

are satisfied.

Lemma 2.2.3 *Let V be an R-finite R-projective C-comodule, and W a C-comodule. Then W is u-acyclic with respect to V if and only if the complex of R-modules $\operatorname{Cobar}_C(W, V^*)$ is u-acyclic.*

Proof. By Lemma I.3.6.16, we have

$$\operatorname{Ext}^i_C(V, W \otimes M) \cong H^i(\operatorname{Cobar}_C(W \otimes M, V^*)) \cong H^i(\operatorname{Cobar}_C(W, V^*) \otimes M)$$

for any R-module M and any $i \geq 0$. □

Lemma 2.2.4 *The following hold:*

1 *Let V and W be C-comodules. If W is u-acyclic with respect to V, then $W \otimes M$ is u-acyclic with respect to V for any R-module M. If V is R-finite projective, then for any commutative R-algebra $R \to R'$, $W' := W \otimes R'$ is u-acyclic with respect to $V' = V \otimes R'$, as a $C' = C \otimes R'$-comodule.*

2 *The class of C-comodules which are u-acyclic with respect to V is closed under direct sums and direct summands.*

3 *Let $0 \to W_1 \to W_2 \to W_3 \to 0$ be an exact sequence of C-comodules. If W_1 is u-acyclic with respect to V and $W_1 \hookrightarrow W_2$ is R-pure, then W_3 is u-acyclic with respect to V if and only if W_2 is u-acyclic with respect to V.*

4 *Assume that V is R-finite and R is noetherian. Then any filtered inductive limit of C-comodules which are u-acyclic with respect to V is again u-acyclic with respect to V.*

5 *For a fixed C-comodule W, the class of C-comodules V with respect to which W is u-acyclic is closed under extensions, direct summands, and epikernels.*

Proof. **1** Let N be an R-module. Then $\mathrm{Ext}^i_C(V, (W \otimes M) \otimes N) = 0$ $(i > 0)$ is obvious. The canonical map

$$\rho : \mathrm{Hom}_C(V, W \otimes M) \otimes N \to \mathrm{Hom}_C(V, W \otimes M \otimes N)$$

agrees with the composite

$$\mathrm{Hom}_C(V, W \otimes M) \otimes N \xrightarrow{\rho^{-1} \otimes 1_N} \mathrm{Hom}_C(V, W) \otimes (M \otimes N) \xrightarrow{\rho} \mathrm{Hom}_C(V, W \otimes M \otimes N),$$

which is an isomorphism.

We prove the second statement of **1**. By Lemma 2.2.3, $\mathrm{Cobar}_C(W, V^*)$ is u-acyclic. By Lemma 2.1.4, $\mathrm{Cobar}_C(W, V^*) \otimes R'$ is a u-acyclic complex over R'. By Lemma I.3.6.15 and Lemma 2.2.3, we are done.

Statement **2** is obvious.

We prove **3**. Let M be an arbitrary R-module. As $W_1 \hookrightarrow W_2$ is R-pure,

$$0 \to W_1 \otimes M \to W_2 \otimes M \to W_3 \otimes M \to 0$$

is exact. Then by assumption, we have

$$\mathrm{Ext}^i_C(V, W_2 \otimes M) \cong \mathrm{Ext}^i_C(V, W_3 \otimes M)$$

for $i > 0$. Now consider the commutative diagram:

$$\begin{array}{ccccccc} & \mathrm{Hom}_C(V, W_1) \otimes M & \to & \mathrm{Hom}_C(V, W_2) \otimes M & \to & \mathrm{Hom}_C(V, W_3) \otimes M & \to 0 \\ & \downarrow\cong & & \downarrow & & \downarrow & \\ 0 \to & \mathrm{Hom}_C(V, W_1 \otimes M) & \to & \mathrm{Hom}_C(V, W_2 \otimes M) & \to & \mathrm{Hom}_C(V, W_3 \otimes M) & \to 0 \end{array}$$

The rows are exact and the first vertical arrow is an isomorphism, since W_1 is u-acyclic. By the five lemma, the second vertical arrow is an isomorphism if and only if the third is. Hence, we have W_3 is u-acyclic with respect to V if and only if W_2 is.

Statement **4** follows immediately from Proposition I.3.7.4.

5 It is clearly closed under direct summands. Now let

$$0 \to V_1 \to V_2 \to V_3 \to 0$$

be an exact sequence of C-comodules such that W is u-acyclic with respect to V_3. Then for any R-module M, we have

$$\operatorname{Ext}^i_C(V_2, W \otimes M) \cong \operatorname{Ext}^i_C(V_1, W \otimes M) \qquad (i > 0),$$

as W is u-acyclic with respect to V_3. Moreover, we have a commutative diagram:

$$\begin{array}{ccccccc} & \operatorname{Hom}_C(V_3, W) \otimes M & \to & \operatorname{Hom}_C(V_2, W) \otimes M & \to & \operatorname{Hom}_C(V_1, W) \otimes M & \to 0 \\ & \downarrow \cong & & \downarrow & & \downarrow & \\ 0 \to & \operatorname{Hom}_C(V_3, W \otimes M) & \to & \operatorname{Hom}_C(V_2, W \otimes M) & \to & \operatorname{Hom}_C(V_1, W \otimes M) & \to 0 \end{array}$$

The rows are exact and the first vertical map is an isomorphism, as W is u-acyclic with respect to V_3. By the five lemma again, we have that W is u-acyclic with respect to V_2 if and only if it is so with respect to V_1. That is, the class in question is closed under extensions and epikernels. □

Lemma 2.2.5 *Let R be a noetherian ring, V an R-finite R-projective C-comodule, and W an R-projective C-comodule. Assume that for any $\mathfrak{p} \in$* Spec R*, we have*

$$\operatorname{Ext}^i_{C(\mathfrak{p})}(V(\mathfrak{p}), W(\mathfrak{p})) = 0 \qquad (i > 0)$$

and $\dim_{\kappa(\mathfrak{p})} \operatorname{Hom}_{C(\mathfrak{p})}(V(\mathfrak{p}), W(\mathfrak{p}))$ *is a locally constant function on* Spec R *with finite values. Then W is u-acyclic with respect to V, and* $\operatorname{Hom}_C(W, V)$ *is R-finite projective.*

Proof. By assumption, $\mathbb{F} := \operatorname{Cobar}(W, V^*)$ is an R-flat complex, and $F^0 = W \otimes V^*$ is R-projective. For $\mathfrak{p} \in \operatorname{Spec} R$, we have

$$\begin{aligned} H^i(\operatorname{Cobar}_C(W, V^*)(\mathfrak{p})) &\cong H^i(\operatorname{Cobar}_{C(\mathfrak{p})}(W(\mathfrak{p}), (V^*)(\mathfrak{p}))) \\ &\cong \operatorname{Ext}^i_{C(\mathfrak{p})}(V(\mathfrak{p}), W(\mathfrak{p})) \end{aligned}$$

by Lemma I.3.6.15 and Lemma I.3.6.16. By assumption and (2.1.17), we have that $\operatorname{Cobar}_C(W, V^*)$ is a u-acyclic complex. By Lemma 2.2.3, W is u-acyclic with respect to V. □

Definition 2.2.6 We say that a triple (X^+, Δ, ∇) is a *semisplit highest weight theory* over C, if the following are satisfied:

D1 X^+ is an ordered set which satisfies the conditions in Lemma 1.1.2, and $\Delta = (\Delta_C(\lambda))_{\lambda\in X^+}$ and $\nabla = (\nabla_C(\lambda))_{\lambda\in X^+}$ are families of R-finite R-projective C-comodules, indexed by X^+.

D2 For any $\mathfrak{p} \in \operatorname{Spec} R$, $(C(\mathfrak{p}), X^+, \Delta(\mathfrak{p}), \nabla(\mathfrak{p}))$ is a split highest weight coalgebra over $\kappa(\mathfrak{p})$.

We also say that C, or better (C, X^+, Δ, ∇), is a *semisplit highest weight coalgebra* over R. We call $X^+ = X^+_C$ the set of *dominant weights*, $\Delta_C(\lambda)$ the *Weyl module* of highest weight λ, and $\nabla_C(\lambda)$ the *induced module* of highest weight λ. We say that (X^+, Δ, ∇) is *split* if $\Delta_C(\lambda)$, $\nabla_C(\lambda)$ and $R(\lambda) := \operatorname{Hom}_C(\Delta_C(\lambda), \nabla_C(\lambda))$ are R-free modules for all $\lambda \in X^+$. Let (X^+, Δ, ∇) be a semisplit highest weight theory over C. We say that $\mathcal{C} = (C_\pi)$ is a *Donkin system* of C associated with (X^+, Δ, ∇), if (C_π) is a family of R-subcoalgebras of C indexed by all finite poset ideals π of X^+, and the conditions

a $C_\emptyset = 0$,

b $\pi' \subset \pi$ implies $C_{\pi'} \subset C_\pi$,

c If π is a finite poset ideal of X^+ and λ is its maximal element, then

$$C_\pi/C_{\pi'} \cong R(\lambda)^* \otimes \Delta_C(\lambda)^* \otimes \nabla_C(\lambda)$$

as (C, C)-bicomodules, where $\pi' = \pi \setminus \{\lambda\}$

are satisfied. If there is no danger of confusion, then we simply say that $\mathcal{C}$ is a Donkin system of C.

Lemma 2.2.7 *Assume that R is noetherian, and let (X^+, Δ, ∇) be a semisplit highest weight theory over C. Then for $\lambda, \mu \in X^+$, $\nabla_C(\mu)$ is u-acyclic with respect to $\Delta_C(\lambda)$. Moreover,*

$$R(\lambda) = \operatorname{Hom}_C(\Delta_C(\lambda), \nabla_C(\mu))$$

is rank-one R-projective.

Proof. Follows immediately from Lemma 2.2.5. □

Lemma 2.2.8 *If (X^+, Δ, ∇) is a semisplit highest weight theory over C and $R \to R'$ is a homomorphism of noetherian commutative rings, then (X^+, Δ', ∇') is a semisplit highest weight theory over C'. If, moreover, (C_π) is a Donkin system over C, then (C'_π) is a Donkin system of C', where $(?)'$ denotes the functor $? \otimes R'$.*

Proof. Condition **D1** is obvious. Condition **D2** follows from Corollary 1.4.8. Note that the construction of $R(\lambda)^*$ is compatible with the base change because of Lemma 2.2.7. Hence, the last assertion is obvious. □

Lemma 2.2.9 *Let (X^+, Δ, ∇) be a semisplit highest weight theory over C. Then $(X^+, \nabla^*, \Delta^*)$ is a semisplit highest weight theory over C^{op}. If, moreover, (C_π) is a Donkin system of C, then (C_π^{op}) is a Donkin system of C^{op}.*

Proof. Condition **D1** is obvious. Condition **D2** follows from Theorem 1.4.7. The last assertion is checked easily, as we have

$$R(\lambda)_{C^{\mathrm{op}}} = \mathrm{Hom}_{C^{\mathrm{op}}}(\nabla_C(\lambda)^*, \Delta_C(\lambda)^*) \cong \mathrm{Hom}_C(\Delta_C(\lambda), \nabla_C(\lambda)) = R(\lambda)_C.$$

□

(2.2.10) The rest of this subsection is devoted to proving the following.

Theorem 2.2.11 *Let R be a noetherian commutative ring, and C an R-projective R-coalgebra. If (X^+, Δ, ∇) is a semisplit highest weight theory over C, then there is a Donkin system (C_π) of C associated with (X^+, Δ, ∇).*

In the rest of this subsection, let R be a noetherian commutative ring, C an R-projective R-coalgebra, and (X^+, Δ, ∇) a semisplit highest weight theory over C.

We construct (C_π), in order to prove the (existence) theorem. The construction is inductive. For clarity, we state what we are actually going to prove.

Proposition 2.2.12 *Let $n \geq 0$. There exists a family $(C_\pi)_{\#\pi \leq n}$ of R-subcoalgebras of C indexed by all finite poset ideals π of X^+ with $\#\pi \leq n$, such that the conditions* **a**, **b** *and* **c** *in* Definition 2.2.6 *and the condition*

d *For any π and any maximal element λ of π, we have that $C_\pi / C_{\pi'}$ is the image of the canonical evaluation map*

$$\rho' : \mathrm{Hom}_C(\nabla_C(\lambda), C/C_{\pi'}) \otimes \nabla_C(\lambda) \to C/C_{\pi'}.$$

Or equivalently, $C_\pi / C_{\pi'}$ is the sum of images of all C-comodule maps from $\nabla_C(\lambda)$ to $C/C_{\pi'}$, where $\pi' := \pi \setminus \{\lambda\}$.

are satisfied.

Note that a family $(C_\pi)_{\#\pi\leq n}$ which satisfies **a**–**d** is unique, if it exists, and C_π does not depend on n. It follows that the proposition implies Theorem 2.2.11. We prove the proposition using induction on n.

If $n=0$, then we set $C_\emptyset = 0$, and **a**–**d** are satisfied. Next, we assume $n>0$. By induction assumption, the family $(C_\pi)_{\#\pi<n}$ which satisfies **a**–**d** (uniquely) exists.

Let π be a poset ideal of X^+ with $\#\pi = n$. To define C_π, we take a maximal element λ of π, and set $\pi' := \pi\setminus\{\lambda\}$. By induction assumption, $C_{\pi'}$ is already constructed, it is R-finite R-projective, and $C_{\pi'} \hookrightarrow C$ is R-pure (see Definition I.3.6.7). By Lemma I.2.1.11, $C/C_{\pi'}$ is R-projective.

For any $\mathfrak{p}\in \operatorname{Spec} R$, it is obvious that $C_{\pi'}(\mathfrak{p})$ belongs to $\mathbb{M}^{C(\mathfrak{p})}(\pi')$, and we have $C_{\pi'}(\mathfrak{p})\subset C(\mathfrak{p})(\pi')$. Thanks to Lemma 2.2.7, we have $C_{\pi'}(\mathfrak{p}) = C(\mathfrak{p})(\pi')$ by dimension counting.

Lemma 2.2.13 *For any $\mu\in X^+$, C, $C_{\pi'}$, and $C/C_{\pi'}$ are u-acyclic with respect to $\Delta_C(\mu)$. $C/C_{\pi'}$ is also u-acyclic with respect to $\nabla_C(\lambda)$. Moreover,*

$$\operatorname{Hom}_C(\Delta_C(\lambda),C),\quad \operatorname{Hom}_C(\Delta_C(\lambda),C/C_{\pi'}),\quad \textit{and}\quad \operatorname{Hom}_C(\nabla_C(\lambda),C/C_{\pi'})$$

are R-finite projective modules with the same rank functions.

Proof. As C, $C_{\pi'}$, and $C/C_{\pi'}$ are all R-projective, the first assertion follows easily from Lemma 2.2.5. For $\mathfrak{p}\in\operatorname{Spec}R$, $C/C_{\pi'}(\mathfrak{p})\cong C(\mathfrak{p})/C(\mathfrak{p})(\pi')$ is a filtered inductive limit of objects of $\mathcal{F}((\nabla_{C(\mathfrak{p})}(\lambda)_{\lambda\in X^+\setminus\pi'}))$ by Proposition 1.3.2. We have by Lemma 1.3.8,

$$\operatorname{Ext}^i_{C(\mathfrak{p})}(\nabla_C(\lambda)(\mathfrak{p}),\nabla_C(\mu)(\mathfrak{p}))=0$$

for $i>0$ and $\mu\not<\lambda$. It follows that

$$\operatorname{Ext}^i_{C(\mathfrak{p})}(\nabla_C(\lambda)(\mathfrak{p}),(C/C_{\pi'})(\mathfrak{p}))=0\qquad(i>0).$$

On the other hand, by Proposition 1.3.2, the dimensions of

$$\begin{aligned}\operatorname{Hom}_{C(\mathfrak{p})}(\nabla_C(\lambda)(\mathfrak{p}),(C/C_{\pi'})(\mathfrak{p}))&\cong \operatorname{Hom}_{C(\mathfrak{p})}(\Delta_C(\lambda)(\mathfrak{p}),C(\mathfrak{p}))\\&\cong \operatorname{Hom}_{\kappa(\mathfrak{p})}(\Delta_C(\lambda)(\mathfrak{p}),\kappa(\mathfrak{p}))\end{aligned}$$

are finite and locally constant on $\operatorname{Spec}R$.

By Lemma 2.2.5, we have that $C/C_{\pi'}$ is u-acyclic with respect to $\nabla_C(\lambda)$. The last assertion is obvious by Lemma 2.2.5 and Proposition 1.3.2. □

Lemma 2.2.14 *The canonical maps*

$$\begin{aligned}&\Delta_C(\lambda)^*\cong\operatorname{Hom}_C(\Delta_C(\lambda),C)\\&\qquad\to\operatorname{Hom}_C(\Delta_C(\lambda),C/C_{\pi'})\leftarrow\operatorname{Hom}_C(\nabla_C(\lambda),C/C_{\pi'})\otimes R(\lambda)\end{aligned}$$

are all isomorphisms of left C-comodules, where the $\leftarrow$ is the map obtained by the composition $\Delta_C(\lambda)\to\nabla_C(\lambda)\to C/C_{\pi'}$.

Proof. It is obvious that these maps are left C-comodule maps. By Lemma 2.2.13, the modules in question are R-finite projective modules with the same rank functions, and so these maps are isomorphisms, if they are surjective. This is checked after tensoring with $\kappa(\mathfrak{p})$. The assertion is now obvious by Lemma 2.2.7, Lemma 2.2.13 and Proposition 1.3.2. □

Lemma 2.2.15 *The canonical pairing*

$$\rho' : \operatorname{Hom}_C(\nabla_C(\lambda), C/C_{\pi'}) \otimes \nabla_C(\lambda) \to C/C_{\pi'}$$

is an R-pure (in particular, injective) (C, C)-bicomodule map.

Proof. As the modules in question are R-projective, it suffices to check that the map in question is injective after tensoring with $\kappa(\mathfrak{p})$ for all $\mathfrak{p} \in \operatorname{Spec} R$, by Lemma I.2.1.4. As we have $(C/C_{\pi'})(\mathfrak{p}) \cong C(\mathfrak{p})/C(\mathfrak{p})(\pi')$, this is clear by Lemma 2.2.13 and Proposition 1.3.2. □

Now we define C_π to be the pull-back of $\operatorname{Im} \rho'$ by the projection $C \to C/C_{\pi'}$. By Lemma 2.2.15, we have $C/C_\pi \cong \operatorname{Coker} \rho'$ is R-flat. Hence, C_π is an R-pure (C, C)-subbicomodule of C, namely, an R-subcoalgebra. By Lemma 2.2.14 and Lemma 2.2.15, the isomorphism $C_\pi/C_{\pi'} \cong R(\lambda)^* \otimes \Delta_C(\lambda)^* \otimes \nabla_C(\lambda)$ is obvious.

Now we claim that the definition of C_π is independent of the choice of a maximal element λ of π. Then conditions **a**–**d** are obvious by construction, and the proof of Proposition 2.2.12 is complete.

So let $\mu \neq \lambda$ be another maximal element of π. We keep on using λ to define C_π, to avoid ambiguity. Set $\pi'_0 := \pi \setminus \{\lambda, \mu\}$, $\pi_0 := \pi \setminus \{\mu\}$, and $\pi' := \pi \setminus \{\lambda\}$ (as above). These are finite poset ideals of π.

By Lemma 2.2.5, it is easy to see that $\operatorname{Ext}^i_C(\nabla_C(\lambda), \nabla_C(\mu)) = 0$ for $i \geq 0$. As we have $C_{\pi'}/C_{\pi'_0} \in \operatorname{add} \nabla_C(\mu)$ by induction assumption, we have

$$\operatorname{Ext}^i_C(\nabla_C(\lambda), C_{\pi'}/C_{\pi'_0}) = 0$$

for $i = 0, 1$. Consider the exact sequence

$$0 \to C_{\pi'}/C_{\pi'_0} \to C/C_{\pi'_0} \xrightarrow{p} C/C_{\pi'} \to 0.$$

By an easy long exact sequence argument, the canonical map

$$\operatorname{Hom}_C(\nabla_C(\lambda), C/C_{\pi'_0}) \to \operatorname{Hom}_C(\nabla_C(\lambda), C/C_{\pi'})$$

induced by p is an isomorphism. By the definition of C_π and the induction assumption, we have that the image $p(C_{\pi_0}/C_{\pi'_0})$ of $C_{\pi_0}/C_{\pi'_0}$ by p agrees with $C_\pi/C_{\pi'}$. This is equivalent to saying that $C_{\pi_0} + C_{\pi'} = C_\pi$.

Hence, even if we replace λ by μ and π' by π_0, the same construction implies $C_{\pi'} + C_{\pi_0} = C_\pi$. Hence, the definition of C_π is independent of choice of λ, and the proof of Proposition 2.2.12 is complete. The proof of Theorem 2.2.11 is also complete. □

2.3 Basic properties of the Donkin system

(2.3.1) Let R be a noetherian commutative ring, C an R-flat coalgebra, (X^+, Δ, ∇) a semisplit highest weight theory over C, and (C_π) a Donkin system of C associated with (X^+, Δ, ∇).

For a (possibly infinite) poset ideal π of X, we define $C_\pi := \varinjlim C_\rho$, where ρ runs through all finite poset ideals of π. It is easy to see that C_π is an R-flat R-subcoalgebra of C.

Lemma 2.3.2 *For any poset ideal π of X^+ and any $\mathfrak{p} \in \operatorname{Spec} R$, we have $C_\pi(\mathfrak{p}) = C(\mathfrak{p})(\pi)$.*

Proof. By Lemma 2.2.8, $(C_\pi(\mathfrak{p}))$ is a Donkin system of $C(\mathfrak{p})$. By the uniqueness Lemma 1.3.6, the assertion is obvious if π is finite. As the base change preserves inductive limits, the lemma is also true for general π. □

Lemma 2.3.3 $C_{X^+} = C$.

Proof. As the canonical injection $C_{X^+} \hookrightarrow C$ is R-pure and C_{X^+} is R-flat, we have that C/C_{X^+} is R-flat. By Lemma 2.3.2 and Lemma 1.3.6, **d**, we have $(C/C_{X^+})(\mathfrak{p}) = 0$ for any $\mathfrak{p} \in \operatorname{Spec} R$. Hence, by Corollary I.2.1.6, we have $C/C_{X^+} = 0$. □

Lemma 2.3.4 *For any poset ideal π of X^+, we have that C_π is R-countable and IFP, and hence it is R-projective. In particular, C is R-countable, IFP, and R-projective.*

Proof. It is clear that $C_\pi = \varinjlim C_\rho$ is IFP, and it is R-countable, since π is countable. It is R-projective by Lemma I.3.11.3. The last assertion is obvious by Lemma 2.3.3. □

An object of the category $\operatorname{add}\mathcal{F}(\Delta)$ (resp. $\operatorname{add}\mathcal{F}(\nabla)$) is called a Δ-*good* (resp. ∇-*good*) C-comodule, see (I.1.11). Note that Δ-good comodules and ∇-good comodules are R-finite R-projective. If π is a finite poset ideal of X^+, then C_π is ∇-good, and hence is R-finite R-projective. The dual algebra C_π^* of C_π is denoted by S_π. Note that S_π is Δ-good. By Lemma 2.2.4, if V is Δ-good and W is a filtered inductive limit of ∇-good comodules, then W is u-acyclic with respect to V.

Lemma 2.3.5 *Let π be a poset ideal of X^+. For $V \in \mathbb{M}^C$, the following are equivalent.*

1 *V is a C_π-comodule.*

2 *For any $\lambda \in X^+ \setminus \pi$, we have $\operatorname{Hom}_C(\Delta_C(\lambda), V) = 0$.*

Proof. **1⇒2** As $V \hookrightarrow V \otimes C_\pi$ is injective, we may assume that $V = V_0 \otimes C_\pi$ for some R-module V_0, after replacing V by $V \otimes C_\pi$. As C_π is u-acyclic with respect to $\Delta_C(\lambda)$ and it is easy to see that $\operatorname{Hom}_C(\Delta_C(\lambda), C_\pi) = 0$ for $\lambda \notin \pi$, this direction is obvious.

We prove **2⇒1**. Any (R-finite) C-subcomodule of V satisfies condition **2**, and if any R-finite C-subcomodule of V is a C_π-comodule, then V itself is a C_π-comodule. Hence, we may assume that V is R-finite. By Lemma 2.3.3, V is a $C_{\pi'}$-comodule for some finite poset ideal π' of X^+ containing π. As

$$0 \to \operatorname{Hom}_C(C_\pi^*, V) \to \operatorname{Hom}_C(C_{\pi'}^*, V) \to \operatorname{Hom}_C((C_{\pi'}/C_\pi)^*, V)$$

is exact and $\operatorname{Hom}_C((C_{\pi'}/C_\pi)^*, V) = 0$ by assumption, we have

$$\operatorname{ind}_C^{C_\pi} V = \operatorname{Hom}_C(C_\pi^*, V) \cong \operatorname{Hom}_C(C_{\pi'}^*, V) = V$$

by Corollary I.3.6.12. Hence, V is a C_π-comodule. □

Corollary 2.3.6 *We have* $\nabla_C(\lambda), \Delta_C(\lambda) \in \mathbb{M}^{C_{(-\infty,\lambda]}}$.

Proof. The assertion for $\nabla_C(\lambda)$ is obvious by the lemma. Hence, by Lemma 2.2.9, we have $\Delta_C(\lambda)^* \in \mathbb{M}^{C^{\mathrm{op}}_{(-\infty,\lambda]}}$. This shows $\Delta_C(\lambda) = \Delta_C(\lambda)^{**} \in \mathbb{M}^{C_{(-\infty,\lambda]}}$. □

Corollary 2.3.7 *If π is a poset ideal of X^+, then $\mathbb{M}^{C_\pi}$ is closed under extensions in $\mathbb{M}^C$.*

Proof. Follows immediately from the lemma. □

Proposition 2.3.8 *Let V be a C-comodule. Then the following are equivalent.*

1 *For any $\lambda \in X^+$ and $i > 0$, we have* $\operatorname{Ext}^i_C(\Delta_C(\lambda), V) = 0$.

2 *For any poset ideal π of X^+ and $i > 0$, we have* $R^i \operatorname{ind}_C^{C_\pi} V = 0$.

3 *For any $\lambda \in X^+$, we have* $\operatorname{Ext}^1_C(\Delta_C(\lambda), V) = 0$.

4 *For any finite poset ideal π of X^+, we have* $R^1 \operatorname{ind}_C^{C_\pi} V = 0$.

5 *For any bijective order-preserving map $f : X^+ \to [X^+]$, there exists a filtration $0 = V_0 \subset V_1 \subset \cdots \subset V$ such that the following conditions are satisfied:*

a $\varinjlim V_i = V$.

b *For $i \geq 1$, there exists an isomorphism of C-comodules*

$$V_i/V_{i-1} \cong R(\lambda(i))^* \otimes \operatorname{Hom}_C(\Delta_C(\lambda(i)), V) \otimes \nabla_C(\lambda(i)),$$

where $\lambda(i) := f^{-1}(i)$ and $R(\lambda(i)) := \operatorname{Hom}_C(\Delta_C(\lambda(i)), \nabla_C(\lambda(i)))$.

If the conditions are satisfied, then a filtration which satisfies **a,b** *in* **5** *is unique. More precisely, if $f : X^+ \to [X^+]$ is an order-preserving bijective map and*

$$0 = V_0 \subset V_1 \subset V_2 \subset \cdots \subset V$$

is a filtration of V which satisfies **a** *and*

b′ *For $i \geq 1$, $V_i/V_{i-1} \cong M_i \otimes \nabla_C(\lambda(i))$ for some R-module M_i,*

then we have $V_i = \operatorname{ind}_C^{C_{\pi(i)}} V$, where $\pi(i) := f^{-1}[1, i]$.

A C-comodule is called *good* if the conditions above are satisfied.

Proof. For any finite poset ideal π of X^+,

$$R^i \operatorname{ind}_C^{C_\pi}(?) = \operatorname{Cotor}_C^i(?, C_\pi) \cong \operatorname{Ext}_C^i(S_\pi, ?)$$

by Lemma I.3.6.16. As $S_\pi \in \operatorname{add} \mathcal{F}(\Delta)$, **1⇒2** and **3⇒4** are obvious.

1⇒3 and **2⇒4** are trivial.

4⇒5. To verify the existence, set for $i \geq 0$, $\pi(i) := f^{-1}([1, i])$ and

$$V_i := \operatorname{res}_C^{C_{\pi(i)}} \operatorname{ind}_C^{C_{\pi(i)}}(V).$$

It is obvious that $0 = V_0 \subset V_1 \subset V_2 \subset \cdots \subset V$. When we set $\lambda(i) := f^{-1}(i)$ for $i \in [X^+]$, then we have an exact sequence

$$0 \to R(\lambda) \otimes \nabla_C(\lambda(i))^* \otimes \Delta_C(\lambda(i)) \to S_{\pi(i)} \to S_{\pi(i-1)} \to 0.$$

Set $N_i := \operatorname{Hom}_C(R(\lambda) \otimes \nabla_C(\lambda(i))^* \otimes \Delta_C(\lambda(i)), V)$. Then

$$0 \to V_{i-1} \to V_i \to N_i \to R^1 \operatorname{ind}_C^{C_{\pi(i-1)}}(V) = 0$$

is exact. As

$$V_i/V_{i-1} \cong N_i \cong R(\lambda)^* \otimes \operatorname{Hom}_C(\Delta_C(\lambda(i)), V) \otimes \nabla_C(\lambda(i))$$

as C-comodules, condition **b** is satisfied.

We prove condition **a**. Let W be an R-finite C-subcomodule of V. Then the image of $\omega : W \to W \otimes C$ is contained in $W \otimes C_\pi$ for some finite poset ideal π of X^+, by Lemma 2.3.3. As π is contained in $\pi(i)$ for some i, we have $W = \operatorname{ind}_C^{C_{\pi(i)}} W \subset V_i$ for some i. Hence, $\varinjlim V_i$ contains any R-finite

C-subcomodule of V, and it agrees with V. Now the conditions **a** and **b** are verified for $V_i := \operatorname{res}_C^{C_{\pi(i)}} \operatorname{ind}_C^{C_{\pi(i)}}(V)$, and the existence has been proved.

5⇒1 As $\nabla_C(\mu)$ is u-acyclic with respect to $\Delta_C(\lambda)$ for any $\lambda, \mu \in X^+$, we have

$$\operatorname{Ext}^i_C(\Delta_C(\lambda), V_j/V_{j-1}) = 0 \qquad (i, j > 0).$$

Hence, by Proposition I.3.7.4,

$$\operatorname{Ext}^i_C(\Delta_C(\lambda), V) \cong \varinjlim \operatorname{Ext}^i_C(\Delta_C(\lambda), V_j) = 0 \qquad (i > 0).$$

Now we have that **1**–**5** are equivalent. We prove the uniqueness of the filtration in **5**. To verify this, assuming that (V_i) is a filtration which satisfies conditions **a**, **b**$'$ and $V_0 = 0$, it suffices to prove $V_i = \operatorname{ind}_C^{C_{\pi(i)}} V$ by induction on i.

If $i = 0$, then the assertion is obvious. Let $i > 0$, and assume that the uniqueness holds up to $i - 1$. By induction assumption, V_{i-1} is a $C_{\pi(i)}$-comodule, and by assumption **b**$'$ and Corollary 2.3.6, V_i/V_{i-1} is also a $C_{\pi(i)}$-comodule. By Corollary 2.3.7, V_i is a $C_{\pi(i)}$-comodule. Hence, $V_i \subset \operatorname{ind}_C^{C_{\pi(i)}} V$.

If $j \le i$, then $\operatorname{Hom}_C(\Delta_C(\lambda(j)), V/V_i) = 0$. This is because

$$\operatorname{Hom}_C(\Delta_C(\lambda(j)), V_l/V_i) = 0$$

for $l \ge i$ (proved easily by induction on l) and $\varinjlim V_l = V$. On the other hand, if $j > i$, then

$$\operatorname{Hom}_C(\Delta_C(\lambda(j)), \operatorname{ind}_C^{C_{\pi(i)}} V/V_i) = 0.$$

This is because $\operatorname{Ext}^1_C(\Delta_C(\lambda(j)), V_i) = 0$ by u-acyclicity of $\nabla_C(l)$ for $1 \le l \le i$ with respect to $\Delta_C(\lambda(j))$, and $\operatorname{Hom}_C(\Delta_C(\lambda(j)), \operatorname{ind}_C^{C_{\pi(i)}} V) = 0$ because $\lambda(j) \notin \pi$.

By Lemma 2.3.5, $\operatorname{ind}_C^{C_{\pi(i)}} V/V_i$ is a $C(\emptyset)$-comodule, and $V_i = \operatorname{ind}_C^{C_{\pi(i)}} V$. This completes the proof of the uniqueness. □

Remark 2.3.9 When we set $\Delta'_C(\lambda) := R(\lambda) \otimes \Delta_C(\lambda)$, then it is easy to see that (X^+, Δ', ∇) is a semisplit highest weight theory over C, and we have

$$R'(\lambda) := \operatorname{Hom}_C(R(\lambda) \otimes \Delta_C(\lambda), \nabla_C(\lambda)) \cong R(\lambda)^* \otimes R(\lambda) \cong R,$$

for any $\lambda \in X^+$. As we have

$$R(\lambda)^* \otimes \Delta_C(\lambda)^* \otimes \nabla_C(\lambda) \cong \Delta'_C(\lambda)^* \otimes \nabla_C(\lambda),$$

(C_π) is a Donkin system associated with (X^+, Δ, ∇) if and only if it is a Donkin system associated with (X^+, Δ', ∇).

Hence, making the additional assumption $\operatorname{Hom}_C(\Delta_C(\lambda), \nabla_C(\lambda)) \cong R$ is not essential.

Corollary 2.3.10 *A Donkin system of C associated with (X^+, Δ, ∇) is unique. Namely, if (C'_π) is a Donkin system associated with (X^+, Δ, ∇), then for any finite poset ideal π of X^+, we have $C'_\pi = C_\pi$.*

Proof. We have an order-preserving bijective map $f : X^+ \to [X^+]$ such that $f(\pi) = [1, \#\pi]$ by Lemma 1.1.2 (applied to π and $X^+ \setminus \pi$). As $C'_i := C'_{\pi(i)}$ ($\pi(i) := f^{-1}[1, i]$) satisfies the conditions **a** and **b**$'$ in Proposition 2.3.8 by the definition of Donkin system and Lemma 2.3.3, we have $C'_i = \operatorname{ind}_C^{C_{\pi(i)}} C = C_{\pi(i)}$. Hence, $C'_\pi = C'_{\#\pi} = C_{\pi(\#\pi)} = C_\pi$. □

(2.3.11) Now we have the following by Theorem 2.2.11, Lemma 2.3.4, and Corollary 2.3.10.

Theorem 2.3.12 *Let R be a noetherian commutative ring, C an R-flat coalgebra, and (X^+, Δ, ∇) a semisplit highest weight theory over C. Then the following are equivalent.*

1 *There is a Donkin system of C associated with (X^+, Δ, ∇).*

2 *There is a unique Donkin system of C associated with (X^+, Δ, ∇).*

3 *C is an R-projective module.*

□

(2.3.13) Let R be a noetherian commutative ring, C an R-projective R-coalgebra, and (X^+, Δ, ∇) a semisplit highest weight theory over C. Let (C_π) be the Donkin system of C associated with (X^+, Δ, ∇). For a (possibly infinite) poset ideal of π of X^+, we denote $\varinjlim C_\rho$ by C_π, where ρ runs through all finite poset ideals of π. We call C_π the *Donkin subcoalgebra* of C with respect to π. For a finite poset ideal π of X^+, we call the dual algebra S_π the *Schur algebra* of C with respect to π. These are determined only by (C, X^+, Δ, ∇).

Lemma 2.3.14 *Let V be a C-comodule. Then the following are equivalent.*

1 *For any $\lambda \in X^+$, V is u-acyclic with respect to $\Delta_C(\lambda)$.*

2 *V is good, and $V_\pi := \operatorname{ind}_C^{C_\pi} V \hookrightarrow V$ is R-pure for any finite poset ideal π of X^+.*

3 *There exists a filtration*

$$0 = V_0 \subset V_1 \subset V_2 \subset \cdots \subset V$$

of V such that each V_i is an R-pure submodule of V_{i+1}, $\varinjlim V_i = V$, and for each $i \geq 1$, there exist some $\lambda \in X^+$ and an R-module M such that $V_i/V_{i-1} \cong \nabla_C(\lambda) \otimes M$.

We say that V is *u-good* if the conditions above are satisfied.

Proof. **1⇒2** It is obvious that V is good. To prove the R-purity of V_π in V, we may assume that π is finite. By Lemma 2.2.4, V is u-acyclic with respect to S_π. Hence, for any R-module M,

$$V_\pi \otimes M = \operatorname{Hom}_C(S_\pi, V) \otimes M \cong \operatorname{Hom}_C(S_\pi, V \otimes M) = (V \otimes M)_\pi$$

is a C-subcomodule of $V \otimes M$ in a natural way. Hence, $V_\pi \hookrightarrow V$ is R-pure.

2⇒3 We take and fix an order-preserving bijective map $f : X^+ \to [X^+]$. Then $V_i = \operatorname{ind}_C^{C_{f^{-1}[1,i]}} V$ is an R-pure submodule of V. Hence, $V_i \hookrightarrow V_{i+1}$ is R-pure. By Proposition 2.3.8, the filtration $0 = V_0 \subset V_1 \subset \cdots$ is the desired one.

3⇒1 For any $\lambda \in X^+$, V_i/V_{i-1} is u-acyclic with respect to $\Delta_C(\lambda)$. As V_{i-1} is an R-pure submodule of V_i, we have that V_i is u-acyclic with respect to $\Delta_C(\lambda)$ by Lemma 2.2.4 and induction on i. Hence, $V = \varinjlim V_i$ is also u-acyclic with respect to $\Delta_C(\lambda)$, again by Lemma 2.2.4. □

The next is obvious.

Lemma 2.3.15 *The set of good C-comodules is closed under extensions, direct summands, monocokernels, and filtered inductive limits. C and $\nabla_C(\lambda)$ $(\lambda \in X^+)$ are u-good. The set of u-good C-comodules is closed under direct summands and filtered inductive limits. If*

$$0 \to W_1 \xrightarrow{i} W_2 \to W_3 \to 0$$

is an exact sequence of C-comodules, W_1 is u-good and i is R-pure, then W_2 is u-good if and only if W_3 is u-good. If V is a good (resp. u-good) C-comodule and π is a (possibly infinite) poset ideal of X^+, then $\operatorname{ind}_C^{C_\pi} V$ and $V/\operatorname{ind}_C^{C_\pi} V$ are good (resp. u-good). If V is u-good, then $V \otimes M$ is u-good for any R-module M. If V is u-good and $R \to R'$ is a noetherian commutative R-algebra, then $V \otimes R'$ is a u-good $C \otimes R'$-comodule.

Lemma 2.3.16 *For any poset ideal π of X^+ and $V \in \mathbb{M}^{C_\pi}$, we have $R^i \operatorname{ind}_C^{C_\pi} V = 0$ $(i > 0)$.*

Proof. By Corollary I.3.6.19, Lemma I.3.6.17 and Lemma I.3.6.16, we have isomorphisms

$$R^i \operatorname{ind}_C^{C_\pi} V \cong \operatorname{Cotor}^i_C(V, C_\pi) \cong \varinjlim \operatorname{Cotor}^i_C(V(\rho), C_\rho) \cong \varinjlim R^i \operatorname{ind}_C^{C_\rho} V(\rho),$$

where ρ runs through finite poset ideals of π. So replacing π by each ρ, we may assume that π is finite.

As C_π is u-good, it is u-acyclic with respect to S_π. Hence, the cobar resolution $\mathrm{Cobar}_{C_\pi} V$ of V as a C_π-comodule is an $\mathrm{ind}_C^{C_\pi}$-acyclic resolution of V as a C-comodule. On the other hand, we have

$$\mathrm{ind}_C^{C_\pi}(\mathrm{Cobar}_{C_\pi} V) = \mathrm{Cobar}_{C_\pi} V,$$

and hence we have $R^i \mathrm{ind}_C^{C_\pi} V = 0$ $(i > 0)$. □

Theorem 2.3.17 *Let π be a poset ideal of X^+, and V and W be C_π-comodules. Then the canonical map*

$$\mathrm{Ext}^i_{C_\pi}(V, W) \to \mathrm{Ext}^i_C(V, W)$$

is an isomorphism for $i \geq 0$.

Proof. Consider the isomorphism $\mathrm{Hom}_C(V, ?) \cong \mathrm{Hom}_{C_\pi}(V, ?) \circ \mathrm{ind}_C^{C_\pi}$ of functors on $\mathbb{M}^C$. As $\mathrm{ind}_C^{C_\pi}$ has an exact left adjoint $\mathrm{res}_C^{C_\pi}$, it preserves injectives. Hence, we have an isomorphism of derived functors

$$\underline{R}^+ \mathrm{Hom}_C(V, ?) \cong \underline{R}^+ \mathrm{Hom}_{C_\pi}(V, ?) \circ \underline{R}^+ \mathrm{ind}_C^{C_\pi}.$$

As we have $\underline{R}^+ \mathrm{ind}_C^{C_\pi} W = W$ by Lemma 2.3.16, we have the desired isomorphism, taking the cohomology. □

Corollary 2.3.18 *If π is a poset ideal of X^+, then $(\pi, \Delta(\pi), \nabla(\pi))$ is a semisplit highest weight theory over C_π, and its associated Donkin system is $(C_\rho)_\rho$, where $\Delta(\pi) = (\Delta_C(\lambda))_{\lambda\in\pi}$ and $\nabla(\pi) = (\nabla_C(\lambda))_{\lambda\in\pi}$, and ρ runs through all finite poset ideals of π.*

Corollary 2.3.19 *If V and W are R-finite C-comodules, then $\mathrm{Ext}^i_C(V, W)$ is R-finite for $i \geq 0$.*

Proof. As we have $C = C_{X^+}$, there exists a finite poset ideal π of X^+ such that both V and W are C_π-comodules. By the theorem and the R-linear equivalence ${}_{S_\pi}\mathbb{M} \cong \mathbb{M}^{C_\pi}$, we have R-isomorphisms

$$\mathrm{Ext}^i_C(V, W) \cong \mathrm{Ext}^i_{C_\pi}(V, W) \cong \mathrm{Ext}^i_{S_\pi}(V, W).$$

As S_π is an R-finite algebra, $\mathrm{Ext}^i_{S_\pi}(V, W)$ is R-finite, and so is $\mathrm{Ext}^i_C(V, W)$. □

Lemma 2.3.20 *If V and W are R-finite C-comodules, then the canonical map*

$$\varphi^i : \mathrm{Ext}^i_{\mathbb{M}^C_f}(V, W) \to \mathrm{Ext}^i_C(V, W)$$

is an isomorphism for $i \geq 0$.

Proof. We may assume that $i \geq 1$. There exists some finite poset ideal π of X^+ such that both V and W are C_π-comodules. For a finite poset ideal ρ of X^+ containing π, we have

$$\operatorname{Ext}^i_{\mathbb{M}_f^{C_\rho}}(V, W) \cong \operatorname{Ext}^i_{C_\rho}(V, W) \cong \operatorname{Ext}^i_C(V, W)$$

since $\mathbb{M}_f^{C_\rho} \cong {}_{S_\rho}\mathbb{M}_f$ contains enough S_ρ-projectives and by Theorem 2.3.17.

When we denote the canonical map $\varinjlim \operatorname{Ext}^i_{\mathbb{M}_f^{C_\rho}}(V, W) \to \operatorname{Ext}^i_{\mathbb{M}_f^C}(V, W)$ by ψ^i, then the last paragraph shows that $\varphi^i \circ \psi^i$ is an isomorphism. So it suffices to show that ψ^i is surjective. Let $\alpha \in \operatorname{Ext}^i_{\mathbb{M}_f^C}(V, W)$, and

$$0 \to W \to M_i \to \cdots \to M_1 \to V \to 0$$

be a representative of α. Then as there is some $\rho \supset \pi$ such that $\bigoplus_j M_j$ is a C_ρ-comodule, α is already realized as an exact sequence of C_ρ-comodules. Hence, ψ^i is surjective. □

3 Ringel's theory over a field

3.1 Ringel's approximation over a field

(3.1.1) Let k be a field, and $(C, X^+, \Delta, \nabla, L)$ a highest weight coalgebra over k.

Lemma 3.1.2 *If* $\lambda \in X^+$, *then we have* $\operatorname{inj.dim}_C \nabla_C(\lambda) \leq \operatorname{coht} \lambda$.

Proof. We may assume that $\operatorname{coht} \lambda < \infty$, and we prove the lemma by induction on $n := \operatorname{coht} \lambda$. Consider the exact sequence

$$0 \to \nabla_C(\lambda) \to Q_C(\lambda) \to Q_C(\lambda)/\nabla_C(\lambda) \to 0.$$

Then by Lemma 1.4.6, $Q_C(\lambda)/\nabla_C(\lambda)$ has a good filtration, and we have

$$[Q_C(\lambda)/\nabla_C(\lambda) : \nabla_C(\mu)] = 0$$

if $\mu \not> \lambda$. In particular, we have $\nabla_C(\lambda) = Q_C(\lambda)$ if $n = 0$, and the assertion holds.

Assume $n > 0$. If $\mu > \lambda$, then $\operatorname{coht} \mu < n$. Hence, for any finite dimensional C-comodule V, we have $\operatorname{Ext}^n_C(V, Q_C(\lambda)/\nabla_C(\lambda)) = 0$ by induction assumption. As $Q_C(\lambda)$ is C-injective, $\operatorname{Ext}^{n+1}_C(V, \nabla_C(\lambda)) = 0$. As $\mathbb{M}^C$ is locally noetherian, $\operatorname{inj.dim}_C \nabla_C(\lambda) \leq n$ by Lemma I.1.9.4. □

Lemma 3.1.3 *If* $\lambda \in X^+$, *then* $\operatorname{inj.dim}_C L_C(\lambda) \leq \operatorname{rank} X^+ + \operatorname{ht} \lambda$. *In particular,* $\operatorname{gl.dim} C \leq 2 \operatorname{rank} X^+$.

Proof. We may assume that $\operatorname{rank} X^+ < \infty$. We prove the first assertion by induction on $n = \operatorname{ht}\lambda$. The sequence

$$0 \to L_C(\lambda) \to \nabla_C(\lambda) \to \nabla_C(\lambda)/L_C(\lambda) \to 0$$

is exact, and any composition factor of $\nabla_C(\lambda)/L_C(\lambda)$ is of the form $L_C(\mu)$, with $\mu < \lambda$. Hence, $L_C(\lambda) = \nabla_C(\lambda)$ if $n = 0$, and the assertion holds by Lemma 3.1.2. If $n > 0$, then $\operatorname{inj.dim}_C L_C(\mu) \le \operatorname{rank} X^+ + n - 1$ for $\mu < \lambda$ by induction assumption. Combining this with $\operatorname{inj.dim}_C \nabla_C(\lambda) \le \operatorname{rank} X^+$, we are done.

We prove the last assertion. By the first part and Proposition I.3.7.4,

$$\operatorname{Ext}_C^{2\operatorname{rank} X^+ + 1}(V, W) = 0$$

for any finite dimensional C-comodule V and any C-comodule W. By Lemma I.1.9.4, $\operatorname{inj.dim}_C W \le 2\operatorname{rank} X^+$. □

We set $\mathcal{A}_C := \mathbb{M}_f^C$, $\mathcal{Y}_C := \mathcal{F}(\nabla)$, $\mathcal{X}_C := \mathcal{F}(\Delta)$, and $\omega_C := \mathcal{X}_C \cap \mathcal{Y}_C$.

Lemma 3.1.4 *A C-comodule V belongs to $\mathcal{Y}_C = \mathcal{F}(\nabla)$ if and only if V is good and finite dimensional.*

Proof. Obvious by Proposition 1.3.2. □

Lemma 3.1.5 *$\mathcal{Y}_C$ is closed under extensions, monocokernels, and direct summands. $\mathcal{X}_C$ is closed under extensions, epikernels, and direct summands.*

Proof. The first assertion is obvious by Lemma 3.1.4. We prove the second assertion. By Theorem 1.4.7, $(C^{\mathrm{op}}, X^+, \nabla^*, \Delta^*, L^*)$ is a highest weight coalgebra, and for $V \in \mathcal{A}_C$, we have $V \in \mathcal{X}_C$ if and only if $V^* \in \mathcal{Y}_{C^{\mathrm{op}}}$ by definition. By the first part, we are done. □

Lemma 3.1.6 *If $V \in \mathcal{A}_C$, then $\mathcal{X}_C$-$\operatorname{resol.dim} V < \infty$. Hence, we have $\hat{\mathcal{X}}_C = \mathcal{A}_C$.*

Proof. There exists a finite poset ideal π of X^+ such that $V \in \mathbb{M}^C(\pi)$. By Lemma 3.1.3 applied to $C(\pi)$, there is an $S_C(\pi)$-finite $S_C(\pi)$-projective resolution $\mathbb{P} \to V$ of V of length at most $2\operatorname{rank}\pi$. As we have $S_C(\pi) \in \mathcal{X}_C$, each term of $\mathbb{P}$ also belongs to $\mathcal{X}_C$ by Lemma 3.1.5, and we are done. □

Lemma 3.1.7 *Let $V \in \mathcal{A}_C$ and $\lambda \in X^+$. Set*

$$\mathcal{X} := \{W \in \mathcal{A}_C \mid \operatorname{Ext}_C^1(\Delta_C(\lambda), W) = 0\}.$$

Then there exists an exact sequence

$$\alpha : 0 \to V \xrightarrow{i} U \to D \to 0 \qquad (3.1.8)$$

in $\mathcal{A}_C$ such that $U \in \mathcal{X}$, $D \in \operatorname{add}\{\Delta_C(\lambda)\}$, and i is left minimal. If α is such an exact sequence, then i is the left minimal $\mathcal{X}$-approximation of V, and in particular, such an α is unique up to isomorphisms of complexes.

For the left minimality and left minimal $\mathcal{X}$-approximation, see (I.1.12). We call the exact sequence α in the lemma the *minimal λ-extension* of V.

Proof. We denote the finite dimensional division k-algebra $\mathrm{End}_C(\Delta_C(\lambda))$ by E, see Lemma 1.2.20.

By Corollary 2.3.19, $\mathrm{Ext}^1_C(\Delta_C(\lambda), V)$ is finite dimensional over k, and is also finite dimensional as a right E-vector space. We take a basis $\alpha_1, \ldots, \alpha_r$ of it as a right E-space. We set $D = \bigoplus_{i=1}^r \Delta_C(\lambda)$, and we define the exact sequence (3.1.8) to be the element of $\mathrm{Ext}^1_C(D, V)$ which corresponds to $\alpha := (\alpha_1, \ldots, \alpha_r)$ by the canonical isomorphism

$$\mathrm{Ext}^1_C(D, V) \cong \bigoplus_i \mathrm{Ext}^1_C(\Delta_C(\lambda), V).$$

As the induced map $\alpha' : \mathrm{Hom}_C(\Delta_C(\lambda), D) \to \mathrm{Ext}^1_C(\Delta_C(\lambda), V)$ is right E-linear, α' is surjective by the choice of α (hence it is also an isomorphism by dimension counting). As $\mathrm{Ext}^1_C(\Delta_C(\lambda), D) = 0$, we have $\mathrm{Ext}^1_C(\Delta_C(\lambda), U) = 0$, in other words, $U \in \mathcal{X}$. By the definition of $\mathcal{X}$, we have $\mathrm{Ext}^1_C(D, \mathcal{X}) = 0$, and hence i is a left $\mathcal{X}$-approximation by the dual assertion of Lemma I.1.12.1.

We show that i is left minimal. Let $\varphi \in \mathrm{End}_C U$ with $\varphi i = i$, and $\psi : D \to D$ be the map induced by φ. By the five lemma, it suffices to show that ψ is an isomorphism. As the functor $\mathrm{Hom}_C(\Delta_C(\lambda), ?) : \mathrm{add}(\Delta_C(\lambda)) \to \mathrm{add}\, E_E$ is an equivalence [12, Proposition II.2.1], it suffices to show that $\psi' := \mathrm{Hom}_C(\Delta_C(\lambda), \psi)$ is an isomorphism. As we have a map of exact sequences

$$\begin{array}{cccccccc} \alpha: & 0 \to & V & \xrightarrow{i} & U & \longrightarrow & D & \to 0 \\ & & \downarrow 1_V & & \downarrow \varphi & & \downarrow \psi & \\ \alpha: & 0 \to & V & \xrightarrow{i} & U & \longrightarrow & D & \to 0, \end{array}$$

we have a commutative diagram

$$\begin{array}{ccc} \mathrm{Hom}_C(\Delta_C(\lambda), D) & \xrightarrow{\alpha'} & \mathrm{Ext}^1_C(\Delta_C(\lambda), V) \\ \downarrow \psi' & & \downarrow \mathrm{id} \\ \mathrm{Hom}_C(\Delta_C(\lambda), D) & \xrightarrow{\alpha'} & \mathrm{Ext}^1_C(\Delta_C(\lambda), V). \end{array}$$

As $\alpha'\psi' = \alpha'$ and α' is an isomorphism, we have $\psi' = \mathrm{id}$, and in particular, ψ' is an isomorphism. □

Lemma 3.1.9 *Let $V \in \mathcal{A}_C$, and $\lambda \in X^+$. We set*

$$\mathcal{X} := \{W \in \mathcal{A}_C \mid \mathrm{Ext}^1_C(\Delta_C(\lambda), W) = 0\}.$$

Assume that

(3.1.10) $$\alpha : 0 \to V \xrightarrow{i} U \to D \to 0$$

is an exact sequence in $\mathcal{A}_C$ with $U \in \mathcal{X}$, and $D \cong \Delta_C(\lambda)^r$, where

$$r := \dim \mathrm{Ext}^1_C(\Delta_C(\lambda), V) / \dim \mathrm{End}_C \Delta_C(\lambda).$$

Then (3.1.10) *is a minimal λ-extension.*

Proof. Left to the reader as an exercise. □

Lemma 3.1.11 *If $V \in \mathcal{A}_C$, then there is an exact sequence*

$$0 \to V \xrightarrow{i} Y \to X \to 0$$

in $\mathcal{A}_C$ such that $Y \in \mathcal{Y}_C$, $X \in \mathcal{X}_C$, and i is a left minimal $\mathcal{Y}_C$-approximation, uniquely. More precisely, such an exact sequence is obtained as follows.

0 *There exists a finite poset ideal π of X^+ such that $\operatorname{Ext}^1_C(\Delta_C(\mu), V) = 0$ for $\mu \notin \pi$. Fix such a finite poset ideal π, and set $n := \#\pi$.*

1 *Fix an order-preserving bijective map $f : \pi \to [1, n]$. Set $\lambda(i) := f^{-1}(i)$.*

2 *Define exact sequences*

$$\alpha(i) : 0 \to V \xrightarrow{\rho(i)} Y(i) \to X(i) \to 0$$

for $i \in [1, n+1]$ by descending induction on i. Set $\alpha(n+1)$ to be the exact sequence

$$0 \to V \xrightarrow{\mathrm{id}} V \to 0 \to 0.$$

Assume that $i \in [1, n]$, and $\alpha(i+1)$ is already defined. Let

$$\beta(i) : 0 \to Y(i+1) \xrightarrow{\eta(i)} Y(i) \to D(i) \to 0$$

be the minimal $\lambda(i)$-extension of $Y(i+1)$. Set $\rho(i) := \eta(i) \circ \rho(i+1)$, and $X(i) := \operatorname{Coker} \rho(i)$. Now $\alpha(i)$ is defined.

3 *For $i \in [1, n+1]$, we have $X(i) \in \mathcal{F}(\{\Delta_C(\lambda(j)) \mid i \le j \le n\})$ and $\rho(i)$ is left minimal.*

4 *If $i \in [1, n+1]$ and $\lambda \in X^+ \setminus f^{-1}([1, i-1])$, then $\operatorname{Ext}^1_C(\Delta_C(\lambda), Y(i)) = 0$.*

5 *$\alpha(1)$ is the left minimal $\mathcal{Y}_C$-approximation, and $X(1) \in \mathcal{F}(\Delta(\pi))$.*

Proof. **0** Take π so that $V \in \mathbb{M}^{C(\pi)}$. Then the desired property is satisfied. **1** is always possible. **2** makes sense, by the uniqueness of the minimal $\lambda(i)$-extension.

We prove **3** by descending induction on i. If $i = n+1$, then the assertion is trivial, so assume that $i \in [1, n]$. Note that $D(i) \in \operatorname{add} \Delta_C(\lambda(i))$. As there is an exact sequence

$$0 \to X(i+1) \to X(i) \to D(i) \to 0,$$

the first assertion follows.

We are to prove that $\rho(i)$ is minimal. So let $\varphi : Y(i) \to Y(i)$ be a C-comodule map such that $\varphi\rho(i) = \rho(i)$. As we have $\lambda(j) \notin (-\infty, \lambda(i)]$

for $j > i$, we have $\mathrm{Hom}_C(X(i+1), D(i)) = 0$ by the first assertion. The composite

$$Y(i+1) \xrightarrow{\eta(i)} Y(i) \xrightarrow{\varphi} Y(i) \to D(i)$$

is zero on $V = \mathrm{Im}\,\rho(i+1)$. The induced map $X(i+1) = \mathrm{Coker}\,\rho(i+1) \to D(i)$ is also zero, so the composite above is zero. It follows that φ maps $Y(i+1)$ to $Y(i+1)$. As $\rho(i+1)$ is left minimal, the restriction $\psi := \varphi|_{Y(i+1)}$ is an isomorphism.

As $\beta(i)$ is a minimal $\lambda(i)$-extension, we have $\mathrm{Ext}^1_C(D(i), Y(i+1)) = 0$, and hence ψ^{-1} is extended to a C-comodule map $\varphi' : Y(i) \to Y(i)$. By the left minimality of $\eta(i)$, both $\varphi \circ \varphi'$ and $\varphi' \circ \varphi$ are isomorphisms. Hence, φ is an isomorphism, and hence $\rho(i)$ is left minimal.

4 We prove the assertion by descending induction on i. If $i = n+1$, then the assertion is trivial. Assume that $i \leq n$. If $\lambda = \lambda(i)$, then as $\beta(i)$ is the left minimal λ-extension, the assertion is obvious. So we may assume that $\lambda \in X^+ \setminus f^{-1}[1, i]$. By induction assumption, we have $\mathrm{Ext}^1_C(\Delta_C(\lambda), Y(i+1)) = 0$. On the other hand, since $D(i) \in \mathrm{add}\,\Delta_C(\lambda(i))$ and $\lambda \not< \lambda(i)$, we have $\mathrm{Ext}^1_C(\Delta_C(\lambda), D(i)) = 0$. Thanks to the exact sequence $\beta(i)$, we have the desired assertion.

5 is now obvious, and the lemma follows. □

For $V \in \mathcal{A}_C$, we denote X and Y in the lemma, which are uniquely determined only by V, by X^V and Y^V, respectively.

Theorem 3.1.12 (Ringel) *$(\mathcal{X}_C, \mathcal{Y}_C, \omega_C)$ is an Auslander–Buchweitz context in $\mathcal{A}_C$.*

Proof. As we know the condition $\mathcal{A}_C = \hat{\mathcal{X}}_C$ by Lemma 3.1.6, it suffices to check that $(\mathcal{X}_C, \mathcal{Y}_C, \omega_C)$ is a weak Auslander–Buchweitz context. Namely, it suffices to check the conditions **AB1**–**AB3** in Theorem I.1.12.10.

AB1 is contained in Lemma 3.1.5. **AB2** is also trivial by Lemma 3.1.5 and $\mathcal{A}_C = \hat{\mathcal{X}}_C$. As $\omega_C = \mathcal{X}_C \cap \mathcal{Y}_C$ is the definition, it remains to prove that ω_C is an injective cogenerator of $\mathcal{X}_C$. Let $V \in \mathcal{X}_C$. Then $Y^V \in \mathcal{Y}_C$. On the other hand, as $X^V \in \mathcal{X}_C$ and $\mathcal{X}_C$ is closed under extensions, $Y^V \in \mathcal{X}_C$. Hence, $Y^V \in \omega_C$ with $X^V \in \mathcal{X}_C$, and ω_C is a cogenerator of $\mathcal{X}_C$. As $\mathrm{Ext}^i_C(\mathcal{X}_C, \mathcal{Y}_C) = 0$ $(i \geq 1)$ and $\omega_C \subset \mathcal{Y}_C$, ω_C is $\mathcal{X}_C$-injective. □

Note that as $\mathrm{End}_C\,V$ is a finite dimensional k-algebra (hence is semiperfect) for any $V \in \mathcal{A}_C$, we have that any $V \in \mathcal{A}_C$ admits a unique minimal $\mathcal{X}_C$-approximation and a minimal $\mathcal{Y}_C$-hull.

3.2 Tilting modules over a field

(3.2.1) Let k be a field, and $(C, X^+, \Delta, \nabla, L)$ a highest weight coalgebra over k. Set $\mathcal{X}_C := \mathcal{F}(\Delta)$, $\mathcal{Y}_C := \mathcal{F}(\nabla)$, and $\omega_C := \mathcal{X}_C \cap \mathcal{Y}_C$.

Lemma 3.2.2 *Let $\lambda \in X^+$,*

$$0 \to \Delta_C(\lambda) \to T_C(\lambda) \to X_C(\lambda) \to 0$$

be the minimal $\mathcal{Y}_C$-hull of $\Delta_C(\lambda)$, and

$$0 \to Y_C(\lambda) \to T'_C(\lambda) \to \nabla_C(\lambda) \to 0$$

be the minimal $\mathcal{X}_C$-approximation of $\nabla_C(\lambda)$. Then we have:

(3.2.3) *$T_C(\lambda)$ is an indecomposable object of ω_C.*

(3.2.4) *$(T_C(\lambda) : L_C(\lambda)) = 1$, and $(T_C(\lambda) : L_C(\mu)) \neq 0$ implies $\mu \leq \lambda$.*

(3.2.5) *Any indecomposable object of ω_C is isomorphic to $T_C(\lambda)$ for some λ. More precisely, if T is an indecomposable object of ω_C, then the set $\{\mu \in X^+ \,|\, (T : L_C(\mu)) \neq 0\}$ has a maximum element, say λ, and we have $T \cong T_C(\lambda)$.*

(3.2.6) *$T'_C(\lambda) \cong T_C(\lambda)$.*

(3.2.7) *Assume that C is finite dimensional, and let A denote the quasi-hereditary algebra C^*. When we set $T := \bigoplus_{\lambda \in X^+} T_C(\lambda)$, then T is a basic tilting–cotilting module of A, and we have $\operatorname{add} T = \omega_A$.*

Proof. Set $\pi := (-\infty, \lambda)$. Then $\operatorname{Ext}^1_C(\Delta_C(\mu), \Delta_C(\lambda)) = 0$ for $\mu \notin \pi$. By Lemma 3.1.11 $X_C(\lambda) \in \mathcal{F}(\{\Delta_C(\mu) \,|\, \mu \in \pi\})$. Hence, $(X_C(\lambda) : L_C(\mu)) = 0$ for $\mu \not< \lambda$. This proves (3.2.4).

Next, we decompose $T_C(\lambda) \cong T_1 \oplus \cdots \oplus T_r$, with T_i indecomposable. This is possible, see Lemma I.1.12.6. As $(T_C(\lambda) : L_C(\lambda)) = 1$, we may assume $(T_1(\lambda) : L_C(\lambda)) = 1$ and $(T_i(\lambda) : L_C(\lambda)) = 0$ $(i \geq 2)$. Then as

$$\operatorname{Hom}_C(\Delta_C(\lambda), T_i) = 0 \qquad (i \geq 2),$$

we must have $i = 1$, by the left minimality of $\Delta_C(\lambda) \to T_C(\lambda)$. Hence, (3.2.3) holds.

We prove (3.2.5). Set

$$\omega' := \operatorname{add}\{T_C(\lambda) \,|\, \lambda \in X^+\}.$$

Then by Corollary I.1.12.13 applied to $\mathcal{X}_0 := \Delta$, ω' is an injective cogenerator of $\mathcal{X}_C$. By Theorem I.1.12.10, **2**, $\omega' = \operatorname{add} \omega' = \omega$. Now the first assertion follows from the Krull–Schmidt theorem, Lemma I.1.12.6. The last assertion is an immediate consequence of the first assertion and (3.2.4).

We prove (3.2.6). The exact sequence

$$0 \to \nabla_C(\lambda)^* \to T'_C(\lambda)^* \to Y_C(\lambda)^* \to 0$$

is a minimal $\mathcal{X}_C^*$-hull of the Weyl module $\nabla_C(\lambda)^*$ for the highest weight coalgebra

$$(C^{\mathrm{op}}, X^+, \nabla^*, \Delta^*, L^*).$$

Hence, $T'_C(\lambda) \in \omega_C$ is indecomposable, and the set $\{\mu \in X^+ \mid (T'_C(\lambda) : L_C(\mu)) \neq 0\}$ has λ as its maximum element. Hence, we have $T'_C(\lambda) \cong T_C(\lambda)$ by (3.2.5).

We prove (3.2.7). It is clear by Theorem I.4.10.22 that T is the (uniquely determined) basic cotilting module of A which satisfies $\operatorname{add} T = \omega_A$. On the other hand, we have $\operatorname{add} T^* = \omega_{A^{\mathrm{op}}}$, and T is also a tilting module. □

We call $T_C(\lambda)$ the indecomposable *tilting module* of C of highest weight λ.

Proposition 3.2.8 *Let k be a field, C a k-coalgebra, and (X^+, Δ, ∇) a split highest weight theory over C. Let π be a poset ideal of X^+, and*

$$\mathbb{F} : 0 \to V \to X \to Y \to 0$$

a sequence in $\mathbb{M}^C(\pi)_f$. Let k' be an extension field of k, and $?'$ denote the functor $? \otimes k'$. Then the following are equivalent.

1 *$\mathbb{F}$ is a minimal $\mathcal{Y}_{C(\pi)}$-hull of V in $\mathbb{M}^C(\pi)_f$.*

2 *$\mathbb{F}$ is a minimal $\mathcal{Y}_C$-hull of V in $\mathbb{M}^C_f$.*

3 *$\mathbb{F}^*$ is a minimal $\mathcal{X}_{C^{\mathrm{op}}}$-approximation of V^* in $\mathbb{M}^{C^{\mathrm{op}}}_f$.*

4 *$\mathbb{F}'$ is a minimal $\mathcal{Y}_{C'}$-hull of V' in $\mathbb{M}^{C'}_f$.*

Proof. Note that $\mathbb{F}$ is exact if and only if $\mathbb{F}^*$ is exact if and only if $\mathbb{F}'$ is exact. For $X \in \mathbb{M}^C(\pi)_f$, $X \in \mathcal{X}_C$ if and only if $X \in \mathcal{X}_{C(\pi)}$ if and only if $X^* \in \mathcal{Y}_{C^{\mathrm{op}}}$. As we have

$$\operatorname{Ext}^1_{C'}(X', \nabla_{C'}(\lambda)) \cong \operatorname{Ext}^1_C(X, \nabla_C(\lambda))'$$

for any $\lambda \in X^+$, and $(?)'$ is faithful exact, $X \in \mathcal{X}_C$ if and only if $X' \in \mathcal{X}_{C'}$. As we have a similar equivalence for Y, we have the desired equivalence, except for the assertions on the minimality. As for minimality, **1**$\Leftrightarrow$**2**$\Leftrightarrow$**3**$\Leftarrow$**4** is obvious.

We prove **2**$\Rightarrow$**4**. By Lemma 3.1.11, it suffices to prove that, for $V \in \mathcal{A}_C$, if

$$\mathbb{G} : 0 \to V \to W \to D \to 0$$

is a minimal λ-extension of V, then $\mathbb{G}'$ is a minimal λ-extension of V'. By assumption, we have $\operatorname{Ext}^1_C(\Delta_C(\lambda), W) = 0$ and $D \cong \Delta_C(\lambda)^{\oplus r}$, where

$r = \dim_k \mathrm{Ext}^1_C(\Delta_C(\lambda), V)$ (note that $k \cong \mathrm{End}_C \Delta_C(\lambda)$, as we consider a *split* highest weight coalgebra). Then we have

$$\mathrm{Ext}^1_{C'}(\Delta_{C'}(\lambda), W') \cong \mathrm{Ext}^1_C(\Delta_C(\lambda), W)' = 0,$$

and $D' \cong \Delta_{C'}(\lambda)^{\oplus r}$. As we have

$$\dim_{k'} \mathrm{Ext}^1_{C'}(\Delta_{C'}(\lambda), V') = \dim_{k'} \mathrm{Ext}^1_C(\Delta_C(\lambda), V)' = r,$$

$\mathbb{G}'$ is a minimal λ-extension of V' by Lemma 3.1.9. □

Corollary 3.2.9 *Under the assumptions of the proposition, we have the following for $\lambda \in X^+$.*

1 *For any poset ideal π of X^+ such that $\lambda \in \pi$, we have $T_{C(\pi)}(\lambda) \cong T_C(\lambda)$.*

2 $T_C(\lambda)^* \cong T_{C^{\mathrm{op}}}(\lambda)$.

3 *If k' is an extension field of k, then we have $T_{C\otimes k'}(\lambda) \cong T_C(\lambda) \otimes k'$.*

Remark 3.2.10 All the results in (III.1) and (III.3) are well-known as the theory of highest weight category and the theory of quasi-hereditary algebras, see [37], [45], [44], and [52]. See the appendix of [52], which contains substantial historical remarks. We have used thorough coalgebra–comodule approach, which is useful in generalizing the base ring to an arbitrary commutative noetherian ring.

4 Ringel's theory over a commutative ring

4.1 Tilting modules over a commutative ring

(4.1.1) Let R be a noetherian commutative ring, and C an R-projective R-coalgebra. Let (X^+, Δ, ∇) be a semisplit highest weight theory over C, and $(C_\pi)_\pi$ its associated Donkin system. We set $\mathcal{A}_C := \mathbb{M}^C_f$.

Lemma 4.1.2 *Let A be a left noetherian R-algebra, R' an R-flat commutative R-algebra, $M \in {}_A\mathbb{M}_f$, and $N \in {}_A\mathbb{M}$. Then the canonical R'-linear map*

$$\mathrm{Ext}^i_A(M, N)' \to \mathrm{Ext}^i_{A'}(M', N')$$

is an isomorphism for $i \geq 0$, where $(?)'$ denotes the exact functor $? \otimes R'$.

Proof. The case $i = 0$ and $M = A$ is trivial. The case $i = 0$ follows from the five lemma. Let $\mathbb{F}$ be a projective resolution of M with each term A-finite. Then we have

$$H^i(\mathrm{Hom}_A(\mathbb{F}, N))' \cong H^i(\mathrm{Hom}_A(\mathbb{F}, N)') \cong H^i(\mathrm{Hom}_{A'}(\mathbb{F}', N')),$$

and as $\mathbb{F}'$ is an A'-projective resolution of M' by flatness, we are done. □

Lemma 4.1.3 ([38, Lemma 3.3.2]) *Let A be an R-finite projective R-algebra, and J an R-finite projective A-module. If $J(\mathfrak{m})$ is $A(\mathfrak{m})$-projective for any $\mathfrak{m} \in \operatorname{Max} R$, then J is A-projective.*

Proof. By Lemma 4.1.2, we may assume that $(R, \mathfrak{m})$ is a complete local ring. Then as A is semiperfect, there is an A-projective cover $\varphi : P \to J$ of J. As $P(\mathfrak{m}) \to J(\mathfrak{m})$ splits as an $A(\mathfrak{m})$-linear map by assumption, $P(\mathfrak{m}) \to J(\mathfrak{m})$ is an isomorphism, since $\operatorname{top} P(\mathfrak{m}) \to \operatorname{top} J(\mathfrak{m})$ is an isomorphism by Lemma I.1.12.5. By Lemma I.2.1.4, φ is injective, and we have $P \cong J$. □

Corollary 4.1.4 *Let π be a finite poset ideal of X^+, and λ its maximal element. Then $\Delta_C(\lambda)$ is $S_C(\pi)$-projective.*

Proof. The case R is a field is obvious by Definition 1.1.4 and Theorem 2.3.17. The general case follows from the lemma. □

Corollary 4.1.5 *Let π be a finite poset ideal of X^+, let $V \in \mathbb{M}^{C_\pi}$, and assume either λ is a maximal element of π or $\lambda \in X^+ \setminus \pi$. Then we have $\operatorname{Ext}^i_C(\Delta_C(\lambda), V) = 0$ for $i > 0$.*

Proof. We set $\rho := \pi \cup (-\infty, \lambda]$. Then ρ is a finite poset ideal of X^+, and λ is a maximal element of ρ. By Corollary 4.1.4 and Theorem 2.3.17, we have

$$\operatorname{Ext}^i_C(\Delta_C(\lambda), V) \cong \operatorname{Ext}^i_{C_\rho}(\Delta_C(\lambda), V) = 0 \qquad (i > 0).$$

□

Lemma 4.1.6 *Let π be a finite poset ideal of X^+, and V an R-finite $S_C(\pi)$-module, and $r \geq 0$. If $\operatorname{proj.dim}_R V \leq r$, then we have $\operatorname{proj.dim}_{S_C(\pi)} V \leq 2 \operatorname{rank} \pi + r$. In particular, we have $\operatorname{proj.dim}_{S_C(\pi)} V < \infty$.*

Proof. Let

$$\mathbb{F} : \cdots \to F_n \xrightarrow{d_n} F_{n-1} \to \cdots \to F_1 \to F_0 \to V \to 0$$

be an $S_C(\pi)$-projective resolution of V with each term $S_C(\pi)$-finite. We set $\Omega_i := \operatorname{Im} d_i$. Then as Ω_r is R-finite projective, replacing V by Ω_r (and r by 0), we may assume that $r = 0$, namely, V is R-projective. Then for each $\mathfrak{m} \in \operatorname{Max} R$,

$$0 \to \Omega_{2\operatorname{rank}\pi}(\mathfrak{m}) \to F_{2\operatorname{rank}\pi-1}(\mathfrak{m}) \to \cdots \to F_1(\mathfrak{m}) \to F_0(\mathfrak{m}) \to V(\mathfrak{m}) \to 0$$

is exact. By Lemma 3.1.3, $\operatorname{gl.dim} S_C(\pi)(\mathfrak{m}) \leq 2 \operatorname{rank} \pi$, and hence $\Omega_{2\operatorname{rank}\pi}(\mathfrak{m})$ is $S_C(\pi)(\mathfrak{m})$-projective. By Lemma 4.1.3, $\Omega_{2\operatorname{rank}\pi}$ is $S_C(\pi)$-projective, and hence $\operatorname{proj.dim}_{S_C(\pi)} V \leq 2 \operatorname{rank} \pi$. □

Theorem 4.1.7 *Let V and W be R-finite R-projective C-comodules. Then we have the following.*

1 *Let n be a non-negative integer. If $\operatorname{Ext}^i_{C(\mathfrak{m})}(V(\mathfrak{m}), W(\mathfrak{m})) = 0$ $(i \geq n)$ for any $\mathfrak{m} \in \operatorname{Max} R$, then we have $\operatorname{Ext}^i_{C \otimes R'}(V \otimes R', W \otimes M) = 0$ $(i \geq n)$ for any R-algebra R' and any R'-module M.*

2 *The following are equivalent.*

a $\operatorname{Ext}^i_C(V, W) = 0$ $(i > 0)$.

b *W is u-acyclic with respect to V, and $\operatorname{Hom}_C(V, W)$ is R-finite projective.*

c *For any $\mathfrak{m} \in \operatorname{Max} R$, we have $\operatorname{Ext}^i_{C(\mathfrak{m})}(V(\mathfrak{m}), W(\mathfrak{m})) = 0$ $(i > 0)$.*

Proof. Let π be a finite poset ideal of X^+ such that both V and W are C_π-comodules, or $S_C(\pi)$-modules. By Lemma 4.1.6, there is an $S_C(\pi)$-finite projective resolution $\mathbb{F}$ of V. Set $\mathbb{G} := \operatorname{Hom}_{S_C(\pi)}(\mathbb{F}, W)$. Then $\mathbb{G}$ is a perfect R-complex (i.e., a bounded complex of R-modules with each term R-finite projective).

As can be seen easily, utilizing Theorem 2.3.17, we have

$$\operatorname{Ext}^i_{C \otimes R'}(V \otimes R', W \otimes M) \cong H^i(\operatorname{Hom}_{S_C(\pi) \otimes R'}(\mathbb{F} \otimes R', W \otimes M)) \cong H^i(\mathbb{G} \otimes M)$$

for any commutative R-algebra R', any R'-module M, and any i. All the assertions follow immediately from Proposition 2.1.14, applied to $\mathbb{G}$. □

Corollary 4.1.8 *Let V be an R-finite R-projective C-comodule. Then the following are equivalent.*

1 *V is good.*

2 *V is u-good.*

2′ $V \in \mathcal{F}(\operatorname{add} \nabla)$.

2′* $V^* \in \mathcal{F}(\operatorname{add} \Delta_{C^{\mathrm{op}}})$.

3 *V is ∇-good, i.e., $V \in \operatorname{add} \mathcal{F}(\nabla)$.*

3* *V^* is Δ-good, i.e., $V^* \in \operatorname{add} \mathcal{F}(\Delta_{C^{\mathrm{op}}})$.*

4 *For any $\mathfrak{m} \in \operatorname{Max} R$, $V(\mathfrak{m})$ is good.*

4* *For any $\mathfrak{m} \in \operatorname{Max} R$, $V^*(\mathfrak{m}) \in \mathcal{F}(\Delta_{C^{\mathrm{op}}(\mathfrak{m})})$.*

Proof. The equivalence **1⇔2⇔4** is an immediate consequence of the theorem.

2⇒2′ Assume that V is u-good. Then there is a finite filtration

$$0 = W_0 \subset W_1 \subset \cdots \subset W_n = V$$

of V such that each W_i is an R-pure C-subcomodule of V, and $W_i/W_{i-1} \cong \nabla_C(\mu(i)) \otimes \operatorname{Hom}_C(\Delta_C(\mu(i)), V) \otimes R(\mu(i))^*$ for each i. Set

$$M_i := \operatorname{Hom}_C(\Delta_C(\mu(i)), V) \otimes R(\mu(i))^*.$$

Then M_i is R-finite. On the other hand, W_i/W_{i-1} is R-finite projective, because of the finite-projectivity of V and the purity of W_i and W_{i-1}. As $\nabla_C(\mu(i))_{\mathfrak{m}}$ is a *non-zero* $R_{\mathfrak{m}}$-free module for any $\mathfrak{m} \in \operatorname{Max} R$, M_i is also R-finite projective. Hence, **2′** is proved.

We prove **2′ ⇒3**. As $\nabla \subset \mathcal{F}(\nabla)$, it follows that $\operatorname{add} \nabla \subset \operatorname{add} \mathcal{F}(\nabla)$. As $\operatorname{add} \mathcal{F}(\nabla)$ is closed under extensions (I.1.11), we have $\mathcal{F}(\operatorname{add} \nabla) \subset \operatorname{add} \mathcal{F}(\nabla)$.

3⇒1 This is clear, since objects of ∇ are good, and good comodules are closed under extensions and direct summands.

3⇔3*, **2′ ⇔2′*** and **4⇔4*** are obvious. □

Lemma 4.1.9 *Let $V \in \mathcal{A}_C$ and $\lambda \in X^+$. We set*

$$\mathcal{X} := \{W \in \mathcal{A}_C \mid \operatorname{Ext}^1_C(\Delta_C(\lambda), W) = 0\}.$$

Then

1 *There exists an exact sequence*

$$\alpha : 0 \to V \xrightarrow{i} U \to D \to 0 \qquad (4.1.10)$$

in $\mathcal{A}_C$ such that $U \in \mathcal{X}$ and D is a finite direct sum of copies of $\Delta_C(\lambda)$.

2 *If, moreover, $(R, \mathfrak{m})$ is local, then we can take i to be left minimal in* **1**. *Then i is a left minimal $\mathcal{X}$-approximation of V, and in particular, such an α is unique up to isomorphisms of complexes.*

We call an exact sequence α which satisfies condition **1** a λ-*extension* of V. If, moreover, i in the exact sequence is left minimal, then it is called the *minimal λ-extension* of V.

Proof. Set $E := \operatorname{End}_C(\Delta_C(\lambda))$. By Theorem 4.1.7, $\Delta_C(\lambda)$ is u-acyclic with respect to itself, and the canonical map $R \to E = \operatorname{End}_C(\Delta_C(\lambda))$ is an isomorphism.

Note that $\operatorname{Ext}^1_C(\Delta_C(\lambda), V)$ is R-finite by Corollary 2.3.19. Let $\alpha_1, \ldots, \alpha_r$ be a generating set of $\operatorname{Ext}^1_C(\Delta_C(\lambda), V)$ as an R-module. If $(R, \mathfrak{m})$ is local, we take this as a minimal set of generators.

We set $D = \bigoplus_{i=1}^r \Delta_C(\lambda)$, and we define the extension (4.1.10) to be the element corresponding to $\alpha = (\alpha_1, \ldots, \alpha_r)$ by the canonical isomorphism

$$\operatorname{Ext}^1_C(D, V) \cong \bigoplus_i \operatorname{Ext}^1_C(\Delta_C(\lambda), V).$$

As the Yoneda product from the left

$$\alpha' : \operatorname{Hom}_C(\Delta_C(\lambda), D) \to \operatorname{Ext}^1_C(\Delta_C(\lambda), V)$$

is E-linear, it is surjective by the choice of α. As we have $\operatorname{Ext}^1_C(\Delta_C(\lambda), D) = 0$, we have $\operatorname{Ext}^1_C(\Delta_C(\lambda), U) = 0$, in other words, $U \in \mathcal{X}$. Now **1** has been proved.

To prove **2**, we keep the assumption that $(R, \mathfrak{m})$ is local. As we have $\operatorname{Ext}^1_C(D, \mathcal{X}) = 0$, it follows that i is a left $\mathcal{X}$-approximation by the dual of Lemma I.1.12.1.

It remains to prove i is left minimal. Let $\varphi \in \operatorname{End}_C U$ and $\varphi i = i$, and let $\psi : D \to D$ be the map induced by φ. By the five lemma, it suffices to prove ψ is an isomorphism. As $\operatorname{Hom}_C(\Delta_C(\lambda), ?) : \operatorname{add}(\Delta_C(\lambda)) \to \operatorname{add} E_E$ is an equivalence [12, Proposition II.2.1], it suffices to show that $\psi' := \operatorname{Hom}_C(\Delta_C(\lambda), \psi) \in \operatorname{End}_E(\operatorname{Hom}_C(\Delta_C(\lambda), D))$ is an isomorphism. As in the proof of Lemma 3.1.7, we have a commutative diagram

$$\begin{array}{ccc} \operatorname{Hom}_C(\Delta_C(\lambda), D) & \xrightarrow{\alpha'} & \operatorname{Ext}^1_C(\Delta_C(\lambda), V) \\ \downarrow \psi' & & \downarrow \text{id} \\ \operatorname{Hom}_C(\Delta_C(\lambda), D) & \xrightarrow{\alpha'} & \operatorname{Ext}^1_C(\Delta_C(\lambda), V). \end{array}$$

By the definition of minimal basis and by the fact $R \cong E$,

$$\alpha' : R^r \cong \operatorname{Hom}_C(\Delta_C(\lambda), D) \to \operatorname{Ext}^1_C(\Delta_C(\lambda), V)$$

is a projective cover (as an R-module, or an E-module). Hence, ψ' is an isomorphism. This proves **2**. □

Utilizing Lemma 4.1.9 and Corollary 4.1.5, we can verify the following, as in the proof of Lemma 3.1.11.

Proposition 4.1.11 *Let π be a finite poset ideal of X^+, and $V \in \mathcal{A}_C$. Assume that $\operatorname{Ext}^1_C(\Delta_C(\mu), V) = 0$ for $\mu \notin \pi$. Then there exists an exact sequence*

$$\alpha : 0 \to V \xrightarrow{i} Y \to X \to 0 \tag{4.1.12}$$

in $\mathcal{A}_C$ such that Y is good and $X \in \mathcal{F}(\{\Delta_C(\mu) \mid \mu \in \pi\})$. If, moreover, $(R, \mathfrak{m})$ is local, then we can take i to be a left minimal good approximation. Such an exact sequence is unique, up to isomorphisms of exact sequences.

□

When $(R, \mathfrak{m})$ is local, we denote the left minimal good approximation (4.1.12) of V, which uniquely exists by the proposition, by

$$0 \to V \to Y^V \to X^V \to 0$$

for $V \in \mathcal{A}_C$. Moreover, for $\lambda \in X^+$, we denote $Y^{\Delta_C(\lambda)}$ by $T_C(\lambda)$.

Lemma 4.1.13 *Let $V \in \mathcal{A}_C$, and π be a finite poset ideal of X^+. Assume $\operatorname{Ext}^1_C(\Delta_C(\mu), V) = 0$ for $\mu \in X^+ \setminus \pi$. Then we have $\operatorname{Ext}^i_C(\Delta_C(\nu), V) = 0$ $(i \geq 2)$ for $\nu \in X^+$, provided $\nu \notin \pi$ or ν is a maximal element of π.*

Proof. By Proposition 4.1.11, there exists an exact sequence

$$0 \to V \to Y \to X \to 0$$

such that Y is good, and $X \in \mathcal{F}(\{\Delta_C(\lambda) \mid \lambda \in \pi\})$. As $X \in \mathbb{M}^{C_\pi}$, it follows that $\operatorname{Ext}^i_C(\Delta_C(\nu), X) = 0$ $(i \geq 1)$ by Corollary 4.1.5. On the other hand, $\operatorname{Ext}^i_C(\Delta_C(\nu), Y) = 0$ $(i \geq 1)$, as Y is good. Hence, we have the required property by an obvious long exact sequence argument. □

Proposition 4.1.14 *Assume that $(R, \mathfrak{m})$ is local. Let V be an R-finite R-projective C-comodule, and π a finite poset ideal of X^+ such that*

$$\operatorname{Ext}^1_C(\Delta_C(\mu), V) = 0 \qquad (\mu \in X^+ \setminus \pi).$$

If $\nu \notin \pi$ or ν is a maximal element of π, then for the minimal ν-extension

$$\alpha : 0 \to V \to V' \to D \to 0$$

of V, we have $\alpha(\mathfrak{m})$ is a minimal ν-extension of $V(\mathfrak{m})$.

Proof. By Lemma 3.1.9, it suffices to show that the number of minimal generators of $\operatorname{Ext}^1_C(\Delta_C(\nu), V)$ as an R-module agrees with

$$\dim_{\kappa(\mathfrak{m})} \operatorname{Ext}^1_{C(\mathfrak{m})}(\Delta_{C(\mathfrak{m})}(\nu), V(\mathfrak{m})).$$

Let ρ be a finite poset ideal of X^+ such that both $\nu \in \rho$ and V is an $S_C(\rho)$-module. Let $\mathbb{F}$ be an $S_C(\rho)$-finite projective resolution of V, which exists by Lemma 4.1.6.

As $\operatorname{Hom}_{S_C(\pi)}(\mathbb{F}, V)$ is an R-perfect complex, there is a spectral sequence

$$E_2^{p,q} = \operatorname{Tor}^R_{-p}(\operatorname{Ext}^q_C(\Delta_C(\nu), V), \kappa(\mathfrak{m})) \Rightarrow \operatorname{Ext}^{p+q}_{C(\mathfrak{m})}(\Delta_C(\nu)(\mathfrak{m}), V(\mathfrak{m})).$$

By Lemma 4.1.13, $E_2^{p,q} = 0$ $(q \neq 0, 1)$. Moreover, obviously $E_2^{p,q} = 0$ for $p > 0$. Hence, $E_2^{0,1} \cong E_\infty^{0,1}$. As $p + q = 1$ and $E_2^{p,q} \neq 0$ imply $(p, q) = (0, 1)$, we have

$$\operatorname{Ext}^1_{C(\mathfrak{m})}(\Delta_C(\nu)(\mathfrak{m}), V(\mathfrak{m})) \cong E_\infty^{0,1} \cong E_2^{0,1} \cong \operatorname{Ext}^1_C(\Delta_C(\nu), V) \otimes \kappa(\mathfrak{m}).$$

This proves the lemma. □

Corollary 4.1.15 *Let $(R, \mathfrak{m})$ be a local ring, and V an R-finite projective C-comodule. Then we have $Y^V(\mathfrak{m}) \cong Y^{V(\mathfrak{m})}$ and $X^V(\mathfrak{m}) \cong X^{V(\mathfrak{m})}$.*

Corollary 4.1.16 *Let $(R, \mathfrak{m})$ be a local ring, and $\lambda \in X^+$. Then we have $T_C(\lambda)(\mathfrak{m}) \cong T_{C(\mathfrak{m})}(\lambda)$. In particular, $T_C(\lambda)$ is indecomposable.*

Proof. The first assertion is an immediate consequence of Corollary 4.1.15. As $T_{C(\mathfrak{m})}(\lambda)$ is indecomposable by (3.2.3), $T_C(\lambda)$ must be also indecomposable by the first assertion and Nakayama's lemma. □

(4.1.17) Now we set

$$\mathcal{Y}_C^{\mathrm{pro}} := \{V \in \mathcal{A}_C \mid \operatorname{proj.dim}_R V < \infty \text{ and } V \text{ is good}\},$$

$\mathcal{X}_C^{\mathrm{pro}} := \operatorname{add} \mathcal{F}(\Delta)$, and $\omega_C^{\mathrm{pro}} := \mathcal{X}_C^{\mathrm{pro}} \cap \mathcal{Y}_C^{\mathrm{pro}}$. By Corollary 4.1.8, we have $\mathcal{X}_C^{\mathrm{pro}} = \mathcal{F}(\operatorname{add} \Delta)$.

Lemma 4.1.18 *Let $f : R \to R'$ be a homomorphism of commutative noetherian rings. Then we have the following.*

1 *$\mathcal{X}_C^{\mathrm{pro}} \otimes R' \subset \mathcal{X}_{C\otimes R'}^{\mathrm{pro}}$ and $\omega_C^{\mathrm{pro}} \otimes R' \subset \omega_{C\otimes R'}^{\mathrm{pro}}$.*

2 *If f is flat, then $\mathcal{Y}_C^{\mathrm{pro}} \otimes R' \subset \mathcal{Y}_{C\otimes R'}^{\mathrm{pro}}$.*

3 *If f is faithfully flat, then for $V \in \mathcal{A}_C$, we have $V \in \mathcal{X}_C^{\mathrm{pro}}$ (resp. $V \in \mathcal{Y}_C^{\mathrm{pro}}$, $V \in \omega_C^{\mathrm{pro}}$) if and only if $V \otimes R' \in \mathcal{X}_{C\otimes R'}^{\mathrm{pro}}$ (resp. $V \otimes R' \in \mathcal{Y}_{C\otimes R'}^{\mathrm{pro}}$, $V \otimes R' \in \omega_{C\otimes R'}^{\mathrm{pro}}$).*

Proof. Follows easily from Corollary 4.1.8 and Corollary 1.4.8. □

Lemma 4.1.19 *We have $\hat{\mathcal{X}}_C^{\mathrm{pro}} = \{V \in \mathcal{A}_C \mid \operatorname{proj.dim}_R V < \infty\}$.*

Proof. As any object of $\mathcal{X}_C^{\mathrm{pro}}$ is R-projective, any object of $\hat{\mathcal{X}}_C^{\mathrm{pro}}$ is of finite R-projective dimension. Conversely, if $V \in \mathcal{A}_C$ and $\operatorname{proj.dim}_R V < \infty$, then there exists a finite poset ideal π of X^+ such that $V \in \mathbb{M}^{C_\pi}$. By Lemma 4.1.6, $\operatorname{proj.dim}_{S_C(\pi)} V < \infty$. As $\operatorname{add} S_C(\pi) \subset \mathcal{X}_C^{\mathrm{pro}}$, it follows that $\mathcal{X}_C^{\mathrm{pro}}$-resol.dim $V < \infty$. □

Theorem 4.1.20 *Let R be noetherian, C an R-projective R-coalgebra, and (X^+, Δ, ∇) a semisplit highest weight theory over C. Then $(\mathcal{X}_C^{\mathrm{pro}}, \mathcal{Y}_C^{\mathrm{pro}}, \omega_C^{\mathrm{pro}})$ is a weak Auslander–Buchweitz context in $\mathcal{A}_C$. When we take an exact sequence*

$$0 \to \Delta_C(\lambda) \to T_\lambda \to X_\lambda \to 0$$

with $X_\lambda \in \mathcal{X}_C^{\mathrm{pro}}$ and T_λ good for each $\lambda \in X^+$, then we have $\omega_C^{\mathrm{pro}} = \operatorname{add}\{T_\lambda \mid \lambda \in X^+\}$.

Proof. It is obvious that $\mathcal{X}_C^{\text{pro}}$ is closed under extensions and direct summands. We prove that $\mathcal{X}_C^{\text{pro}}$ is closed under epikernels. Let

$$0 \to V \to X \to X_1 \to 0$$

be an exact sequence in $\mathcal{A}_C$ such that $X, X_1 \in \mathcal{X}_C^{\text{pro}}$. As X_1 is R-projective, we have that

$$0 \to X_1^* \to X^* \to V^* \to 0$$

is exact, and X_1^* and X^* are u-good C^{op}-comodules by Corollary 4.1.8. As V^* is R-finite projective, we have that V^* is a u-good C^{op}-comodule by Lemma 2.3.15. By Corollary 4.1.8 again, $V \in \mathcal{X}_C^{\text{pro}}$.

Good comodules are closed under extensions, direct summands, and monocokernels. On the other hand, comodules with finite R-projective dimension are also closed under extensions, direct summands, and monocokernels. Hence, $\mathcal{Y}_C^{\text{pro}}$ is closed under extensions, direct summands, and monocokernels.

By Lemma 4.1.19, we have $\mathcal{Y}_C^{\text{pro}} \subset \hat{\mathcal{X}}_C^{\text{pro}}$. Hence, conditions **AB1** and **AB2** in Theorem I.1.12.10 hold.

We verify condition **AB3**. Let $\lambda \in X^+$. Then by Proposition 4.1.11, there is an exact sequence

$$0 \to \Delta_C(\lambda) \to T_\lambda \to X_\lambda \to 0$$

with T_λ good and $X_\lambda \in \mathcal{X}_C^{\text{pro}}$. In the sequel, we fix such exact sequences for all $\lambda \in X^+$.

We set $T = \{T_\lambda \mid \lambda \in X^+\}$. Then we have $\operatorname{add} T \subset \omega_C^{\text{pro}}$. In fact, as $X_\lambda, \Delta_C(\lambda) \in \mathcal{X}_C^{\text{pro}}$, we have $T_\lambda \in \mathcal{X}_C^{\text{pro}}$. As T_λ is good and R-finite projective, $T_\lambda \in \mathcal{Y}_C^{\text{pro}}$.

Next, we show that $\operatorname{add} T$ is an injective cogenerator of $\mathcal{X}_C^{\text{pro}}$. In fact, as $\operatorname{add} T \subset \mathcal{Y}_C^{\text{pro}}$, it follows that $\operatorname{add} T$ is $\mathcal{X}_C^{\text{pro}}$-injective. As we have $\mathcal{X}_C^{\text{pro}} = \operatorname{add} \mathcal{F}(\Delta)$, we conclude that $\operatorname{add} T$ is an injective cogenerator of $\mathcal{X}_C^{\text{pro}}$, by Corollary I.1.12.13.

By Lemma I.1.12.14, $(\mathcal{X}_C^{\text{pro}}, \mathcal{Y}_C^{\text{pro}}, \operatorname{add} T)$ is a weak Auslander–Buchweitz context, and we have $\operatorname{add} T = \omega_C^{\text{pro}}$. □

Corollary 4.1.21 *Let (X^+, Δ, ∇) be a semisplit highest weight theory over an R-projective R-coalgebra C. If R is a regular ring, then we have that the triple $(\mathcal{X}_C^{\text{pro}}, \mathcal{Y}_C^{\text{pro}}, \omega_C^{\text{pro}})$ is an Auslander–Buchweitz context in $\mathcal{A}_C$.*

Proof. Follows immediately from Lemma 4.1.19, Theorem 4.1.20 and Lemma I.4.10.17. □

Corollary 4.1.22 *If $(R, \mathfrak{m})$ is a local ring, then we have*

$$\omega_C^{\text{pro}} = \operatorname{add}\{T_C(\lambda) \mid \lambda \in X^+\}.$$

Proof. Trivial. □

4.2 Minimal Ringel's approximations over local rings

(4.2.1) Let $(R, \mathfrak{m})$ be a noetherian local commutative ring. Let C be an R-projective R-coalgebra, (X^+, Δ, ∇) a semisplit highest weight theory over C, and (C_π) its associated Donkin system.

Lemma 4.2.2 *Let R be Henselian local. Then any object of ω_C^{pro} is uniquely a finite direct sum of copies of $T_C(\lambda)$ for $\lambda \in X^+$. In particular, any indecomposable object T of ω_C^{pro} is isomorphic to $T_C(\lambda)$ for a unique λ. An object T of ω_C^{pro} is isomorphic to $T_C(\lambda)$ if and only if $T(\mathfrak{m}) \cong T_{C(\mathfrak{m})}(\lambda)$.*

Proof. Any object of $\mathcal{A}_C$ has a semiperfect endomorphism ring by Corollary 2.3.19. The first assertion follows immediately from Corollary 4.1.22, Corollary 4.1.16, and Lemma I.1.12.6. The rest of the assertions are trivial. □

Proposition 4.2.3 *Any object of ω_C^{pro} is uniquely a finite direct sum of $T_C(\lambda)$ $(\lambda \in X^+)$. In particular, any indecomposable object of ω_C^{pro} is isomorphic to $T_C(\lambda)$ for a unique λ. An object T of ω_C^{pro} is isomorphic to $T_C(\lambda)$ if and only if $T(\mathfrak{m}) \cong T_{C(\mathfrak{m})}(\lambda)$.*

Proof. It suffices to prove the first assertion.

Since $T_C(\lambda)(\mathfrak{m}) \cong T_{C(\mathfrak{m})}(\lambda)$ is indecomposable for $\lambda \in X^+$ by Corollary 4.1.16 and (3.2.3), the uniqueness assertion is obvious by the Krull–Schmidt theorem (Lemma I.1.12.6).

Next, we prove that any object V in $\omega_C^{\mathrm{pro}} = \mathrm{add}\{T_C(\lambda) \mid \lambda \in X^+\}$ is a direct sum of $T_C(\lambda)$.

We prove the existence of a decomposition. It suffices to prove that if $V \in \omega_C$ with $V \neq 0$, then there is some $\lambda \in X^+$ and a split mono $T_C(\lambda) \to V$. If so, then there is a decomposition $V \cong V_1 \oplus T_C(\lambda_0)$. And if $V_1 \neq 0$, then there is a decomposition $V_1 \cong T_C(\lambda_1) \oplus V_2$, and if $V_2 \neq 0$, then $V_2 \cong T_C(\lambda_2) \oplus V_3$. Continuing thus, we have a sequence of C-subcomodules

$$0 \subset T_C(\lambda_0) \subset T_C(\lambda_0) \oplus T_C(\lambda_1) \subset T_C(\lambda_0) \oplus T_C(\lambda_1) \oplus T_C(\lambda_2) \subset \cdots$$

of V. This is an increasing sequence of C-subcomodules of V, and eventually we have $V_{r+1} = 0$ for some r. This shows that we have a decomposition $V = T_C(\lambda_0) \oplus \cdots \oplus T_C(\lambda_r)$ for some r, and completes the proof.

Let $\hat{R}$ be the $\mathfrak{m}$-adic completion of R, and set $\hat{?} = ? \otimes \hat{R}$.

By Lemma 4.2.2, $\widehat{T_C(\lambda)} \cong T_{\hat{C}}(\lambda)$, since $\widehat{T_C(\lambda)} \in \omega_{\hat{C}}^{\mathrm{pro}}$ by Lemma 4.1.18 and

$$\widehat{T_C(\lambda)}(\mathfrak{m}\hat{R}) = T_C(\lambda) \otimes \hat{R} \otimes (\hat{R}/\mathfrak{m}\hat{R}) \cong T_C(\lambda)(\mathfrak{m}).$$

As $\hat{V} \neq 0$ and $\hat{V} \in \omega_{\hat{C}}^{\text{pro}}$, there is a split mono $\hat{\varphi} : \widehat{T_C(\lambda)} \hookrightarrow \hat{V}$ for some $\lambda \in X^+$ by Lemma 4.2.2. Set $T := T_C(\lambda)$, and $\hat{T} := \widehat{T_C(\lambda)}$.

Let $\hat{\psi} : \hat{V} \to \hat{T}$ be a $\hat{C}$-comodule map such that $\hat{\psi}\hat{\varphi} = \text{id}$. As V is u-acyclic with respect to T and vice versa, we can write $\varphi = \sum_i \varphi_i \otimes c_i$, with $\varphi_i \in \text{Hom}_C(T, V)$ and $c_i \in \hat{R}$, and $\psi = \sum_j \psi_j \otimes c'_j$ with $\psi_j \in \text{Hom}_C(V, T)$ and $c'_j \in \hat{R}$. As $\text{End}_{\hat{C}} \hat{T}$ is local, there are some i and j such that $\hat{\psi}_j \hat{\varphi}_i$ is an isomorphism. As $\hat{?}$ is faithful exact, $\psi_j \varphi_i$ is also an isomorphism, and hence φ_i is a split mono. □

Corollary 4.2.4 *For any local homomorphism $(R, \mathfrak{m}) \to (R', \mathfrak{n})$ of commutative noetherian local rings, we have $T_C(\lambda)' \cong T_{C'}(\lambda)$, where $(?)'$ is the functor $? \otimes R'$.*

Proof. By Lemma 4.1.18, we have that $T_C(\lambda)' \in \omega_{C'}^{\text{pro}}$. We have

$$T_C(\lambda)' \otimes_{R'} R'/\mathfrak{n} \cong T_C(\lambda) \otimes R'/\mathfrak{n} \cong T_C(\lambda) \otimes R/\mathfrak{m} \otimes_{R/\mathfrak{m}} R'/\mathfrak{n}$$
$$\cong T_{C \otimes R/\mathfrak{m}}(\lambda) \otimes_{R/\mathfrak{m}} R'/\mathfrak{n} \cong T_{C \otimes R'/\mathfrak{n}}(\lambda)$$

by Corollary 4.1.16 and Corollary 3.2.9. By the proposition, we have the isomorphism. □

(4.2.5) Let $(R, \mathfrak{m})$ and C be as in (4.2.1). Let (X^+, Δ, ∇) be a semisplit highest weight theory over C.

Theorem 4.2.6 *Let*

$$\mathbb{F} : 0 \to Y \xrightarrow{i} X \xrightarrow{p} V \to 0$$

be a sequence in $\mathcal{A}_C$. Let $\hat{R}$ be the $\mathfrak{m}$-adic completion of R, and $\hat{?}$ denote the functor $? \otimes \hat{R}$. Then the following are equivalent.

1 *$\mathbb{F}$ is an $\mathcal{X}_C^{\text{pro}}$-approximation of V, and p is right minimal.*

2 *$\mathbb{F}$ is an $\mathcal{X}_C^{\text{pro}}$-approximation of V, and there is no non-zero direct summand of X contained in $\text{Im}\, i$.*

3 *$\hat{\mathbb{F}}$ is an $\mathcal{X}_{\hat{C}}^{\text{pro}}$-approximation of $\hat{V}$, and $\hat{p}$ is right minimal.*

4 *$\hat{\mathbb{F}}$ is an $\mathcal{X}_{\hat{C}}^{\text{pro}}$-approximation of $\hat{V}$, and there is no non-zero direct summand of $\hat{X}$ contained in $\text{Im}\, \hat{\imath}$.*

In particular, if $V \in \hat{\mathcal{X}}_C$, then there is a unique minimal $\mathcal{X}_C^{\text{pro}}$-approximation. Similarly, there is a unique minimal $\mathcal{Y}_C^{\text{pro}}$-hull of V, and it agrees with the left minimal good approximation.

Proof. First note that $\mathbb{F}$ is exact if and only if $\hat{\mathbb{F}}$ is. By Lemma 4.1.18, $\mathbb{F}$ is an $\mathcal{X}^{\text{pro}}$-approximation of V if and only if $\hat{\mathbb{F}}$ is an $\mathcal{X}_{\hat{C}}^{\text{pro}}$-approximation of $\hat{V}$.

Hence, the only problem is the minimality. The implications **1⇒2** and **4⇔3** follow from Proposition I.1.12.8. We prove **3⇒1**. Assume that $\varphi : X \to X$ is a C-comodule map such that $p\varphi = p$. Then as we have $\hat{p}\hat{\varphi} = \hat{p}$, it follows from the assumption on $\hat{p}$ that $\hat{\varphi}$ is an isomorphism. Hence, φ is also an isomorphism.

We prove **2⇒4**. Assume the contrary, and let T be a non-zero direct summand of X contained in $\operatorname{Im} i$. As $T \in \omega_{\hat{C}}^{\text{pro}}$, replacing T by its direct summand, we may assume that $T \cong T_{\hat{C}}(\lambda) \cong T_C(\lambda) \otimes \hat{R}$. By assumption, there exist some $j \in \operatorname{Hom}_{\hat{C}}(T_C(\lambda) \otimes \hat{R}, \hat{Y})$ and $q \in \operatorname{Hom}_{\hat{C}}(\hat{X}, T_C(\lambda) \otimes \hat{R})$ such that $q\hat{i}j = \text{id}$. As $\hat{R}$ is R-flat, we can write

$$j = \sum_s j'_s \otimes c_s \qquad (j'_s \in \operatorname{Hom}_C(T_C(\lambda), Y),\ c_s \in \hat{R})$$

and

$$q = \sum_t q'_t \otimes c'_t \qquad (q'_t \in \operatorname{Hom}_C(X, T_C(\lambda)),\ c'_t \in \hat{R}).$$

Then we have $\sum_{s,t} q'_t i j'_s \otimes c_s c'_t = \text{id}$. As $\operatorname{End}_{\hat{C}}(T_C(\lambda) \otimes \hat{R})$ is local, there exist some s, t such that $q'_t i j'_s$ is an isomorphism. Then $\operatorname{Im} i j'_s$ is a non-zero direct summand of X contained in $\operatorname{Im} i$, which is a contradiction.

The last assertion is proved easily, and the proof is left to the reader. □

4.3 Cohen–Macaulay analogue of u-good module

(4.3.1) Let R be a commutative noetherian ring.

Definition 4.3.2 Let M and N be R-modules. We say that an R-linear map $\varphi : M \to N$ is *semipure* if $\varphi \otimes 1 : M \otimes L \to N \otimes L$ is injective for any R-module L which is locally of finite flat dimension, see (I.4.10.18).

Note that an R-semipure map is injective, and any R-pure map is R-semipure. If R is a regular ring, then an R-semipure map is R-pure.

Lemma 4.3.3 *Let R be a Cohen–Macaulay ring, N a maximal Cohen–Macaulay R-module, M an R-finite module, and $\varphi : M \to N$ an R-linear map. Then the following are equivalent.*

1 *φ is semipure.*

2 *For any* $\mathfrak{m} \in \operatorname{Max} R$, *there exists some parameter ideal (i.e., an ideal generated by a system of parameters)* J *of* $R_{\mathfrak{m}}$ *such that*

$$\varphi \otimes 1 : M \otimes R_{\mathfrak{m}}/J \to N \otimes R_{\mathfrak{m}}/J$$

is injective.

3 φ *is injective, and* $\operatorname{Coker} \varphi$ *is maximal Cohen–Macaulay.*

Proof. **1⇒2** is obvious. **2⇒3** We may assume that $(R, \mathfrak{m})$ is local. Set $d := \dim R$, $C := \operatorname{Coker} \varphi$, $I := \operatorname{Im} \varphi$, and $K := \operatorname{Ker} \varphi$. As $\varphi \otimes R/J : M/JM \to N/JN$ is injective and factors through the induced surjective map $M/JM \to I/JI$, the map $M/JM \to I/JI$ is an isomorphism, and the induced map $I/JI \to N/JN$ is also injective. Consider the short exact sequence

$$0 \to I \to N \to C \to 0,$$

and its derived long exact sequence

$$\cdots \to \operatorname{Tor}_1^R(N, R/J) \to \operatorname{Tor}_1^R(C, R/J) \to I/JI \to N/JN \to C/JC \to 0.$$

As N is maximal Cohen–Macaulay, we have

$$\operatorname{Tor}_1^R(N, R/J) \cong \operatorname{Ext}_R^{d-1}(R/J, N) = 0$$

by Corollary I.2.9.2. As $I/JI \to N/JN$ is injective, $\operatorname{Tor}_1^R(C, R/J) = 0$. By Lemma I.2.9.6, $\operatorname{depth} C = d$, and C is maximal Cohen–Macaulay. Hence, I is also maximal Cohen–Macaulay. Hence, $\operatorname{Tor}_1^R(I, R/J) = 0$, and

$$0 \to K/JK \to M/JM \to I/JI \to 0$$

is exact. As $M/JM \to I/JI$ is an isomorphism, $K/JK = 0$. By Nakayama's lemma, $K = 0$.

3⇒1 If L is an R-module locally of finite flat dimension, then $\operatorname{Tor}_1^R(C, L) = 0$ by Theorem I.4.10.19. The assertion follows immediately. □

Definition 4.3.4 We say that an R-complex

$$\mathbb{F} : 0 \to F^0 \xrightarrow{d^0} F^1 \xrightarrow{d^1} \cdots \tag{4.3.5}$$

is *su-acyclic* if for any R-module L locally of finite flat dimension, $H^i(\mathbb{F} \otimes L) = 0$ $(i > 0)$ and the canonical map $H^0(\mathbb{F}) \otimes L \to H^0(\mathbb{F} \otimes L)$ is an isomorphism.

Lemma 4.3.6 *Let* R *be a Cohen–Macaulay ring, and* $\mathbb{F}$ (4.3.5) *be an* R-*complex. Assume that* F^i *is maximal Cohen–Macaulay for* $i \geq 0$. *Then the following are equivalent.*

1 $H^i(\mathbb{F}) = 0$ $(i > 0)$.

2 $\mathbb{F}$ *is su-acyclic, and* $H^0(\mathbb{F})$ *is a maximal Cohen–Macaulay* R*-module.*

3 *For any* $\mathfrak{m} \in \operatorname{Max} R$*, there exists some parameter ideal* J *of* $R_\mathfrak{m}$ *such that* $H^i(\mathbb{F} \otimes R_\mathfrak{m}/J) = 0$ $(i > 0)$.

Proof. **1⇒2** We may assume that $(R, \mathfrak{m})$ is local. Let L be an R-module with $\text{flat.dim}_R L < \infty$. We take a bounded flat resolution $\mathbb{G}$ of L. By Theorem I.4.10.19, we have $\operatorname{Tor}_i^R(F_j, L) = 0$ $(i > 0,\ j \geq 0)$, so $\mathbb{F} \otimes \mathbb{G}$ is quasi-isomorphic to $\mathbb{F} \otimes L$. Hence, there exists a spectral sequence

$$E_2^{p,q} = \operatorname{Tor}_{-p}^R(H^q(\mathbb{F}), L) \Rightarrow H^{p+q}(\mathbb{F} \otimes L).$$

By assumption, this spectral sequence degenerates. Hence, we have $H^i(\mathbb{F} \otimes L) = 0$ for $i > 0$, the canonical map

$$H^0(\mathbb{F}) \otimes L \to H^0(\mathbb{F} \otimes L)$$

is an isomorphism, and $\operatorname{Tor}_i^R(H^0(\mathbb{F}), L) = 0$ for $i > 0$. Considering the special case where $L = R/J$ with J a parameter ideal of R, we conclude that $H^0(\mathbb{F})$ is maximal Cohen–Macaulay.

2⇒3 is obvious.

3⇒1 We may assume $(R, \mathfrak{m})$ is local. We use induction on $d = \dim R$. If $d = 0$, then we have $J = 0$, and the assertion is obvious. Let $d > 0$, and $J = (x_1, \ldots, x_d)$, and set $x := x_1$. Then by induction assumption (applied to R/xR and $\mathbb{F}/x\mathbb{F}$), we have $H^i(\mathbb{F}/x\mathbb{F}) = 0$ $(i > 0)$. Hence, for $i > 0$, the multiplication by x

$$x : H^i(\mathbb{F}) \to H^i(\mathbb{F})$$

is surjective. By Nakayama's lemma, we have $H^i(\mathbb{F}) = 0$ $(i > 0)$. □

(4.3.7) Let R be a noetherian commutative ring, C an R-projective coalgebra, (X^+, Δ, ∇) a semisplit highest weight theory over C, and (C_π) its associated Donkin system.

Definition 4.3.8 Let V and W be C-comodules. We say that W is *su-acyclic* with respect to V if

1 $\operatorname{Ext}_C^i(V, W \otimes M) = 0$ $(i > 0)$,

2 The canonical map $\operatorname{Hom}_C(V, W) \otimes M \to \operatorname{Hom}_C(V, W \otimes M)$ is an isomorphism

hold for any R-module M locally of finite flat dimension.

Lemma 4.3.9 *The following hold:*

1 *Let V and W be C-comodules. If W is su-acyclic with respect to V, then $W \otimes M$ is su-acyclic with respect to V for any R-flat module M. If, moreover, V is R-finite projective, then $W' = W \otimes R'$ is su-acyclic with respect to $V' = V \otimes R'$ as a $C' = C \otimes R'$-comodule, for any homomorphism of commutative noetherian rings locally of finite flat dimension.*

2 *The class of C-comodules su-acyclic with respect to V is closed under direct sums and direct summands.*

3 *Let $0 \to W_1 \to W_2 \to W_3 \to 0$ be an exact sequence of C-comodules, and assume that W_1 is su-acyclic with respect to V and the map $W_1 \hookrightarrow W_2$ is R-semipure. Then W_3 is su-acyclic with respect to V if and only if W_2 is su-acyclic with respect to V.*

4 *Assume that V is R-finite. A filtered inductive limit of C-comodules which are su-acyclic with respect to V is again su-acyclic with respect to V.*

5 *For a C-comodule W, the class of C-comodules V such that W is su-acyclic with respect to V is closed under extensions, direct summands, and epikernels.*

Proof. Similar to Lemma 2.2.4, and is easy. □

Lemma 4.3.10 *Let V be a C-comodule. Then the following are equivalent.*

1 *For any $\lambda \in X^+$, V is su-acyclic with respect to $\Delta_C(\lambda)$.*

2 *V is good, and for any finite poset ideal π of X^+, we have $V_\pi := \operatorname{ind}_C^{C_\pi} V \hookrightarrow V$ is R-semipure.*

3 *There exists a filtration*

$$0 = V_0 \subset V_1 \subset V_2 \subset \cdots \subset V$$

of V such that $\varinjlim V_i = V$, and for any $i \geq 1$, V_i is an R-semipure submodule of V_{i+1}, and there exists $\lambda \in X^+$ and an R-module M such that $V_i/V_{i-1} \cong \nabla_C(\lambda) \otimes M$.

A C-comodule V is called *su-good* if the equivalent conditions above are satisfied. The proof of the lemma is similar to Lemma 2.3.14, and we omit it. If R is a regular ring, then su-good and u-good are the same thing.

Lemma 4.3.11 *Assume that R is Cohen–Macaulay. Let V and W be C-comodules, and assume that V is R-finite R-projective, and W is a maximal Cohen–Macaulay R-module. If $\operatorname{Ext}^i_C(V, W) = 0$ $(i > 0)$, then W is su-acyclic with respect to V, and $\operatorname{Hom}_C(V, W)$ is maximal Cohen–Macaulay.*

Proof. Let π be a finite poset ideal of X^+ such that both V and W are C_π-comodules. Then $\mathrm{Cobar}_{C_\pi}(W, V^*)$ is an R-complex with each term maximal Cohen–Macaulay. By Lemma 4.3.6, we have the desired assertions. □

Corollary 4.3.12 *Assume R is Cohen–Macaulay. Let V be a C-comodule which is maximal Cohen–Macaulay as an R-module. Then the following are equivalent.*

1 *V is good.*

2 *V is su-good.*

3 *$V \in \mathcal{F}(\nabla \otimes \mathcal{X}_R)$, where $\mathcal{X}_R$ denotes the full subcategory of ${}_R\mathbb{M}$ consisting of maximal Cohen–Macaulay R-modules.*

4.4 Cohen–Macaulay Ringel's approximation

(4.4.1) Let R be a commutative noetherian ring with a pointwise dualizing module K_R. In particular, R is Cohen–Macaulay by Bass's conjecture (I.2.5.1). We set $\mathcal{A}_R := {}_R\mathbb{M}_f$, and define $\mathcal{X}_R$ to be the full subcategory of $\mathcal{A}_R$ consisting of maximal Cohen–Macaulay R-modules, and $\mathcal{Y}_R$ to be the full subcategory of $\mathcal{A}_R$ consisting of $V \in \mathcal{A}_R$ such that $\mathrm{inj.dim}_{R_\mathfrak{p}} V_\mathfrak{p} < \infty$ for any $\mathfrak{p} \in \operatorname{Spec} R$. We set $\omega_R := \operatorname{add} K_R$. By Proposition I.4.10.11, $(\mathcal{X}_R, \mathcal{Y}_R, \omega_R)$ is an Auslander–Buchweitz context in $\mathcal{A}_R$. We denote the full subcategory of $\mathcal{A}_R$ consisting of finite projective R-modules by $\mathcal{X}_R^{\mathrm{pro}}$. We set $\mathcal{P}_R := \hat{\mathcal{X}}_R^{\mathrm{pro}}$, i.e., $\mathcal{P}_R$ denotes the full subcategory of $\mathcal{A}_R$ consisting of R-finite modules of finite projective dimension.

Let C be an R-projective R-coalgebra, and (X^+, Δ, ∇) a semisplit highest weight theory over C, and (C_π) its associated Donkin system. We define $\mathcal{X}_C$ to be the full subcategory of $\mathcal{A}_C$ consisting of $V \in \mathcal{A}_C$ such that $V \in \mathcal{X}_R$ (as an R-module) and $\mathrm{Hom}(V, K_R)$ is a good C^{op}-comodule. We also define $\mathcal{Y}_C$ to be the full subcategory of $\mathcal{A}_C$ consisting of $V \in \mathcal{A}_C$ such that $V \in \mathcal{Y}_R$ (as an R-module) and $\mathrm{Hom}(K_R, V)$ is good. We set $\omega_C := \mathcal{X}_C \cap \mathcal{Y}_C$. We define $\mathcal{P}_C$ to be the full subcategory of $\mathcal{A}_C$ consisting of $V \in \mathcal{A}_C$ such that $V \in \mathcal{P}_R$ (as an R-module) and V is good.

The following is trivial by Proposition I.4.10.11.

Lemma 4.4.2 *The functor* $\mathrm{Hom}(?, K_R)$ *is an exact equivalence from the exact category $\mathcal{X}_C$ to the exact category of maximal Cohen–Macaulay good C^{op}-comodules, and* $\mathrm{Hom}(?, K_R)$ *itself is its quasi-inverse. The functor* $\mathrm{Hom}(K_R, ?)$ *is an exact equivalence from $\mathcal{Y}_C$ to $\mathcal{P}_C$, and $? \otimes K_R$ is its quasi-inverse.*

Hence by Corollary 4.3.12, we have:

Corollary 4.4.3 *$\mathcal{X}_C = \mathcal{F}(\Delta \otimes \mathcal{X}_R)$, and $\mathcal{X}_C$ is closed under extensions, direct summands, and epikernels in $\mathcal{A}_C$. The full subcategory $\mathcal{Y}_C$ is closed under extensions, direct summands, and monocokernels in $\mathcal{A}_C$.*

Lemma 4.4.4 *$\mathcal{A}_C = \hat{\mathcal{X}}_C$.*

Proof. Let $V \in \mathcal{A}_C$. Then there exists a finite poset ideal π of X^+ such that $V \in \mathcal{A}_{C_\pi}$. We set $S := S_C(\pi)$. As $V \in \hat{\mathcal{X}}_R$, we have $h < \infty$, where $h := \omega_R$-proj.dim V. Take an exact sequence of S-modules

$$0 \to V' \to F_{h-1} \to \cdots \to F_1 \to F_0 \to V \to 0$$

such that each F_i S-finite projective. Then each F_i is also R-finite R-projective, and hence V' is maximal Cohen–Macaulay. As each F_i belongs to $\mathcal{X}_C$, it suffices to show $V' \in \hat{\mathcal{X}}_C$.

Set $r := \operatorname{rank} \pi$, and take an exact sequence

$$0 \to V'' \to F_{2r-1} \to \cdots \to F_1 \to F_0 \to V' \to 0$$

of S-modules with each F_i S-finite projective. By Lemma 4.1.6, we have

$$\text{proj.dim}_{S^{\text{op}}} \nabla_C(\lambda)^* \leq 2r$$

for $\lambda \in \pi$. Hence $\operatorname{Ext}^i_{C^{\text{op}}}(\nabla_C(\lambda)^*, \operatorname{Hom}(V', K_R)) = 0$ for $\lambda \in \pi$ and $i \geq 2r+1$. As

$$0 \to \operatorname{Hom}(V', K_R) \to \operatorname{Hom}(F_0, K_R) \to \operatorname{Hom}(F_1, K_R) \to \\ \cdots \to \operatorname{Hom}(F_{2r-1}, K_R) \to \operatorname{Hom}(V'', K_R) \to 0$$

is exact and each $\operatorname{Hom}(F_j, K_R)$ is a good C_π^{op}-comodule, we have

$$\operatorname{Ext}^i_{C^{\text{op}}}(\nabla_C(\lambda)^*, \operatorname{Hom}(V'', K_R)) = 0$$

for $\lambda \in \pi$ and $i \geq 1$. Hence, $\operatorname{Hom}(V'', K_R)$ is a good C_π^{op}-comodule. Hence, $V'' \in \mathcal{X}_C$, and the lemma is proved. □

Theorem 4.4.5 *Let R be a Cohen–Macaulay ring with a pointwise dualizing module K_R, C an R-projective R-coalgebra, and (X^+, Δ, ∇) a semisplit highest weight theory over C. Then $(\mathcal{X}_C, \mathcal{Y}_C, \omega_C)$ is an Auslander–Buchweitz context of $\mathcal{A}_C$, and $\omega_C = K_R \otimes \omega_C^{\text{pro}}$.*

Proof. The conditions **AB1,AB2** in Theorem I.1.12.10 are obvious by Corollary 4.4.3 and Lemma 4.4.4.

Next, we prove $\omega_C = K_R \otimes \omega_C^{\text{pro}}$. First, we show $K_R \otimes \omega_C^{\text{pro}} \subset \omega_C$. If $V \in \omega_C^{\text{pro}}$, then as $\operatorname{Hom}(K_R \otimes V, K_R) \cong \operatorname{Hom}(V, \operatorname{Hom}(K_R, K_R)) \cong V^*$, we have $K_R \otimes V \in \mathcal{X}_C$. On the other hand, as V is R-finite R-projective, we have $\operatorname{Hom}(K_R, K_R \otimes V) \cong V$, and hence $K_R \otimes V \in \mathcal{Y}_C$. Hence $K_R \otimes \omega_C^{\text{pro}} \subset \omega_C$.

We show $\omega_C \subset K_R \otimes \omega_C^{\text{pro}}$. If $T \in \omega_C$, then as $T \in \mathcal{Y}_C$, we have $T \cong K_R \otimes \operatorname{Hom}(K_R, T)$. So it suffices to show that $\operatorname{Hom}(K_R, T) \in \omega_C^{\text{pro}}$. In fact, $\operatorname{Hom}(K_R, T)$ is R-projective and good, and hence $\operatorname{Hom}(K_R, T) \in \mathcal{Y}_C^{\text{pro}}$. On the other hand, as

$$\begin{aligned}\operatorname{Hom}(K_R, T)^* &\cong \operatorname{Hom}(\operatorname{Hom}(K_R, T), \operatorname{Hom}(K_R, K_R)) \\ &\cong \operatorname{Hom}(K_R \otimes \operatorname{Hom}(K_R, T), K_R) \cong \operatorname{Hom}(T, K_R),\end{aligned}$$

it follows that $\operatorname{Hom}(K_R, T)^*$ is an R-finite R-projective good C^{op}-comodule. By Corollary 4.1.8, $\operatorname{Hom}(K_R, T) \in \mathcal{X}_C^{\text{pro}}$. Now $\omega_C = K_R \otimes \omega_C^{\text{pro}}$ is proved.

It remains to prove the condition **AB3**. As $\omega_C = \mathcal{X}_C \cap \mathcal{Y}_C$ by definition, it suffices to prove that ω_C is an injective cogenerator of $\mathcal{X}_C$.

In fact, if $\lambda \in X^+$, $V \in \mathcal{X}_R$, and $M \in \omega_C^{\text{pro}}$, then there is a spectral sequence

$$E_2^{p,q} = \operatorname{Ext}_C^p(\Delta_C(\lambda), \operatorname{Ext}_R^q(V, K_R \otimes M)) \Rightarrow \operatorname{Ext}_C^{p+q}(\Delta_C(\lambda) \otimes V, K_R \otimes M)$$

by Proposition I.3.5.13. As $\operatorname{Ext}_R^q(V, K_R \otimes M) = 0$ $(q > 0)$, we have isomorphisms

$$\begin{aligned}\operatorname{Ext}_C^i(\Delta_C(\lambda) \otimes V, K_R \otimes M) &\cong \operatorname{Ext}_C^i(\Delta_C(\lambda), \operatorname{Hom}(V, K_R \otimes M)) \\ &\cong \operatorname{Ext}_C^i(\Delta_C(\lambda), \operatorname{Hom}(V, K_R) \otimes M).\end{aligned}$$

As M is u-good, $\operatorname{Ext}_C^i(\Delta_C(\lambda) \otimes V, K_R \otimes M) = 0$ for $i > 0$. As $\mathcal{X}_C = \mathcal{F}(\Delta \otimes \mathcal{X}_R)$ and $\omega_C = K_R \otimes \omega_C^{\text{pro}}$, it follows that $\operatorname{Ext}_C^i(\mathcal{X}_C, \omega_C) = 0$ $(i > 0)$.

Let $\lambda \in X^+$ and $V \in \mathcal{X}_R$, and consider a $\mathcal{Y}_R$-hull

$$0 \to V \xrightarrow{i} Y^V \to X^V \to 0$$

of V. Let

$$0 \to \Delta_C(\lambda) \xrightarrow{j} T_\lambda \to X_\lambda \to 0$$

be a $\mathcal{Y}_C^{\text{pro}}$-hull of $\Delta_C(\lambda)$. Then the map $j \otimes i : \Delta_C(\lambda) \otimes V \to T_\lambda \otimes Y^V$ is injective. When we set $L := \operatorname{Coker} j \otimes i$, then there is an exact sequence

$$0 \to \Delta_C(\lambda) \otimes X^V \to L \to X_\lambda \otimes Y^V \to 0.$$

Hence, $L \in \mathcal{X}_C$. On the other hand, as $\omega_C = \omega_C^{\text{pro}} \otimes K_R = \omega_C^{\text{pro}} \otimes \omega_R$, it follows that $T_\lambda \otimes Y^V \in \omega_C$. By Corollary I.1.12.13, ω_C is an injective cogenerator of $\mathcal{X}_C$. □

4.5 Applications to split reductive groups

(4.5.1) Let R be a noetherian commutative ring, and $G = \operatorname{Spec} H$ a split reductive group over R. We use the notation, the terminology and the conventions in (I.4.5) and (I.4.7).

Lemma 4.5.2 *If R is a field, then $(H, X_G^+, \Delta_G, \nabla_G)$ is a split highest weight coalgebra over R, where $\Delta_G := (\Delta_G(\lambda))_{\lambda \in X_G^+}$ and $\nabla_G := (\nabla_G(\lambda))_{\lambda \in X_G^+}$.*

Proof. First, consider the case that $R = k$ is an algebraically closed field. We set $L_G := (L_G(\lambda))_{\lambda \in X_G^+}$. It suffices to show the conditions **a–h** (**g** is for $i = 1, 2$) in Proposition 1.2.2. By Lemma I.4.5.10, the condition **a** is obvious. The conditions **b–g** are obvious from (I.4.7). Hence, $(X_G^+, \Delta_G, \nabla_G, L_G)$ is a highest weight theory over H. By Lemma 1.2.20, $\operatorname{End}_G L_G(\lambda)$ is a finite dimensional division k-algebra, and hence it agrees with k. So **h** is also true.

We prove the case that R is a general field. Let R' be the algebraic closure of R. By Proposition I.3.6.20, **2**, we have $\nabla_G(\lambda) \otimes_R R' \cong \nabla_{G \otimes R'}(\lambda)$, and $\Delta_G(\lambda) \otimes_R R' \cong \Delta_{G \otimes R'}(\lambda)$. By Corollary 1.4.8, we are done. □

As is described in [90, p. 248], this fact, Serre's duality, and Borel–Bott–Weil formula together yield a simple proof of the following well-known fact.

Corollary 4.5.3 *If R is a field of characteristic zero, then any G-module is a direct sum of $L_G(\lambda)$. $\operatorname{Ext}_G^1(V, W) = 0$ for any G-modules V and W in this case.*

Proof. Recall (I.4.7.11) that we have $\Delta_G(\lambda) \cong \nabla_G(\lambda)$ for any dominant weight λ. The composite map

$$L_G(\lambda) \cong \operatorname{soc}(\nabla_G(\lambda)) \hookrightarrow \nabla_G(\lambda) \cong \Delta_G(\lambda) \to \operatorname{top}(\Delta_G(\lambda)) \cong L_G(\lambda)$$

is non-zero, because $(\operatorname{rad}(\Delta_G(\lambda)) : L_G(\lambda)) = 0$. This shows $\nabla_G(\lambda) \cong \Delta_G(\lambda) \cong L_G(\lambda)$. Hence, we have $\operatorname{Ext}_G^1(L_G(\lambda), L_G(\mu)) = 0$ for all $\lambda, \mu \in X_G^+$. As any simple G-module is isomorphic to some $L_G(\lambda)$, we have $\operatorname{Ext}_G^1(V, W) = 0$ for finite dimensional G-modules V and W. By Proposition I.3.7.4, this is true even if W is not finite dimensional. By Lemma I.1.9.4, any G-module is injective. Hence, the last assertion is now clear. As is proved by induction on length, any finite dimensional G-module is semisimple. By Lemma I.1.10.5, any G-module is semisimple, which completes the proof of the corollary. □

Lemma 4.5.4 *For $\lambda \in X_G^+$, we have $R^i \operatorname{ind}_B^G(R_\lambda) = 0$ $(i > 0)$, and $\operatorname{ind}_B^G(R_\lambda)$ is R-finite R-free. For any commutative R-algebra R', we have*

$$R' \otimes \nabla_G(\lambda) \cong \nabla_{R' \otimes G}(\lambda).$$

Proof. It suffices to prove that the complex $\mathbb{F} := \mathrm{Cobar}_{R[B]}(R_\lambda, H)$ is u-acyclic with $H^0(\mathbb{F})$ R-finite free. As the construction of $\mathbb{F}$ is compatible with the base change, we may assume that $R = \mathbb{Z}$. Note that $\mathbb{F}$ is an R-flat complex. Note also that $F^0 = R_\lambda \otimes H \cong H$ is R-projective by Theorem II.2.2.8. Hence, we can apply (2.1.17) to $\mathbb{F}$. It suffices to prove that $H^i(\mathbb{F} \otimes \kappa(\mathfrak{p})) = 0$ $(i > 0)$ for $\mathfrak{p} \in \operatorname{Spec} R$, and that $\dim_{\kappa(\mathfrak{p})} H^0(\mathbb{F} \otimes \kappa(\mathfrak{p}))$ is finite and constant.

Let $\mathfrak{p} \in \operatorname{Spec} R$, and k be the algebraic closure of $\kappa(\mathfrak{p})$. By Kempf's vanishing (I.4.7.6), we have

$$H^i(\mathbb{F} \otimes \kappa(\mathfrak{p})) \otimes_{\kappa(\mathfrak{p})} k \cong H^i(\mathbb{F} \otimes k) \cong R^i \operatorname{ind}_{B\otimes k}^{G\otimes k} k_\lambda = 0$$

for $i > 0$. Hence, we have $H^i(\mathbb{F} \otimes \kappa(\mathfrak{p})) = 0$ $(i > 0)$.

By (I.4.7.9),

$$\dim_{\kappa(\mathfrak{p})} H^0(\mathbb{F} \otimes \kappa(\mathfrak{p})) = \dim_k H^0(\mathbb{F} \otimes k) = \dim_k \nabla_{G\otimes k}(\lambda)$$

is independent of k and is finite. □

Hence, we have the following.

Lemma 4.5.5 *$(H, X_G^+, \Delta_G, \nabla_G)$ is a split highest weight coalgebra defined over $\mathbb{Z}$.*

By Theorem II.2.2.8 and Theorem 2.2.11, we have:

Lemma 4.5.6 *There is a unique Donkin system (H_π) of H associated with $(X_G^+, \Delta_G, \nabla_G)$.*

Let π be a poset ideal of X_G^+. We set $C_G(\pi) := H_\pi = \varinjlim_\rho H_\rho$, where ρ runs through all finite poset ideals of π. We call $C_G(\pi)$ the *Donkin subcoalgebra* of G with respect to π. We say that a G-module V belongs to π if V is a $C_G(\pi)$-comodule.

If π is finite, then we denote $C_G(\pi)^*$ by $S_G(\pi)$ and call it the *Schur algebra* of G with respect to π.

We denote the weak Auslander–Buchweitz context $(\mathcal{X}_H^{\mathrm{pro}}, \mathcal{Y}_H^{\mathrm{pro}}, \omega_H^{\mathrm{pro}})$ of ${}_G\mathbb{M}_f$ by

$$(\mathcal{X}_G^{\mathrm{pro}}, \mathcal{Y}_G^{\mathrm{pro}}, \omega_G^{\mathrm{pro}}).$$

Definition 4.5.7 We call an object of ω_G^{pro} a *tilting* G-module.

The name 'tilting' here is not the same as the tilting module explained in (I.4.9), but the origin of this word is in (3.2.7), and these are connected in some sense, see [50].

If R has a pointwise dualizing module K_R, then we denote the Auslander–Buchweitz context $(\mathcal{X}_H, \mathcal{Y}_H, \omega_H)$ of ${}_G\mathbb{M}_f$ by $(\mathcal{X}_G, \mathcal{Y}_G, \omega_G)$.

(4.5.8) The following theorem was proved by O. Mathieu [107].

Theorem 4.5.9 *Assume that R is an algebraically closed field. If V and W are finite dimensional good G-modules, then $V \otimes W$ is good.*

Corollary 4.5.10 *Let R be an arbitrary noetherian commutative ring. Let V and W be G-modules. Then the following hold:*

1 *If V is R-finite R-projective good and W is good, then $V \otimes W$ is good.*

2 *If $V, W \in \mathcal{X}_G^{\text{pro}}$, then $V \otimes W \in \mathcal{X}_G^{\text{pro}}$.*

3 *If $V, W \in \omega_G^{\text{pro}}$, then $V \otimes W \in \omega_G^{\text{pro}}$.*

Proof. **1** If W is also R-finite R-projective, then the problem is reduced to the case that R is an algebraically closed field, using Corollary 4.1.8 and Lemma 4.1.18. By Corollary 4.1.8, **2** and **3** are also obvious from this.

We prove the general case of **1**. Let $\lambda \in X_G^+$. Then we have

$$\operatorname{Ext}_G^1(\Delta_G(\lambda), V \otimes W) \cong \operatorname{Ext}_G^1(\Delta_G(\lambda) \otimes V^*, W)$$

by Proposition II.1.1.14, **2**. The right-hand side vanishes as $\Delta_G(\lambda) \otimes V^* \in \mathcal{X}_G^{\text{pro}}$ by the first paragraph, and W is good. Hence, $V \otimes W$ is good. □

We also remark the following.

Proposition 4.5.11 ([47, Proposition 3.2.7]) *Let R be an algebraically closed field. For a G-module V, V is good if and only if $\operatorname{res}_{[G,G]}^G V$ is good as a $[G,G]$-module, where $[G,G]$ is the derived subgroup of G.*

By the proposition, we have the following.

Corollary 4.5.12 *Any rank-one R-projective G-module is tilting.*

Proof. We may assume that R is an algebraically closed field. If V is one-dimensional, then $GL(V) = \mathbb{G}_m$ is commutative. Hence, we have $\operatorname{res}_{[G,G]}^G V \cong R = \nabla_{[G,G]}(0)$. By the proposition, we have V is also good as a G-module. As V^* is also one-dimensional, we have V^* is also good, and hence V is tilting. □

For more about good modules and tilting modules, see [91, 47, 49, 6] and references therein.

4.6 Good modules of a general linear group

In this subsection, we give some remarks on good modules of $G = GL(n, R)$.

(4.6.1) We follow the notation, terminology and conventions of (I.4.6).

We say that a sequence of non-negative integers $\lambda = (\lambda_1, \lambda_2, \ldots)$ is a *partition* if $\lambda_1 \geq \lambda_2 \geq \cdots$, and $\lambda_n = 0$ for $n \gg 0$. For a partition λ, the *transpose* $\tilde{\lambda} = (\tilde{\lambda}_1, \tilde{\lambda}_2, \ldots)$ of λ is defined by

$$\tilde{\lambda}_i := \#\{j \in \mathbb{N} \mid i \leq \lambda_j\}.$$

It is easy to see that $\tilde{\lambda}$ is a partition, and we have $\tilde{\tilde{\lambda}} = \lambda$. The sum $\sum_i \lambda_i$ is called the *degree* of λ.

The minimum non-negative integer r such that $\lambda_{r+1} = 0$ is called the *length* of λ. By definition, the length of λ agrees with $\tilde{\lambda}_1$. The set of partitions of length at most n, viewed as a subset of $\mathbb{Z}^n = X(T)$, is denoted by $\Gamma(G)$. Note that $\Gamma(G)$ is a poset ideal of X^+. A G-module which belongs to $\Gamma = \Gamma(G)$ is called a *polynomial representation*. We have $C_G(\Gamma) = R[\mathrm{End}(R^n)]$, and $C_G(\Gamma)$ is a subbialgebra of H.

Example 4.6.2 Set $\lambda := (r, 0, 0, \ldots, 0) \in \mathbb{Z}^n = X(T)$, and

$$\pi(r) := (-\infty, \lambda] = \{\mu \in X_G^+ \mid \mu \leq \lambda\}.$$

Then $\pi(r)$ agrees with the set of partitions of degree r with length at most n. Moreover, we have $\Gamma = \coprod_{r \geq 0} \pi(r)$. The Schur algebra $S_G(\pi(r))$ with respect to $\pi(r)$ is nothing but the classical Schur algebra studied by I. Schur. See [62] for more.

(4.6.3) For a partition λ, Akin–Buchsbaum–Weyman [4] defined the *Schur functor* L_λ. It is a universally free functor of type $(1, 0)$ defined over $\mathbb{Z}$. The module $L_\lambda(R^n)$ is a G-module by (I.4.8). If $\lambda_1 > n$, then $L_\lambda(R^n) = 0$, while if $\lambda_1 \leq n$, then we have $L_\lambda(R^n) \cong \nabla_G(\tilde{\lambda})$. More generally, if X is a scheme and $\mathcal{Q}$ a locally free coherent $\mathcal{O}_X$-module, then $L_\lambda(\mathcal{Q})$ is a $GL(\mathcal{Q})$-module. Note also that $L_\lambda(\mathcal{Q})$ is a locally free coherent $\mathcal{O}_X$-module.

Akin–Buchsbaum–Weyman [4] defined the Schur map $d_\lambda : \bigwedge_\lambda \to L_\lambda$, which is a universal map between universally free functors, where $\bigwedge_\lambda$ is defined by

$$\textstyle\bigwedge_\lambda V := \bigwedge^{\lambda_1} V \otimes \cdots \otimes \bigwedge^{\lambda_n} V.$$

Note that d_λ is surjective. Note also that the following holds.

Proposition 4.6.4 *Set* $\bigwedge := \{\bigwedge_\lambda R^n \mid \lambda \in \tilde{\Gamma}\}$. *Then we have* $\mathrm{add} \bigwedge = \omega^{\mathrm{pro}}_{C_G(\Gamma)}$.

We show the outline of the proof of the proposition. First we must prove that $\bigwedge_\lambda R^n$ is tilting. As $\bigwedge^i V$ is R-finite free for any i, $\bigwedge_\lambda R^n$ is also R-finite free. So we may assume that R is an algebraically closed field. By Corollary 4.5.10, it suffices to show that $\bigwedge^i V$ is tilting for each i. In fact, we have $\bigwedge^i V = \nabla_G(\omega_i) = \Delta_G(\omega_i)$, where ω_i is the partition $\varepsilon_1 + \cdots + \varepsilon_i$. As ω_i is minimal in X_G^+, the proof is reduced to proving $\bigwedge^i V = L_G(\omega_i)$, which is proved easily using (I.4.7.10).

As $\bigwedge_\lambda R^n$ is tilting and is a polynomial representation, we have add $\bigwedge \subset \omega^{\text{pro}}_{C_G(\Gamma)}$. Hence, it suffices to show that add $\bigwedge$ is an injective cogenerator of $\mathcal{X}_G^{\text{pro}}$, to prove the proposition. By Corollary I.1.12.13, it suffices to show that for any $\lambda \in \tilde{\Gamma}$, there is an exact sequence

$$0 \to \Delta_G(\tilde{\lambda}) \to \textstyle\bigwedge_\lambda R^n \to D_\lambda \to 0$$

with $D_\lambda \in \mathcal{X}^{\text{pro}}_{C_G(\Gamma)}$.

We claim that $\operatorname{Hom}_G(\Delta_G(\tilde{\lambda}), \bigwedge_\lambda R^n) \cong R$, and $\operatorname{Hom}_G(\Delta_G(\mu), \bigwedge_\lambda R^n) = 0$ for $\mu \not\leq \tilde{\lambda}$. As we know that $\bigwedge_\lambda R^n$ is u-good, the proof of the claim is reduced to the case where $R = \mathbb{C}$, and this is well-known and checked at the character level, see [104].

If R is a field, then $\bigwedge_\lambda R^n \cong T_G(\tilde{\lambda}) \oplus T'_\lambda$, where T'_λ is a finite direct sum of $T_G(\mu)$ with $\mu < \tilde{\lambda}$. Note that it follows, if R is a field, that any non-zero map from $\Delta_G(\tilde{\lambda})$ to $\bigwedge_\lambda R^n$ factors through $T_G(\tilde{\lambda})$, and is a $\mathcal{Y}^{\text{pro}}_{C_G(\Gamma)}$-hull (hence is injective).

Now take a free generator d'_λ of $\operatorname{Hom}_G(\Delta_G(\tilde{\lambda}), \bigwedge_\lambda R^n) \cong R$ as an R-module. Then for any $\mathfrak{m} \in \operatorname{Max} R$, the base change $d'_\lambda(\mathfrak{m})$ is injective, and hence $D_\lambda := \operatorname{Coker} d'_\lambda$ is R-projective. As $D_\lambda(\mathfrak{m})$ is Δ-good for any $\mathfrak{m}$, D_λ is also Δ-good. As D_λ is a polynomial representation, $D_\lambda \in \mathcal{X}^{\text{pro}}_{C_G(\Gamma)}$. □

Remark 4.6.5 In [4], an explicit construction of $d'_\lambda : \Delta_G(\tilde{\lambda}) \to \bigwedge_\lambda R^n$ is given. In fact, d'_λ is constructed as a universal map of universally free functors there (but the notation is different).

Example 4.6.6 As we have seen, an R-finite R-projective good G-module is u-good, see Corollary 4.1.8. However, this is not true without the R-projectivity assumption. Set $R := \mathbb{Z}$, $G := GL_2$, and $V := \mathbb{Z}^2$. Let $\bar{?}$ denote the functor $\otimes_{\mathbb{Z}} \mathbb{Z}/2\mathbb{Z}$. There is a short exact sequence of G-modules

$$0 \to \bar{V}^{(1)} \xrightarrow{i} S_2\bar{V} \xrightarrow{p} \textstyle\bigwedge^2 \bar{V} \to 0,$$

where $p(\bar{v}_1\bar{v}_2) = \bar{v}_1 \wedge \bar{v}_2$, and $V^{(1)}$ is the first Frobenius twisting of $\bar{V}$, see [90, (I.9.10)]. The image $\operatorname{Im} i$ is the $\mathbb{Z}/2\mathbb{Z}$-span $\langle \bar{x}_1^2, \bar{x}_2^2 \rangle$, where $\bar{x}_1, \bar{x}_2$ is a basis of $\bar{V}$. Taking the pull-back of this exact sequence by the canonical projection $\pi : \bigwedge^2 V \to \bigwedge^2 \bar{V}$, we have a commutative diagram of G-modules

$$\begin{array}{ccccccccc}
 & & 0 & & 0 & & 0 & & \\
 & & \uparrow & & \uparrow & & \uparrow & & \\
0 & \to & \bar{V}^{(1)} & \xrightarrow{i} & S_2\bar{V} & \xrightarrow{p} & \bigwedge^2 \bar{V} & \to & 0 \\
 & & \uparrow & & \uparrow \rho & & \uparrow \pi & & \\
0 & \to & \bar{V}^{(1)} & \to & M & \to & \bigwedge^2 V & \to & 0 \\
 & & \uparrow & & \uparrow & & \uparrow 2 & & \\
 & & 0 & \to & \bigwedge^2 V & \to & \bigwedge^2 V & \to & 0 \\
 & & & & \uparrow & & \uparrow & & \\
 & & & & 0 & & 0 & &
\end{array}$$

all of whose rows and columns are exact. As $S_2\bar{V} = \nabla_{\bar{G}}(2)$, the exactness of the middle column shows that M is good. Note that $\operatorname{Im} i = \operatorname{soc} \nabla_{\bar{G}}(2)$. It follows that

$$\operatorname{Hom}_G(\textstyle\bigwedge^2 V, S_2\bar{V}) \cong \operatorname{Hom}_{\bar{G}}(\bigwedge^2 \bar{V}, S_2\bar{V}) = \operatorname{Hom}_{\bar{G}}(\bigwedge^2 \bar{V}, \operatorname{soc}(S_2\bar{V})) = 0.$$

Hence,

$$\operatorname{Hom}_G(\textstyle\bigwedge^2 V, M)^- \cong \operatorname{Hom}_G(\bigwedge^2 V, \bigwedge^2 V)^- \cong \mathbb{Z}/2\mathbb{Z}.$$

On the other hand, $\bar{\rho}$ is an isomorphism, and

$$\operatorname{Hom}_G(\textstyle\bigwedge^2 V, \bar{M}) \cong \operatorname{Hom}_G(\bigwedge^2 V, S_2\bar{V}) = 0.$$

Hence, M is good, but is not u-good.

Chapter IV

Approximations of Equivariant Modules and their Applications

1 Approximations of (G, A)-modules

Throughout this section, R is a noetherian commutative ring, $G = \operatorname{Spec} H$ is an R-flat R-group scheme, and A is an R-flat noetherian commutative G-algebra.

1.1 Graded G-algebras

(1.1.1) Let us assume that G has a fixed closed subgroup $\mathbb{G}_m \subset Z(G)$, where $Z(G)$ is the center of G. Then any G-module is a $\mathbb{G}_m$-module by restriction, and hence it is a $\mathbb{Z}$-graded R-module by (I.4.3). In the sequel, the grading of a G-module is the one obtained in this way, unless otherwise specified. For a G-module V, V_i denotes the degree i component of V, so that $V = \bigoplus_{i\in\mathbb{Z}} V_i$ as an R-module. Any G-linear map preserves the grading of G-modules.

Lemma 1.1.2 *Let $V \in {}_G\mathbb{M}$ and $i \in \mathbb{Z}$. Then V_i is a G-submodule of V, and $V = \bigoplus_{i\in\mathbb{Z}} V_i$ is a direct sum decomposition of V as a G-module.*

Proof. For any commutative R-algebra A, $g \in G(A)$, $v \in V_i$ and $t \in \mathbb{G}_m(A) = A^\times$, we have

$$t(g(v \otimes 1)) = g(t(v \otimes 1)) = g(v \otimes t^i) = t^i(g(v \otimes 1))$$

as we have $t \in Z(G(A))$. In particular, when we set $A = H \otimes R[\mathbb{G}_m]$, and define $g \in G(A) = \operatorname{Hom}_{R\text{-alg}}(H, H \otimes R[\mathbb{G}_m])$ by $h \mapsto h \otimes 1$ and $t \in \mathbb{G}_m(A) = \operatorname{Hom}_{R\text{-alg}}(R[\mathbb{G}_m], H \otimes R[\mathbb{G}_m])$ by $s \mapsto 1 \otimes s$, then the equation above shows $\omega_V(v) \in V_i \otimes H$. Hence, V_i is an H-subcomodule of V, that is, a G-submodule of V. □

(1.1.3) As A is a G-algebra, it is a $\mathbb{G}_m$-algebra, and is a $\mathbb{Z}$-graded R-algebra. From now on, until the end of this subsection, A is assumed to be *positively graded*, i.e., $A_i = 0$ $(i < 0)$ and $R \to A_0$ is an isomorphism. We set $\mathfrak{M} := \bigoplus_{i>0} A_i$. Note that $\mathfrak{M}$ is a G-ideal of A, and we have $A/\mathfrak{M} \cong R$. We also assume that G is R smooth with connected geometric fibers.

We denote by $\mathcal{P}_0$ the full subcategory of ${}_G\mathbb{M}$ consisting of R-finite R-projective G-modules. By $\mathcal{P}$, we denote the full subcategory of ${}_{G,A}\mathbb{M}$ consisting of A-finite A-projective (G, A)-modules.

Lemma 1.1.4 *We have $\mathcal{P} = \mathcal{F}(A \otimes \mathcal{P}_0)$. Namely, for any A-finite A-projective (G, A)-module P, there exists a filtration*

$$0 = P_{(0)} \subset P_{(1)} \subset \cdots \subset P_{(r)} = P \qquad (r \geq 0)$$

of P such that for each $1 \leq i \leq r$, there exists $Q_{(i)} \in \mathcal{P}_0$ such that $P_{(i)}/P_{(i-1)} \cong A \otimes Q_{(i)}$. Moreover, we then have $P/\mathfrak{M}P \cong \bigoplus_{i=1}^{r} Q_{(i)}$ as G-modules.

Proof. Plainly we have that $P/\mathfrak{M}P$ is $(R = A/\mathfrak{M})$-finite R-projective. We prove the existence of such a filtration by induction on

$$s := \sum_{\mathfrak{p} \in \operatorname{Min} R} \operatorname{rank}_{R_\mathfrak{p}} (P/\mathfrak{M}P)_\mathfrak{p}.$$

If $s = 0$, then we have $P/\mathfrak{M}P = 0$. As $\mathfrak{M}$ is contained in the $\mathbb{G}_m$-radical of A, we have $P = 0$ by the equivariant Nakayama, Lemma II.2.3.18. So this case is obvious.

We assume $s > 0$, and $P \neq 0$. When we decompose $P = \bigoplus_{j \in \mathbb{Z}} P_j$ with respect to the grading of P, we have $P_j = 0$ for $j \ll 0$, because P is A-finite and A is positively graded. Let j_0 be the minimum number j such that $P_j \neq 0$. As P is R-flat A-finite and A is of finite type over R, we have that P_j is an R-finite R-projective G-module for any $j \in \mathbb{Z}$. So we have a canonical (G, A)-linear map $\varphi : A \otimes P_{j_0} \to P$ given by $\varphi(a \otimes p) = ap$.

We have that

$$\varphi \otimes A/\mathfrak{M} : P_{j_0} \cong (A \otimes P_{j_0}) \otimes_A A/\mathfrak{M} \to P/\mathfrak{M}P$$

is an isomorphism from P_{j_0} to the degree j_0 component $(P/\mathfrak{M}P)_{j_0} \cong P_{j_0}$ of $P/\mathfrak{M}P$, which obviously G-splits (in particular, R-splits). Hence, when we set $C := \operatorname{Coker} \varphi$, then $C/\mathfrak{M}C$ is $R = A/\mathfrak{M}$-finite R-projective. Another consequence is that $\operatorname{Tor}_1^A(C, A/\mathfrak{M}) = 0$. Hence, by an easy spectral sequence argument, we have $\operatorname{Tor}_1^A(C, V) = 0$ for any $A/\mathfrak{M}$-module V.

Since A is positively graded, any graded maximal ideal $\mathfrak{m}$ of A contains $\mathfrak{M}$. It follows that $\operatorname{Tor}_1^{A_\mathfrak{m}}(C_\mathfrak{m}, \kappa(\mathfrak{m})) = 0$, and hence $C_\mathfrak{m}$ is $A_\mathfrak{m}$-finite free for any graded maximal ideal $\mathfrak{m}$ of A. This shows that for any graded prime ideal $\mathfrak{P}$ of A, $C_\mathfrak{P}$ is $A_\mathfrak{P}$-finite projective. By Corollary II.2.4.3, C is A-finite A-projective. Now it is easy to see that φ is injective (apply the equivariant Nakayama to $\operatorname{Ker}\varphi$). As C has a filtration of the desired form by induction assumption, P also has such a filtration. The last assertion is now obvious. □

Corollary 1.1.5 *If M is a (G, A)-module which is rank-one projective as an A-module, then there exists a rank-one R-projective G-module Λ such that $M \cong A \otimes \Lambda$. We have $\Lambda \cong M/\mathfrak{M}M$, and such a Λ is unique up to isomorphisms.*

Proof. We may assume that $\operatorname{Spec} R$ is connected. Setting $M = \bigoplus_{j \in \mathbb{Z}} M_j$, we define j_0 to be the minimum j such that $M_j \neq 0$. We set $\Lambda := M_{j_0}$. Then the canonical map $\varphi : A \otimes \Lambda \to M$ is A-pure by the proof of the lemma. As M is rank-one A-projective, we must have that Λ is of rank one and $\operatorname{Coker}\varphi = 0$. The last assertion is obvious by $(A \otimes \Lambda) \otimes_A A/\mathfrak{M} \cong \Lambda$. □

As A is R-flat of finite type, each A_i is R-finite R-projective, and hence if we take $r \gg 0$, then $Q := A_1 \oplus \cdots \oplus A_r$ is an R-finite R-projective G-module and Q generates A as an R-algebra. Hence, letting $S := \operatorname{Sym} Q$, we have a surjective G-algebra map $S \to A$, and we have $A \cong S/I$ for some G-ideal I of S. Note that S is R-smooth, because it is R-flat and all its geometric fibers are polynomial rings.

(1.1.6) Let us assume that R has a pointwise dualizing module K_R, and A is Cohen–Macaulay.

Assume that $\operatorname{Spec} R$ is connected. For any base change by $\operatorname{Spec} k \to \operatorname{Spec} R$ with k a field, the Hilbert functions of the base changes of S and A are independent of the field k. Hence, S and A have well-defined relative dimensions over R. We denote the relative dimensions of S and A by n and d, respectively. As can be checked easily, I is perfect of codimension $h := n - d$. Now we define $K_S := K_R \otimes \bigwedge^{\mathrm{top}} \Omega_{S/R}$, and $K_A := \operatorname{Ext}_S^h(A, K_S)$. By Proposition I.2.11.6, we have that K_S is a (G, S)-module, and is pointwise dualizing as an S-module. We also have that K_A is a (G, A)-module, and is pointwise dualizing as an A-module. If $\operatorname{Spec} R$ is disconnected, then we define K_S and K_A componentwise.

(1.1.7) Let R, S and $A = S/I$ be as in the last paragraph. We set $U := \operatorname{Hy}(G)$. As we are assuming that G is R-smooth with connected geometric fibers, U is a generalized hyperalgebra of H by Theorem II.2.2.8. Note also that U is R-projective by Lemma II.2.2.5. Any (G, S)-module is a (U, S)-module in a natural way, see (I.3.14.5).

Lemma 1.1.8 *If A is R-smooth, then the Yoneda product*

$$\operatorname{Ext}^i_S(A,A)\otimes_A \operatorname{Ext}^j_S(A,A)\to \operatorname{Ext}^{i+j}_S(A,A)$$

is (G,A)-linear, and the Yoneda algebra $\operatorname{Ext}_S(A,A)$ is isomorphic to the exterior algebra $\bigwedge(I/I^2)^*$ as a (G,A)-algebra.

Proof. By (I.5.2.11), the Yoneda product is (U,S)-linear, hence it is (G,A)-linear. From the short exact sequence

$$0\to I\to S\to S/I\to 0,$$

we have a long exact sequence of (G,S)-modules

$$\operatorname{Hom}_S(S/I,S/I)\xrightarrow{\cong}\operatorname{Hom}_S(S,S/I)\to\operatorname{Hom}_S(I,S/I)\to\operatorname{Ext}^1_S(S/I,S/I)\to 0,$$

and hence $\operatorname{Ext}^1_S(S/I,S/I)\cong(I/I^2)^*$. As $\operatorname{Hom}_S(S/I,S/I)\cong S/I=A$, it suffices to show that the Yoneda algebra $E=\operatorname{Ext}_S(S/I,S/I)$ is a skew commutative S/I-algebra, and the induced map

$$\bigwedge(I/I^2)^*=\bigwedge(\operatorname{Ext}^1_S(S/I,S/I))\to E$$

is an isomorphism. Now the problem is not related to the G-action any more, and the question is local. Hence, we may forget the G-action, and may assume that $(S,\mathfrak{n})$ is a noetherian local ring, and I is generated by a regular sequence $I=(a_1,\dots,a_h)$ by assumption and Theorem I.2.7.10, **2**.

Although the assertion is now more or less well-known, we briefly illustrate how it is proved. Let $\mathbb{F}$ be the Koszul complex $\operatorname{Kos}(a_1,\dots,a_h;S)=\bigwedge F$, where $F:=R^h$. Then $\mathbb{F}$ is a finite-free S-resolution of $A=S/I$. For any $\eta\in\bigwedge^i F^*$, we have a map $\Phi(\eta):\mathbb{F}\to\mathbb{F}[i]$ given by

$$\Phi(g_1\wedge\cdots\wedge g_i)(f_1\wedge\cdots f_r)$$
$$=\begin{cases}\sum_\sigma(-1)^\sigma\det(g_u(f_{\sigma v}))_{u,v}f_{\sigma(i+1)}\wedge\cdots\wedge f_{\sigma(r)} & (r\geq i)\\ 0 & \text{(otherwise)},\end{cases}$$

where the sum is taken over all $\sigma\in\mathfrak{S}_r$ such that $\sigma 1<\cdots<\sigma i$ and $\sigma(i+1)<\cdots<\sigma r$. An easy calculation shows that $\Phi(\eta\wedge\eta')=\Phi(\eta)\circ\Phi(\eta')$. Note also that the boundary map $\mathbb{F}\to\mathbb{F}[1]$ is given by $\Phi(e^*)$, where $e^*=\sum_{i=1}^h a_ie_i^*$, where $e_1^*,\dots,e_h^*$ is the standard basis of F^*. This shows

$$\Phi(\eta)\circ d=\Phi(\eta\wedge e^*)=(-1)^i\Phi(e^*\wedge\eta)=(-1)^i d\circ\Phi(\eta),$$

and hence $\Phi(\eta):\mathbb{F}\to\mathbb{F}[i]$ is a chain map. It is trivial that $\Phi(1)=\mathrm{id}$, so Φ gives a degree-preserving S-algebra map

$$\Phi:\bigwedge F^*\to Z^\bullet(\mathbb{Y}),$$

where $\mathbb{Y} := \operatorname{Hom}_S^\bullet(\mathbb{F}, \mathbb{F})$, and $Z^\bullet(\mathbb{Y})$ denotes the subalgebra of cocycles in the DG algebra $\mathbb{Y}$ (the multiplication of $\mathbb{Y}$ is given by the composition). See [21] for DG algebras.

To prove that the Yoneda algebra is skew commutative and is generated by degree one, it suffices to prove the composite map

$$\bigwedge F^* \to Z^\bullet(\mathbb{Y}) \to H^\bullet(\mathbb{Y}) = E$$

is surjective. Let $\bar{\eta} \in \operatorname{Ext}_S^i(A, A) = H^i(\operatorname{Hom}_S(\mathbb{F}, S/I)) = \bigwedge^i F^* \otimes S/I$. Then we have $\eta \in \bigwedge^i F^*$ which restricts to $\bar{\eta}$. Obviously, $\Phi(\eta) : \mathbb{F} \to \mathbb{F}[i]$ followed by the augmentation $\mathbb{F}[i] \to S/I[i]$ represents $\bar{\eta}$ in

$$Z^i(\operatorname{Hom}_S(\mathbb{F}, S/I)) = Z^0(\operatorname{Hom}_S(\mathbb{F}, S/I[i])).$$

Hence the map in question is surjective. □

Lemma 1.1.9 *The following hold.*

1 *If J is a G-ideal of A and J is perfect of codimension h', then we have $K_{A/J} \cong \operatorname{Ext}_A^{h'}(A/J, K_A)$ as $(G, A/J)$-modules.*

2 *If A is R-smooth, then we have $K_A \cong \omega_{A/R} \otimes K_R$ as (G, A)-modules.*

3 *The (G, A)-module K_A is determined only by A and K_R up to isomorphisms of (G, A)-modules, and is independent of the choice of S and I. That is, if S' is an R-smooth positively graded G-algebra and I' is a G-ideal of S' such that $A \cong S'/I'$, then K_A agrees with the (G, A)-module K'_A constructed in the same way using S' and I'.*

Proof. As the question is componentwise, we may assume that $\operatorname{Spec} R$ is connected. We prove **1**. As R is Cohen–Macaulay and K_R is a maximal Cohen–Macaulay R-module, S is Cohen–Macaulay and K_S is a maximal Cohen–Macaulay S-module. By Corollary I.4.10.20, the spectral sequence

$$E_2^{p,q} = \operatorname{Ext}_A^p(A/J, \operatorname{Ext}_S^q(A, K_S)) \Rightarrow \operatorname{Ext}_S^{p+q}(A/J, K_S)$$

in Proposition II.1.1.14 collapses, and hence we have

$$K_{A/J} = \operatorname{Ext}_S^{h+h'}(A/J, K_S) \cong \operatorname{Ext}_A^{h'}(A/J, K_A).$$

Next, we prove **2**. As A is R-smooth, I is a local complete intersection ideal of codimension h by Theorem I.2.7.10. By Theorem I.2.7.6, we have an exact sequence of (G, A)-modules

$$0 \to I/I^2 \to \Omega_{S/R} \otimes_S A \to \Omega_{A/R} \to 0.$$

Taking the top exterior power and utilizing Lemma 1.1.8, we have an isomorphism

$$K_S \otimes_S A \cong (\omega_{A/R} \otimes K_R) \otimes_A \operatorname{Ext}^h_S(S/I, S/I)^*.$$

When we denote the canonical projection $S \to S/I$ by p, then we have a canonical (G, A)-linear map

$$\operatorname{Ext}^h_S(S/I, p) : \operatorname{Ext}^h_S(S/I, S) \to \operatorname{Ext}^h_S(S/I, S/I).$$

As we have $\operatorname{proj.dim}_S S/I = h$, this is surjective. As both $\operatorname{Ext}^h_S(S/I, S)$ and $\operatorname{Ext}^h_S(S/I, S/I)$ are rank-one S/I-projective, the map $\operatorname{Ext}^h_S(S/I, p)$ is an isomorphism. Moreover, the Yoneda product

$$K_S \otimes_S \operatorname{Ext}^h_S(S/I, S) \to \operatorname{Ext}^h_S(S/I, K_S) = K_A$$

is an isomorphism, as we have

$$\operatorname{Ext}^i_S(S/I, S) = 0 \quad (i \neq h) \quad \text{and} \quad \operatorname{Tor}^S_i(\operatorname{Ext}^h_S(S/I, S), K_S) = 0 \quad (i \neq 0).$$

Hence, we have

$$K_A \cong \operatorname{Ext}^h_S(S/I, S) \otimes_S K_S \cong (K_S \otimes_S A) \otimes_A \operatorname{Ext}^h_S(S/I, S/I) \cong \omega_{A/R} \otimes K_R.$$

3 follows from **1,2**. □

We call K_A the *G-equivariant pointwise dualizing module* of A.

Corollary 1.1.10 *Assume that* Spec R *is connected, and A has relative dimension d over R. Then as a G-module, we have $\operatorname{Ext}^d_A(R, K_A) \cong K_R$, where $R = A/\mathfrak{M}$.*

Lemma 1.1.11 *If K is a (G, A)-module whose underlying A-module is pointwise dualizing, then there is a unique rank-one R-projective G-module Λ such that $K \cong K_A \otimes \Lambda$. Conversely, for any rank-one R-projective G-module Λ, $K_A \otimes \Lambda$ is pointwise dualizing as an A-module.*

Proof. The converse is trivial. As $\operatorname{Hom}_A(K_A, K)$ is a rank-one A-projective (G, A)-module, there is a rank-one R-projective G-module Λ such that $\operatorname{Hom}_A(K_A, K) \cong A \otimes \Lambda$. Hence,

$$K_A \cong \operatorname{Hom}_A(\operatorname{Hom}_A(K_A, K), K) \cong \operatorname{Hom}(A \otimes \Lambda, K) \cong K \otimes \Lambda^*,$$

and we have $K \cong K_A \otimes \Lambda$. Such a Λ is unique, because $\Lambda \cong \operatorname{Hom}_A(K_A, K) \otimes_A A/\mathfrak{M}$. □

1.2 Reductive group actions on graded algebras

Throughout this subsection, let R be a noetherian commutative ring, G a split reductive R-group scheme over R, and let us consider a closed subgroup $\mathbb{G}_m$ of G contained in the center $Z(G)$ of G. Let A be a noetherian positively graded R-flat G-algebra. As in the last subsection, $\mathfrak{M}_A = \mathfrak{M}$ stands for the G-ideal $\bigoplus_{i>0} A_i$ of A.

(1.2.1) We fix $T \subset B \subset G$ and $\Delta \subset \Sigma$ as in (I.4.5.6), and follow the notation in (I.4.5).

Then as we have $\mathbb{G}_m \subset Z(G)^\circ \subset T$ at each geometric fiber, we have $\mathbb{G}_m \subset T$ over R, where $(?)^\circ$ denotes the connected component containing the unit element. This shows that a canonical map between the character groups $|?| : X(T) \to X(\mathbb{G}_m) \cong \mathbb{Z}$ is defined as a restriction. For any $\lambda \in X(T)$, we call $|\lambda|$ the *degree* of λ. The set $|?|^{-1}(r)$ of weights of T of degree r is denoted by X_r. For $\pi \subset X(T)$, we denote $\pi \cap X_r$ by π_r.

For a poset ideal π of X^+, $\mathcal{A}_{G,A}(\pi)$ denotes the category of A-finite (G, A)-modules belonging to π as G-modules. Let $\mathcal{A}_{G,A}$ stand for $\mathcal{A}_{G,A}(X^+)$. Note that $\mathcal{A}_{G,A}(\pi)$ is a very thick full subcategory of $\mathcal{A}_{G,A} = {}_{G,A}\mathbb{M}_f$.

We say that a poset ideal Γ of X^+ is *t-closed* if $C_G(\Gamma)$ is an R-subalgebra (hence it is a subbialgebra) of $H = R[G]$. If Γ is t-closed and V and W are $C_G(\Gamma)$-comodules, then so is $V \otimes W$. We say that a poset ideal Γ of X^+ is t-closed with respect to A if $A \otimes V$ belongs to Γ for any G-module V belonging to Γ.

By definition, if Γ is a t-closed poset ideal of X^+ and A belongs to Γ, then Γ is t-closed with respect to A.

(1.2.2) Let Γ be a poset ideal of X^+ which is t-closed with respect to A.

Lemma 1.2.3 *For $V \in {}_G\mathbb{M}_f$, the following are equivalent.*

1 $V \in \mathbb{M}_f^{C_G(\Gamma)}$.

2 $A \otimes V \in \mathcal{A}_{G,A}(\Gamma)$.

Proof. **1⇒2** This is obvious, as Γ is assumed to be t-closed with respect to A.

2⇒1 If $A \otimes V \in \mathcal{A}_{G,A}(\Gamma)$, then $V \cong (A \otimes V) \otimes_A A/\mathfrak{M}_A$ is a quotient of $A \otimes V$. Hence, it is an R-finite G-module belonging to Γ. □

(1.2.4) For $V \in \mathcal{A}_{G,A}(\Gamma)$ and $s, t \in \mathbb{Z}$ $(s \leq t)$, we set

$$V_{[s,t]} := V_s \oplus V_{s+1} \oplus \cdots \oplus V_t.$$

For $s, t \in \mathbb{Z}$ $(s \leq t)$ and $\pi \subset X^+$, we denote by ${}_{(\pi,s,t;A)}\mathbb{M}_f(\Gamma)$ the full subcategory of $\mathcal{A}_{G,A}(\Gamma)$ consisting of $M \in \mathcal{A}_{G,A}(\Gamma)$ such that $M_{[s,t]}$ belongs to π. Note that ${}_{(\pi,s,t;A)}\mathbb{M}_f(\Gamma)$ is a very thick full subcategory of $\mathcal{A}_{G,A}(\Gamma)$.

Let π be a finite poset ideal of $X^+ = X^+_G$. The Schur algebra of G with respect to π is denoted by $S_G(\pi) = S(\pi)$. By definition, $(?)_{[s,t]}$ is an exact functor from ${}_{(\pi,s,t;A)}\mathbb{M}_f(\Gamma)$ to ${}_{S(\pi)}\mathbb{M}_f$.

Remark 1.2.5 The conjugate action of G on itself is trivial on $\mathbb{G}_m \subset Z(G)$. Hence, it follows that any root of G is of degree 0. It follows that if $\mu \leq \lambda$, then $|\mu| = |\lambda|$.

Lemma 1.2.6 *Let ρ be a finite poset ideal of Γ, and V an R-finite G-module which is a quotient of a finite direct sum of copies of A. Then there is a finite poset ideal π of Γ such that $V \otimes W$ is an $S(\pi)$-module for any $S(\rho)$-module W.*

Proof. By assumption, $V \otimes S(\rho)$ is a quotient of a finite direct sum of copies of $A \otimes S(\rho)$. As $A \otimes S(\rho)$ belongs to Γ, we have that $V \otimes S(\rho)$ also belongs to Γ. As $V \otimes S(\rho)$ is R-finite, it belongs to π for a poset ideal π of Γ. Let W be an $S(\rho)$-module. Then as $V \otimes W$ is a homomorphic image of a direct sum of copies of $V \otimes S(\rho)$, it belongs to π. □

Definition 1.2.7 Let $s, t \in \mathbb{Z}$ $(s \leq t)$. We say that a subset π of X^+ is $(s, t; A, \Gamma)$-closed if the following conditions are satisfied.

i $\pi = \pi_s \cup \pi_{s+1} \cup \cdots \cup \pi_t$

ii π is a finite poset ideal of Γ

iii If $M \in {}_G\mathbb{M}$ belongs to π, then $(A \otimes M)_{[s,t]}$ belongs to π.

Lemma 1.2.8 *Let $s, t \in \mathbb{Z}$ $(s \leq t)$, and ρ be a finite subset of Γ such that $\rho = \rho_s \cup \rho_{s+1} \cup \cdots \cup \rho_t$. Then there is an $(s, t; A, \Gamma)$-closed set containing ρ.*

Proof. If $\lambda \in \rho$ and $\mu \leq \lambda$, then we have $|\mu| = |\lambda|$ by Remark 1.2.5. Hence, we have $s \leq |\mu| \leq t$. Replacing ρ by $\bigcup_{\lambda \in \rho}(-\infty, \lambda]$, we may assume that ρ is a finite poset ideal of Γ.

We prove the lemma by induction on $w = t - s$. If $w = 0$, then $\pi := \rho$ is $(s, t; A, \Gamma)$-closed, and we are done. In fact, if M belongs to π, then $M = M_s$. As A is positively graded, $(A \otimes M)_{[s,s]} = A_0 \otimes M_s \cong M$ also belongs to π.

So we assume $w > 0$. By induction assumption applied to $\rho_{[s,t-1]} := \rho_s \cup \cdots \cup \rho_{t-1}$, there is an $(s, t-1; A, \Gamma)$-closed subset π' containing $\rho_{[s,t-1]}$. By Lemma 1.2.6, there is a finite poset ideal τ of Γ such that for any $S(\pi')$-module W, $A_{[1,w]} \otimes W$ belongs to τ. We set $\pi_i := \pi'_i$ $(s \leq i < t)$, $\pi_t := \tau_t \cup \rho_t$, and $\pi := \pi_s \cup \cdots \cup \pi_t$. Then π is a finite poset ideal of Γ.

If M is an $S(\pi)$-module, then $M = M_{[s,t]}$. Hence,

$$(A \otimes M)_{[s,t-1]} = (A \otimes M_{[s,t-1]})_{[s,t-1]}.$$

As $M_{[s,t-1]}$ belongs to π', $(A \otimes M)_{[s,t-1]}$ also belongs to π' by the $(s, t-1; A, \Gamma)$-closedness, and hence $(A \otimes M)_{[s,t-1]}$ belongs to π. On the other hand, we have

$$(A \otimes M)_t = (A_{[1,w]} \otimes M_{[s,t-1]})_t \oplus M_t.$$

As M_t is a direct summand of M, it belongs to π. Moreover, $A_{[1,w]} \otimes M_{[s,t-1]}$ belongs to τ by the choice of τ. Hence, $(A_{[1,w]} \otimes M_{[s,t-1]})_t$ belongs to τ_t, and it also belongs to π. Hence, π is $(s, t; A, \Gamma)$-closed, and obviously it contains ρ. □

If π is an $(s, t; A, \Gamma)$-closed set and $M \in {}_{S(\pi)}\mathbb{M}_f$, then we have $A \otimes M \in {}_{(\pi,s,t;A)}\mathbb{M}_f(\Gamma)$. Hence, $(?)_{[s,t]} : {}_{(\pi,s,t;A)}\mathbb{M}_f(\Gamma) \to {}_{S(\pi)}\mathbb{M}_f$ has a left adjoint $A \otimes ?$. Namely, for $V \in {}_{S(\pi)}\mathbb{M}_f$ and $M \in {}_{(\pi,s,t;A)}\mathbb{M}_f(\Gamma)$, the standard map

$$\text{(1.2.9)} \qquad \operatorname{Hom}_{S(\pi)}(V, M_{[s,t]}) \cong \operatorname{Hom}_{{}_{(\pi,s,t;A)}\mathbb{M}_f(\Gamma)}(A \otimes V, M)$$

is an isomorphism.

Lemma 1.2.10 *Let π be an $(s, t; A, \Gamma)$-closed set. Then for $V \in {}_{S(\pi)}\mathbb{M}_f$, $M \in {}_{(\pi,s,t;A)}\mathbb{M}_f(\Gamma)$ and $i \geq 0$, we have*

$$\operatorname{Ext}^i_{S(\pi)}(V, M_{[s,t]}) \cong \operatorname{Ext}^i_{{}_{(\pi,s,t;A)}\mathbb{M}_f(\Gamma)}(A \otimes V, M).$$

Proof. If P is a finite $S(\pi)$-projective module, then $A \otimes P$ is a projective object of ${}_{(\pi,s,t;A)}\mathbb{M}_f(\Gamma)$ by the isomorphism (1.2.9). Hence, if $\mathbb{F}$ is an $S(\pi)$-projective resolution of V, then $A \otimes \mathbb{F}$ is a ${}_{(\pi,s,t;A)}\mathbb{M}_f(\Gamma)$-projective resolution of $A \otimes V$, since A is R-flat.

Hence, we have

$$\operatorname{Ext}^i_{S(\pi)}(V, M_{[s,t]}) \cong H^i(\operatorname{Hom}_{S(\pi)}(\mathbb{F}, M_{[s,t]})) \cong$$
$$H^i(\operatorname{Hom}_{{}_{(\pi,s,t;A)}\mathbb{M}_f(\Gamma)}(A \otimes \mathbb{F}, M)) \cong \operatorname{Ext}^i_{{}_{(\pi,s,t;A)}\mathbb{M}_f(\Gamma)}(A \otimes V, M).$$

□

Lemma 1.2.11 *Let π be an $(s, t; A, \Gamma)$-closed set, $N \in {}_{(\pi,s,t;A)}\mathbb{M}_f(\Gamma)$, and V be an $S(\pi)$-projective module. Then we have $\operatorname{Ext}^i_{\mathcal{A}_{G,A}(\Gamma)}(A \otimes V, N) = 0$ $(i > 0)$.*

Proof. Let α be an element of $\operatorname{Ext}^i_{\mathcal{A}_{G,A}(\Gamma)}(A \otimes V, N)$ represented by an exact sequence

$$\text{(1.2.12)} \qquad 0 \to N \to M_1 \to \cdots \to M_i \to A \otimes V \to 0$$

in $\mathcal{A}_{G,A}(\Gamma)$. By Lemma 1.2.8, there is an $(s, t; A, \Gamma)$-closed set π' such that $\pi' \supset \pi$ and $[M_1 \oplus \cdots \oplus M_i]_{[s,t]}$ belongs to π'.

By Theorem III.2.3.17 and Lemma 1.2.10, we have

$$\operatorname{Ext}^i_{(\pi',s,t;A)\mathbb{M}_f(\Gamma)}(A\otimes V, N) \cong \operatorname{Ext}^i_{S(\pi')}(V, N_{[s,t]})$$
$$\cong \operatorname{Ext}^i_G(V, N_{[s,t]}) \cong \operatorname{Ext}^i_{S(\pi)}(V, N_{[s,t]}) = 0$$

for $i > 0$. Hence, α is already zero in $\operatorname{Ext}^i_{(\pi',s,t;A)\mathbb{M}_f(\Gamma)}(A\otimes V, N)$. Hence, it is zero in

$$\operatorname{Ext}^i_{\mathcal{A}_{G,A}(\Gamma)}(A\otimes V, N).$$

□

Lemma 1.2.13 *For $M, N \in \mathcal{A}_{G,A}(\Gamma)$, there exists an $X \in \mathcal{X}^{\mathrm{pro}}_{C_G(\Gamma)}$ such that*

i *There is a surjective (G,A)-linear map $A\otimes X \to M$;*

ii $\operatorname{Ext}^i_{\mathcal{A}_{G,A}(\Gamma)}(A\otimes X, N) = 0$ $(i>0)$;

iii $\operatorname{Ext}^i_{G,A}(A\otimes X, N) = 0$ $(i>0)$

are satisfied.

Proof. There exists a $V \in \mathbb{M}_f^{C_G(\Gamma)}$ such that there is a surjective (G,A)-linear map $A\otimes V \to M$. As Γ is t-closed with respect to A, we may assume that $M = A\otimes V$.

We take $s,t \in \mathbb{Z}$ $(s\le t)$ so that $V = V_{[s,t]}$. Take an $(s,t;A,\Gamma)$-closed set π to which $V\oplus N_{[s,t]}$ belongs. Take a surjective $S(\pi)$-linear map $X \to V$ such that X is $S(\pi)$-finite projective. Then $X \in \mathcal{X}^{\mathrm{pro}}_{C_G(\Gamma)}$. Hence, it suffices to prove that conditions **i**–**iii** are satisfied for this choice.

Condition **i** is obvious. Condition **ii** is satisfied by Lemma 1.2.11. We check **iii**. By Proposition II.1.1.14 and Theorem III.2.3.17, we have

$$\operatorname{Ext}^i_{G,A}(A\otimes X, N) \cong \operatorname{Ext}^i_G(X, N) \cong \operatorname{Ext}^i_G(X, N_{[s,t]}) \cong \operatorname{Ext}^i_{S(\pi)}(X, N_{[s,t]}) = 0$$

for $i > 0$. □

Theorem 1.2.14 *Let $M, N \in \mathcal{A}_{G,A}(\Gamma)$. Then the canonical map*

$$\operatorname{Ext}^i_{\mathcal{A}_{G,A}(\Gamma)}(M, N) \to \operatorname{Ext}^i_{G,A}(M,N)$$

is an isomorphism for $i \ge 0$.

Proof. We use induction on i. The case $i = 0$ is trivial. If $i > 0$, then we take $A\otimes X \to M$ as in Lemma 1.2.13, and we have a short exact sequence

$$0 \to \Omega \to A\otimes X \to M \to 0.$$

Then we have an associated map between the two long exact sequences of Ext, and by the choice of $A\otimes X \to M$, we are done by induction assumption applied to Ω. □

Lemma 1.2.15 *Let $M, N \in \mathcal{A}_{G,A}$. Then $\operatorname{Ext}^i_{G,A}(M, N)$ is R-finite for $i \geq 0$.*

Proof. By Lemma 1.2.13, there exists an exact sequence

$$\mathbb{X} : \cdots \to A \otimes X_1 \to A \otimes X_0 \to M \to 0$$

such that $X_j \in \mathcal{X}_G^{\mathrm{pro}}$ and $\operatorname{Ext}^i_{G,A}(A \otimes X_j, N) = 0$ for $i > 0$, for $j \geq 0$. So we have $\operatorname{Ext}^i_{G,A}(M, N) \cong H^i(\operatorname{Hom}_{G,A}(\mathbb{X}, N))$. So it suffices to prove that $\operatorname{Hom}_{G,A}(A \otimes X_i, N) \cong \operatorname{Hom}_G(X_i, N)$ is R-finite for each i. If we take $s, t \in \mathbb{Z}$ such that $X_i = (X_i)_{[s,t]}$, then we have $\operatorname{Hom}_G(X_i, N) \cong \operatorname{Hom}_G(X_i, N_{[s,t]})$, which is R-finite. □

1.3 Relative Ringel's approximation

(1.3.1) In this subsection, we follow the notation in the last subsection. So let R be a noetherian commutative ring, G a split reductive R-group scheme with a fixed $\mathbb{G}_m \subset Z(G)$, and A a noetherian R-flat positively graded G-algebra. We set $\mathfrak{M}_A := \bigoplus_{i \geq 1} A_i$. Let Γ be a finite poset ideal of X^+ which is t-closed with respect to A.

Moreover, in this subsection, we assume that A is good as a G-module.

Lemma 1.3.2 *Let V be a good G-module, and M a good (G, A)-module. If V is R-finite R-projective or M is R-flat A-finite, then $M \otimes V$ is good. In particular, $A \otimes V$ is good, if V is a good G-module.*

Proof. It suffices to show that $M_i \otimes V$ is good for each $i \in \mathbb{Z}$. The assertion follows immediately from Corollary III.4.5.10. □

Lemma 1.3.3 *Any rank-one A-projective (G, A)-module is good.*

Proof. Let P be a rank-one A-projective (G, A)-module. Then by Corollary 1.1.5, there exists a rank-one R-projective G-module Λ such that $P \cong A \otimes \Lambda$. By Corollary III.4.5.12, Λ is tilting. By Lemma 1.3.2, P is good. □

(1.3.4) We denote the category of R-finite R-projective G-modules belonging to Γ by $\mathcal{P}_0(\Gamma)$. We denote the category of A-finite A-projective (G, A)-modules belonging to Γ by $\mathcal{P}(\Gamma)$.

Lemma 1.3.5 *We have $\mathcal{P}(\Gamma) = \mathcal{F}(A \otimes \mathcal{P}_0(\Gamma))$.*

Proof. As we are assuming that Γ is t-closed with respect to A, $\mathcal{F}(A \otimes \mathcal{P}_0(\Gamma)) \subset \mathcal{P}(\Gamma)$ is obvious. We prove the opposite direction. If $A \otimes V \in \mathcal{P}(\Gamma)$ and $V \in \mathcal{P}_0$, then we have $V \in \mathbb{M}^{C_G(\Gamma)}$ by Lemma 1.2.3. By Lemma 1.1.4, any element of $\mathcal{P}(\Gamma)$ is contained in $\mathcal{F}(A \otimes \mathcal{P}_0(\Gamma))$ because $\mathcal{F}(A \otimes \mathcal{P}_0(\Gamma))$ is closed under extensions. □

Lemma 1.3.6 *Let $F \in \mathcal{P}(\Gamma)$. Then the following hold:*

1 *F is good $\iff$ $F/\mathfrak{m}F$ is ∇-good $\iff$ $F \in \mathcal{F}(A \otimes (\mathcal{Y}_G^{\text{pro}} \cap \mathcal{P}_0(\Gamma)))$.*

2 *$\operatorname{Hom}_A(F, A)$ is good $\iff$ $F/\mathfrak{m}F$ is Δ-good $\iff$ $F \in \mathcal{F}(A \otimes \mathcal{X}_{C_G(\Gamma)}^{\text{pro}})$.*

Proof. By Lemma 1.1.4 and Lemma 1.2.3, we may assume that $F = A \otimes V$ with $V \in \mathcal{P}_0(\Gamma)$. Then as we have $V = F/\mathfrak{M}_A F$ and $\operatorname{Hom}_A(A \otimes V, A) \cong A \otimes V^*$, assertion **2** is reduced to **1** by Corollary III.4.1.8.

We prove the first equivalence in **1**. The $\Leftarrow$ direction is obvious by Lemma 1.3.2. The $\Rightarrow$ is also obvious, because $V = A_0 \otimes V$ is a direct summand of $A \otimes V$ as a G-module. The second equivalence is easy. □

Now we set

$$\mathcal{X}_{G,A}^{\text{pro}}(\Gamma) := \{M \in \mathcal{A}_{G,A}(\Gamma) \mid M \text{ is } A\text{-projective, and } \operatorname{Hom}_A(M, A) \text{ is good}\},$$
$$\mathcal{Y}_{G,A}^{\text{pro}}(\Gamma) := \{N \in \mathcal{A}_{G,A}(\Gamma) \mid N \text{ is good, and } \operatorname{proj.dim}_A N < \infty\},$$

and $\omega_{G,A}^{\text{pro}}(\Gamma) := \mathcal{X}_{G,A}^{\text{pro}}(\Gamma) \cap \mathcal{Y}_{G,A}^{\text{pro}}(\Gamma)$. If $\Gamma = X_G^+$, then we may drop (Γ), and may write $\mathcal{X}_{G,A}^{\text{pro}}$, $\mathcal{Y}_{G,A}^{\text{pro}}$, $\omega_{G,A}^{\text{pro}}$, $\mathcal{P}$, $\mathcal{P}_0$, and so on.

Lemma 1.3.7 *The full subcategory $\mathcal{X}_{G,A}^{\text{pro}}(\Gamma)$ of $\mathcal{A}_{G,A}(\Gamma)$ is closed under extensions, direct summands, and epikernels. The full subcategory $\mathcal{Y}_{G,A}^{\text{pro}}(\Gamma)$ of $\mathcal{A}_{G,A}(\Gamma)$ is closed under extensions, direct summands, and monocokernels.*

Proof. This is trivial. □

Lemma 1.3.8 *If $X \in \mathcal{X}_{G,A}^{\text{pro}}$ and Y is a good (G, A)-module, then we have $\operatorname{Ext}_{G,A}^i(X, Y) = 0$ $(i > 0)$.*

Proof. By Lemma 1.3.6, we may assume that $X = A \otimes V$, with $V \in \mathcal{X}_G^{\text{pro}} = \operatorname{add} \mathcal{F}(\Delta_G)$. By Proposition II.1.1.14, we have

$$\operatorname{Ext}_{G,A}^i(A \otimes V, Y) \cong \operatorname{Ext}_G^i(V, Y) = 0$$

for $i > 0$. □

Proposition 1.3.9 *We have $\omega_{G,A}^{\text{pro}}(\Gamma) = A \otimes \omega_{C_G(\Gamma)}^{\text{pro}}$, and $\omega_{G,A}^{\text{pro}}(\Gamma)$ is an injective cogenerator of $\mathcal{X}_{G,A}^{\text{pro}}(\Gamma)$.*

Proof. By Lemma 1.3.6, the inclusion $A \otimes \omega_{C_G(\Gamma)}^{\text{pro}} \subset \omega_{G,A}^{\text{pro}}(\Gamma)$ follows. We prove the inclusion in the opposite direction. Let $T \in \omega_{G,A}^{\text{pro}}(\Gamma)$. By Lemma 1.1.4, there is a filtration

$$0 = T_{(0)} \subset \cdots \subset T_{(r)} = T$$

of T such that $T_{(i)}/T_{(i-1)} \cong A \otimes Q_{(i)}$, with $Q_{(i)}$ an R-finite R-projective G-module, and we have $Q := \bigoplus_i Q_{(i)} \cong T/\mathfrak{M}_A T$. By Lemma 1.2.3 and Lemma 1.3.6, Q is a tilting G-module belonging to Γ, and hence so is its direct summand $Q_{(i)}$ for each i. By Lemma 1.3.8, we have

$$\operatorname{Ext}^1_{G,A}(A \otimes Q_{(i)}, A \otimes Q_{(j)}) = 0$$

for $1 \leq i, j \leq r$. Hence, the filtration splits, and we have $T = \bigoplus_i A \otimes Q_{(i)} \cong A \otimes Q$. This shows $T \in A \otimes \omega^{\mathrm{pro}}_{C_G(\Gamma)}$.

Now we prove that $\omega^{\mathrm{pro}}_{G,A}(\Gamma)$ is an injective cogenerator of $\mathcal{X}^{\mathrm{pro}}_{G,A}(\Gamma)$. Let $X = A \otimes V$, with $V \in \mathcal{X}^{\mathrm{pro}}_{C_G(\Gamma)}$. As $\omega^{\mathrm{pro}}_{C_G(\Gamma)}$ is an injective cogenerator of $\mathcal{X}^{\mathrm{pro}}_{C_G(\Gamma)}$ by Theorem III.4.1.20, there is an exact sequence of G-modules

$$0 \to V \to Q \to V' \to 0$$

such that $Q \in \omega^{\mathrm{pro}}_{C_G(\Gamma)}$ and $V' \in \mathcal{X}^{\mathrm{pro}}_{C_G(\Gamma)}$. Then

$$0 \to A \otimes V \to A \otimes Q \to A \otimes V' \to 0$$

is exact, and we have $A \otimes Q \in \omega^{\mathrm{pro}}_{G,A}(\Gamma)$ and $A \otimes V' \in \mathcal{X}^{\mathrm{pro}}_{G,A}(\Gamma)$. In view of Lemma 1.3.6, Lemma 1.3.7 and Lemma 1.3.8, we have that $\omega^{\mathrm{pro}}_{G,A}(\Gamma)$ is an injective cogenerator of $\mathcal{X}^{\mathrm{pro}}_{G,A}(\Gamma)$ by Corollary I.1.12.13. □

Lemma 1.3.10 *We have* $\widehat{\mathcal{X}^{\mathrm{pro}}_{G,A}(\Gamma)} = \{M \in \mathcal{A}_{G,A}(\Gamma) \mid \operatorname{proj.dim}_A M < \infty\} \supset \mathcal{Y}^{\mathrm{pro}}_{G,A}(\Gamma)$.

Proof. By Lemma 1.2.13, for any $M \in \mathcal{A}_{G,A}(\Gamma)$, there is an $A \otimes X \in \mathcal{X}^{\mathrm{pro}}_{G,A}(\Gamma)$ and a surjective (G, A)-linear map $f : A \otimes X \to M$. Then as $\operatorname{Ker} f$ belongs to Γ, we may proceed by induction on $\operatorname{proj.dim}_A M$, and it suffices to prove that if $M \in \mathcal{P}(\Gamma)$, then $M \in \widehat{\mathcal{X}^{\mathrm{pro}}_{G,A}(\Gamma)}$. As we have

$$M/\mathfrak{M}_A M \in \mathbb{M}_f^{C_G(\Gamma)} = \widehat{\mathcal{X}^{\mathrm{pro}}_{C_G(\Gamma)}},$$

we prove this assertion by induction on

$$s := \mathcal{X}^{\mathrm{pro}}_{C_G(\Gamma)}\text{-resol.dim}\, M/\mathfrak{M}_A M = \mathcal{Y}^{\mathrm{pro}}_{C_G(\Gamma)}\text{-proj.dim}\, M/\mathfrak{M}_A M.$$

If $s = 0$, then $M/\mathfrak{M}_A M$ is Δ-good, and we have $M \in \mathcal{X}^{\mathrm{pro}}_{G,A}(\Gamma)$ by Lemma 1.3.6. So this case is obvious. Let us consider the case $s > 0$. Then there is an exact sequence

$$0 \to M' \to A \otimes X \to M \to 0$$

such that $X \in \mathcal{X}^{\mathrm{pro}}_{C_G(\Gamma)}$. Then $M' \in \mathcal{P}(\Gamma)$, and

$$0 \to M'/\mathfrak{M}_A M' \to X \to M/\mathfrak{M}_A M \to 0$$

is exact. As $\mathcal{Y}^{\text{pro}}_{CG(\Gamma)}$-proj.dim $M'/\mathfrak{M}_A M' = s-1$, $M' \in \widehat{\mathcal{X}^{\text{pro}}_{G,A}(\Gamma)}$ by induction assumption. As $A \otimes X \in \mathcal{X}^{\text{pro}}_{G,A}(\Gamma)$, we have $M \in \widehat{\mathcal{X}^{\text{pro}}_{G,A}(\Gamma)}$, and the assertion is proved. □

Combining the results above, we conclude the following.

Theorem 1.3.11 *The triple*

$$(\mathcal{X}^{\text{pro}}_{G,A}(\Gamma), \mathcal{Y}^{\text{pro}}_{G,A}(\Gamma), \omega^{\text{pro}}_{G,A}(\Gamma))$$

is a weak Auslander–Buchweitz context of $\mathcal{A}_{G,A}(\Gamma)$. Moreover, we have

$$\begin{aligned} \omega^{\text{pro}}_{G,A}(\Gamma) &= A \otimes \omega^{\text{pro}}_{CG(\Gamma)} \\ \widehat{\mathcal{X}^{\text{pro}}_{G,A}(\Gamma)} &= \{M \in \mathcal{A}_{G,A}(\Gamma) \mid \text{proj.dim}_A M < \infty\}. \end{aligned}$$

In particular, if A is a regular ring, then $(\mathcal{X}^{\text{pro}}_{G,A}(\Gamma), \mathcal{Y}^{\text{pro}}_{G,A}(\Gamma), \omega^{\text{pro}}_{G,A}(\Gamma))$ is an Auslander–Buchweitz context of $\mathcal{A}_{G,A}(\Gamma)$.

Corollary 1.3.12 *Let h be a non-negative integer, $M \in \mathcal{A}_{G,A}$ and $0 \neq M$. Then the following are equivalent.*

1 $\mathcal{X}^{\text{pro}}_{G,A}(\Gamma)$-resol.dim$(M) = h$ *and* $\text{grade}_A M \geq h$.

2 *M is a G-module belonging to Γ, M is perfect of codimension h, and $\text{Ext}^h_A(M, A)$ is good.*

Proof. We prove **1**⇒**2**. Let

$$\mathbb{F} : 0 \to F_h \to \cdots F_1 \to F_0 \to M \to 0$$

be an $\mathcal{X}^{\text{pro}}_{G,A}(\Gamma)$-resolution of M of length h. As this is an A-projective resolution of M, we have

$$h \geq \text{proj.dim}_A M \geq \text{grade}_A M \geq h,$$

and we have that M is perfect of codimension h. As F_0 belongs to Γ, M belongs to Γ. Moreover, as $\text{Hom}_A(\mathbb{F}, A)[h]$ is a resolution of $\text{Ext}^h_A(M, A)$ and $\text{Hom}_A(F_i, A)$ is good for each i, $\text{Ext}^h_A(M, A)$ is good, as good modules are closed under monocokernels.

We prove **2**⇒**1**. By the theorem, we have $M \in \widehat{\mathcal{X}^{\text{pro}}_{G,A}(\Gamma)}$. Hence, there is an exact sequence

$$0 \to N \to F_{h-1} \to \cdots \to F_1 \to F_0 \to M \to 0 \tag{1.3.13}$$

such that $F_i \in \mathcal{X}^{\text{pro}}_{G,A}(\Gamma)$ for $i = 0, 1, \ldots, h-1$. It suffices to prove $N \in \mathcal{X}^{\text{pro}}_{G,A}(\Gamma)$. As we have $\text{proj.dim}_A M = h$, it is clear that $N \in \mathcal{P}(\Gamma)$. Hence, the sequence (1.3.13) is an A-projective resolution of M. Hence, the sequence

$$0 \to F_0^\star \to F_1^\star \to \cdots \to F_{h-1}^\star \to N^\star \to \mathrm{Ext}_A^h(M, A) \to 0$$

is exact, where $(?)^\star = \mathrm{Hom}_A(?, A)$. As each $F_i^\star$ and $\mathrm{Ext}_A^h(M, A)$ is good, $N^\star$ is also good, since good G-modules are closed under extensions and monocokernels. Hence, we have $N \in \mathcal{X}_{G,A}^{\mathrm{pro}}(\Gamma)$. □

Corollary 1.3.14 *Let h be a non-negative integer, $M \in \mathcal{A}_{G,A}$, and $0 \neq M$. Then the following are equivalent.*

1 $\omega_{G,A}^{\mathrm{pro}}(\Gamma)$-resol.dim$(M) = h$ *and* $\mathrm{grade}_A M \geq h$.

2 *M is a good G-module belonging to Γ, M is perfect of codimension h, and* $\mathrm{Ext}_A^h(M, A)$ *is good.*

Proof. **1⇒2** M is perfect of codimension h and $\mathrm{Ext}_A^h(M, A)$ is good by Corollary 1.3.12. On the other hand, as any object in $\omega_{G,A}^{\mathrm{pro}}(\Gamma)$ is good and good modules are closed under monocokernels, it follows that M is good.

2⇒1 As we have $M \in \mathcal{Y}_{G,A}^{\mathrm{pro}}(\Gamma)$,

$$\omega_{G,A}^{\mathrm{pro}}(\Gamma)\text{-resol.dim}(M) = \mathcal{X}_{G,A}^{\mathrm{pro}}(\Gamma)\text{-resol.dim}(M) = h$$

by Corollary 1.3.12 and Theorem I.1.12.10, **7**. □

(1.3.15) Let us denote the full subcategory of good G-modules in ${}_G\mathbb{M}$ by $\mathcal{G}$. For a G-module V, it is easy to see that Δ_G-inj.dim $V = \mathcal{G}$-cores.dim V. Hence, $\check{\mathcal{G}}$ is closed under extensions, epikernels, monocokernels, and direct summands in ${}_G\mathbb{M}$.

Lemma 1.3.16 *If M is an A-finite (G, A)-module with* $\mathrm{proj.dim}_A M < \infty$, *then $M \in \check{\mathcal{G}}$ as a G-module.*

Proof. As $\check{\mathcal{G}}$ is closed under monocokernels and $M \in \widehat{\mathcal{X}_{G,A}^{\mathrm{pro}}}$, we may assume that $M \in \mathcal{P}$. As $\check{\mathcal{G}}$ is also closed under extensions, we may assume that $M = A \otimes V$ with $V \in \mathcal{P}_0$, by Lemma 1.1.4. Then as $V^* \in \widehat{\mathcal{X}_G^{\mathrm{pro}}}$, it follows that $V \in \check{\mathcal{Y}}_G^{\mathrm{pro}}$. By Lemma 1.3.2, $A \otimes V \in \check{\mathcal{G}}$. □

Theorem 1.3.17 *Let $(R, \mathfrak{m})$ be local. Then the following are equivalent for a sequence*

$$\mathbb{F} : 0 \to Y \xrightarrow{i} X \xrightarrow{p} V \to 0$$

in $\mathcal{A}_{G,A}(\Gamma)$, where $\hat{R}$ is the $\mathfrak{m}$-adic completion of R, and $\hat{?} = ? \otimes \hat{R}$.

1 *$\mathbb{F}$ is an $\mathcal{X}^{\mathrm{pro}}_{G,A}(\Gamma)$-approximation of V, and p is right minimal.*

2 *$\mathbb{F}$ is an $\mathcal{X}^{\mathrm{pro}}_{G,A}(\Gamma)$-approximation of V, and there is no non-zero direct summand of X contained in $\operatorname{Im} i$.*

3 *$\hat{\mathbb{F}}$ is an $\mathcal{X}^{\mathrm{pro}}_{\hat{G},\hat{A}}(\Gamma)$-approximation of $\hat{V}$, and $\hat{p}$ is right minimal.*

4 *$\hat{\mathbb{F}}$ is an $\mathcal{X}^{\mathrm{pro}}_{\hat{G},\hat{A}}(\Gamma)$-approximation of $\hat{V}$, and there is no non-zero direct summand of $\hat{X}$ contained in $\operatorname{Im}\hat{i}$.*

In particular, there is a unique minimal $\mathcal{X}^{\mathrm{pro}}_{G,A}(\Gamma)$-approximation of V, if $\operatorname{proj.dim}_A V < \infty$. *Similarly, there is a unique minimal $\mathcal{Y}^{\mathrm{pro}}_{G,A}(\Gamma)$-hull of V, if* $\operatorname{proj.dim}_A V < \infty$.

Proof. As $\operatorname{End}_{\hat{G},\hat{A}} \hat{X}$ is $\hat{R}$-finite by Lemma 1.2.15, it is semiperfect. Hence, the theorem is proved similarly to Theorem III.4.2.6. □

Corollary 1.3.18 *Assume that $(R,\mathfrak{m})$ is local. If $M \in \mathcal{Y}^{\mathrm{pro}}_{G,A}(\Gamma)$, then there is an $\omega^{\mathrm{pro}}_{G,A}(\Gamma)$-resolution $\mathbb{F}$ of M such that for any (possibly infinite) $\omega^{\mathrm{pro}}_{G,A}(\Gamma)$-resolution $\mathbb{G}$ of M, $\mathbb{F}$ is a direct summand of $\mathbb{G}$ as a (G,A)-complex. Such an $\mathbb{F}$ is unique up to isomorphisms of (G,A)-complexes, and the length of $\mathbb{F}$ is finite and is equal to* $\omega^{\mathrm{pro}}_{G,A}(\Gamma)\text{-resol.dim}(M)$.

Proof. As the uniqueness is obvious, we prove the existence. We use induction on $s := \omega^{\mathrm{pro}}_{G,A}(\Gamma)\text{-resol.dim}(M)$.

If $s = 0$, then we set $\mathbb{F} = \cdots \to 0 \to 0 \to M \to 0$. If $s > 0$, then we take a minimal $\mathcal{X}^{\mathrm{pro}}_{G,A}(\Gamma)$-approximation

$$0 \to Y \to X \to M \to 0$$

of M. We set $F_0 := X$, and we denote the unique $\omega^{\mathrm{pro}}_{G,A}(\Gamma)$-resolution with the desired property of Y, which exists by induction assumption, by $\mathbb{F}'$. We define $\mathbb{F}$ to be the augmented complex $\mathbb{F}' \to F_0 \to 0$, which is obviously an $\omega^{\mathrm{pro}}_{G,A}(\Gamma)$-resolution of M, and is of length s by induction assumption.

Let

$$\mathbb{G} : \cdots \to G_1 \xrightarrow{\partial_1} G_0 \to M \to 0$$

be an arbitrary $\omega^{\mathrm{pro}}_{G,A}(\Gamma)$-resolution of M. It suffices to prove that $\mathbb{F}$ is a direct summand of $\mathbb{G}$ as a (G,A)-complex.

We set $I_i := \operatorname{Im}\partial_i$. As each homogeneous component of I_i is R-finite, it belongs to some finite poset ideal π of Γ. As we have $\operatorname{gl.dim} S(\pi) < \infty$ for any finite poset ideal π of Γ and I_i admits a (possibly infinite) good resolution

$$\cdots \to G_{i+1} \to G_i \to I_i \to 0,$$

each homogeneous component of I_i is good, and hence I_i is good. This shows that

$$\alpha : 0 \to I_1 \to G_0 \to M \to 0$$

is an $\mathcal{X}^{\mathrm{pro}}_{G,A}(\Gamma)$-approximation of M. If $s = 0$, then α splits, and hence $\mathbb{F}$ is a direct summand of $\mathbb{G}$.

Now assume that $s > 0$. By the theorem, α is isomorphic to the direct sum $\beta \oplus \mathrm{id}_W$ for some $W \in \omega^{\mathrm{pro}}_{G,A}(\Gamma)$, where

$$\beta : 0 \to J_1 \to F_0 \to M \to 0 \qquad (J_1 := \mathrm{Im}\,\partial^{\mathbb{F}}_1 = Y)$$

and

$$\mathrm{id}_W : 0 \to W \to W \to 0 \to 0.$$

As the kernel G'_1 of the composite map $G_1 \to I_1 \to W$ is an extension of I_2 by J_1, it is good. This shows that $G_1 \to W$ splits, and we have that $G_1 \cong G'_1 \oplus W$, and hence $G'_1 \in \omega^{\mathrm{pro}}_{G,A}(\Gamma)$. Now we know that $\mathbb{G}$ is isomorphic to the direct sum of id_W and

$$\mathbb{G}' : \cdots \to G_2 \to G'_1 \to F_0 \to 0.$$

Hence, it suffices to show that $\mathbb{F}$ is a direct summand of $\mathbb{G}'$. However, this is clear by induction assumption applied to J_1 and our construction. □

1.4 Relative Cohen–Macaulay Ringel's approximation

(1.4.1) Let R be a noetherian commutative ring, and G a split reductive R-group scheme with a fixed closed subgroup $\mathbb{G}_m \subset Z(G)$. Let S be a noetherian R-flat positively graded G-algebra. We assume that S is a regular ring and good (as a G-module).

As S is faithfully flat over R, R is also regular. We set $\mathfrak{M}_S := \bigoplus_{i \geq 1} S_i$. Let I be a G-ideal of S. We assume that I is perfect of codimension h, and we set $A := S/I$. Moreover, we assume that A is faithfully flat over R and good as a G-module. We define $\mathfrak{M}_A := \bigoplus_{i \geq 1} A_i$. Note that $K_R := R$ is a pointwise dualizing module of R. Corresponding to this pointwise dualizing module, the equivariant pointwise dualizing modules K_S and K_A of S and A, respectively, are determined uniquely, see (1.1.7). Moreover, we assume that $K_A = \mathrm{Ext}^h_S(A, K_S)$ is also good.

The settings above will be effective until the end of this subsection. Note that A is positively graded by assumption.

Lemma 1.4.2 *K_S is good.*

Proof. As S is Gorenstein, K_S is rank-one projective as an S-module. Hence, by Lemma 1.3.3, K_S is good. □

Hence, $S = S/0$ also satisfies the general assumption for $A = S/I$.

Let us denote by $\mathcal{X}_A$ the full subcategory of ${}_{G,A}\mathbb{M}$ consisting of (G, A)-modules which are maximal Cohen–Macaulay as A-modules. Let $\mathcal{Y}_A$ denote the full subcategory of ${}_{G,A}\mathbb{M}$ consisting of A-finite (G, A)-modules such that $\text{inj.dim}_{A_\mathfrak{p}} M_\mathfrak{p} < \infty$ for any $\mathfrak{p} \in \operatorname{Spec} A$, and set $\omega_A := \mathcal{X}_A \cap \mathcal{Y}_A$. Note that for $W \in \mathcal{A}_{G,A}$, we have $W \in \omega_A$ if and only if M, as an A-module (forget the (G, A)-module structure), lies in $\operatorname{add} K_A$ (see (I.4.10)). Note that any object in $\mathcal{X}_A$ is R-flat.

Now we define the full subcategories $\mathcal{X}_{G,A}$, $\mathcal{Y}_{G,A}$, and $\omega_{G,A}$ by

$$\begin{aligned} \mathcal{X}_{G,A} &:= \{M \in \mathcal{A}_{G,A} \mid M \in \mathcal{X}_A \text{ and } \operatorname{Hom}_A(M, K_A) \text{ is good}\}, \\ \mathcal{Y}_{G,A} &:= \{N \in \mathcal{A}_{G,A} \mid N \in \mathcal{Y}_A \text{ and } \operatorname{Hom}_A(K_A, N) \text{ is good}\}, \end{aligned}$$

and $\omega_{G,A} := \mathcal{X}_{G,A} \cap \mathcal{Y}_{G,A}$.

Lemma 1.4.3 $? \otimes_A K_A : \hat{\mathcal{P}} \to \mathcal{Y}_A$ *is an equivalence of exact categories, with* $\operatorname{Hom}_A(K_A, ?)$ *its quasi-inverse.*

Proof. Obvious by (I.4.10.16). □

Lemma 1.4.4 $\operatorname{Hom}_A(?, K_A)$ *is a duality of* $\mathcal{X}_A$*. That is,* $\operatorname{Hom}_A(?, K_A)$ *is a contravariant equivalence of exact categories from* $\mathcal{X}_A$ *to itself. The quasi-inverse of* $\operatorname{Hom}(?, K_A)$ *is* $\operatorname{Hom}(?, K_A)$ *itself.*

Proof. Note that the standard isomorphism

$$M \to \operatorname{Hom}_A(\operatorname{Hom}_A(M, K_A), K_A) \qquad (m \mapsto (f \mapsto fm))$$

is (G, A)-linear. By this observation and (I.4.10.15), the lemma is obvious. □

Lemma 1.4.5 *The following hold:*

1 *For* $M \in \mathcal{X}_A$*,* M *is good if and only if* $\operatorname{Hom}_A(M, K_A) \in \mathcal{X}_{G,A}$ *if and only if there exists some* $L \in \mathcal{X}_{G,A}$ *such that* $M \cong \operatorname{Hom}_A(L, K_A)$.

2 *For* $N \in \hat{\mathcal{P}}$*,* N *is good if and only if* $N \otimes_A K_A \in \mathcal{Y}_{G,A}$ *if and only if there exists some* $L \in \mathcal{Y}_{G,A}$ *such that* $N \cong \operatorname{Hom}_A(K_A, L)$.

3 $\mathcal{X}_{G,A}$ *is closed under extensions, direct summands, and epikernels in* ${}_{G,A}\mathbb{M}$.

4 $\mathcal{Y}_{G,A}$ *is closed under extensions, direct summands, and monocokernels in* ${}_{G,A}\mathbb{M}$.

5 *If V is a Δ-good G-module and $X \in \mathcal{X}_{G,A}$, then $X \otimes V \in \mathcal{X}_{G,A}$.*

6 *If W is a ∇-good G-module and $Y \in \mathcal{Y}_{G,A}$, then $Y \otimes W \in \mathcal{Y}_{G,A}$.*

7 *If T is a tilting G-module and $Z \in \omega_{G,A}$, then $Z \otimes T \in \omega_{G,A}$.*

8 $K_A \in \omega_{G,A}$.

9 $A \in \mathcal{X}_{G,A}$.

Proof. **1,2** follow from Lemma 1.4.3 and Lemma 1.4.4. **3,4** follow easily from **1,2**, Lemma 1.4.3, and Lemma 1.4.4.

5 As $X \otimes V$ is maximal Cohen–Macaulay and

$$\mathrm{Hom}_A(X \otimes V, K_A) \cong \mathrm{Hom}(X, K_A) \otimes V^*,$$

we have $X \otimes V \in \mathcal{X}_{G,A}$ by Lemma 1.3.2.

6 We set $Z := \mathrm{Hom}_A(K_A, Y)$. Then we have $Z \in \mathcal{Y}^{\mathrm{pro}}_{G,A} = \hat{\omega}^{\mathrm{pro}}_{G,A}$. Let $\mathbb{F}$ be an $\omega^{\mathrm{pro}}_{G,A}$-resolution of Z of finite length. As $\mathbb{F}$ is an A-finite projective resolution of Z, $\mathbb{F} \otimes W$ is a finite A-projective resolution of $Z \otimes W$. As each term of $\mathbb{F} \otimes W$ is good by Lemma 1.3.2, it follows that $Z \otimes W$ is also good of finite A-projective dimension. As we have

$$\mathrm{Hom}_A(K_A, Y \otimes W) \cong \mathrm{Hom}_A(K_A, Y) \otimes W = Z \otimes W,$$

it follows that $Y \otimes W \in \mathcal{Y}_{G,A}$. **7** follows immediately from **5,6**.

We prove **8**. As we have $K_A \in \mathcal{X}_A \cap \mathcal{Y}_A$ and $A \cong \mathrm{Hom}_A(K_A, K_A)$ is good by our assumption, we have $K_A \in \omega_{G,A}$ by **1** and **2**.

As we assume that A is Cohen–Macaulay and K_A is good, **9** is obvious. □

Proposition 1.4.6 *We have $\omega_{G,A} \cong K_A \otimes \omega^{\mathrm{pro}}_G$. That is, for any $M \in {}_G\mathbb{M}_f$, we have*

$$M \in \omega_{G,A} \iff \text{There exists a tilting } G\text{-module } T \text{ such that } M \cong K_A \otimes T.$$

Proof. The direction $\Leftarrow$ follows from Lemma 1.4.5, **6**.

We prove the direction $\Rightarrow$. We set $F = \mathrm{Hom}_A(K_A, M)$. Then as $M \in \mathcal{Y}_{G,A}$, F is good. As $M \in \mathrm{add}\, K_A$ (as an A-module), F is A-finite A-projective. Moreover, by Lemma 1.4.3, $M = F \otimes_A K_A$. We set $N = \mathrm{Hom}_A(M, K_A)$. Then as $M \in \mathcal{X}_{G,A}$, N is good. As $M \in \mathrm{add}\, K_A$ as an A-module, N is A-finite A-projective. Now we have

$$N = \mathrm{Hom}_A(F \otimes_A K_A, K_A) \cong \mathrm{Hom}_A(F, \mathrm{Hom}_A(K_A, K_A)) \cong \mathrm{Hom}_A(F, A).$$

By Lemma 1.3.6, $T = F/\mathfrak{M}_A F$ is tilting. By Lemma 1.3.9, $F \cong A \otimes T$. It follows that $M \cong (A \otimes T) \otimes_A K_A \cong K_A \otimes T$. □

Corollary 1.4.7 *If R is local and $\lambda \in X^+$, then $K_A \otimes T_G(\lambda)$ is indecomposable. Any object in $\omega_{G,A}$ is uniquely a direct sum of (G, A)-modules of the form $K_A \otimes T_G(\lambda)$.*

Proof. As $T_G(\lambda) \cong \mathrm{Hom}_A(K_A, K_A \otimes T_G(\lambda)) \otimes A/\mathfrak{M}_A$ is indecomposable, we have that $K_A \otimes T_G(\lambda)$ is also indecomposable by the G-equivariant Nakayama's lemma. The second assertion is obvious by the proposition. □

Lemma 1.4.8 *If $X \in \mathcal{X}_{G,A}$ and T is a ∇-good G-module, then we have $\mathrm{Ext}^i_{G,A}(X, K_A \otimes T) = 0$ $(i > 0)$. In particular, if we have $Z \in \omega_{G,A}$, then $\mathrm{Ext}^i_{G,A}(X, Z) = 0$ $(i > 0)$.*

Proof. As we have $\mathrm{Ext}^i_A(X, K_A) = 0$ $(i > 0)$, it follows that

$$\mathrm{Ext}^i_{G,A}(X, K_A \otimes T) \cong \mathrm{Ext}^i_{G,A}(X \otimes T^*, K_A) \cong \mathrm{Ext}^i_G(T^*, \mathrm{Hom}_A(X, K_A))$$

by Proposition II.1.1.14. As T^* is Δ-good and $\mathrm{Hom}_A(X, K_A)$ is good, the last Ext-module vanishes for $i > 0$. □

Proposition 1.4.9 *We have $\mathcal{A}_{G,A} = \hat{\mathcal{X}}_{G,A}$.*

Proof. Let $M \in \mathcal{A}_{G,A}$. By Lemma 1.2.13, for any $s > 0$, there exists an exact sequence

$$(1.4.10) \quad 0 \to X \to A \otimes V_s \to \cdots \to A \otimes V_1 \to A \otimes V_0 \to M \to 0$$

such that each V_i is Δ-good, and hence $A \otimes V_i \in \mathcal{X}_{G,A}$.

To prove $M \in \hat{\mathcal{X}}_{G,A}$, it suffices to prove $X \in \hat{\mathcal{X}}_{G,A}$. As we have $\hat{\mathcal{X}}_A = \mathcal{A}_{G,A}$, we may and shall assume that $M \in \mathcal{X}_A$, replacing M by X for sufficiently large s. As $N := \mathrm{Hom}_A(M, K_A)$ is an S-finite (G, S)-module and S is a regular ring, we have that $\mathrm{proj.dim}_S N < \infty$. By Lemma 1.3.16, we have $N \in \check{\mathcal{G}}$. We prove that $X \in \mathcal{X}_{G,A}$ if $s \geq \mathcal{G}\text{-cores.dim}(N) - 1$ in (1.4.10). Taking the canonical dual of (1.4.10), we get an exact sequence

$$0 \to N \to K_A \otimes V_0^* \to \cdots \to K_A \otimes V_s^* \to \mathrm{Hom}_A(X, K_A) \to 0$$

in $\mathcal{X}_A$. As each $K_A \otimes V_i^*$ is good and $s \geq \mathcal{G}\text{-cores.dim}(N) - 1$, we have that $\mathrm{Hom}_A(X, K_A)$ is good, and hence $X \in \mathcal{X}_{G,A}$. □

Proposition 1.4.11 *$\omega_{G,A}$ is an injective cogenerator of $\mathcal{X}_{G,A}$.*

Proof. There exists a rank-one R-projective G-module Λ such that $K_S \cong S \otimes \Lambda$. Then $\mathrm{Ext}_A^h(A, S) \cong K_A \otimes \Lambda^*$ is good. Applying Corollary 1.3.14 to $A \in {}_{G,S}\mathbb{M}_f$, we have that there exists an exact sequence

$$0 \to T_h \to T_{h-1} \to \cdots \to T_1 \xrightarrow{\partial} T_0 \to A \to 0$$

such that $T_i \in \omega_{G,S}^{\mathrm{pro}}$ $(0 \leq i \leq h)$. We set $J := \mathrm{Im}\,\partial$. Then J is good and R-flat. Hence, $\mathrm{Ext}^1_{G,S}(S, J) \cong H^1(G, J) = 0$, and we have a commutative diagram of (G, S)-modules

$$\begin{array}{ccccccccc} 0 & \to & I & \to & S & \to & A & \to & 0 \\ & & \downarrow j' & & \downarrow j & & \downarrow = & & \\ 0 & \to & J & \to & T_0 & \to & A & \to & 0. \end{array}$$

We set $Y' := \mathrm{Coker}\, j \cong \mathrm{Coker}\, j'$. As $j \otimes S/\mathfrak{M}_S$ is an R-split mono, we have $\mathrm{Tor}_1^S(S/\mathfrak{M}_S, Y') = 0$ and $Y'/\mathfrak{M}_S Y'$ is R-flat. Hence, j is injective and Y' is S-finite S-projective, by an argument similar to the proof of Lemma 1.1.4. As S and T_0 are good, Y' is good. Note also that j' is injective.

For $M \in \mathcal{X}_{G,A}$, we set $N = \mathrm{Hom}_A(M, K_A)$. As S is regular, the triple $(\mathcal{X}_{G,S}^{\mathrm{pro}}, \mathcal{Y}_{G,S}^{\mathrm{pro}}, \omega_{G,S}^{\mathrm{pro}})$ is an Auslander–Buchweitz context in $\mathcal{A}_{G,S}$. Hence, the (G, S)-module N has an $\mathcal{X}_{G,S}^{\mathrm{pro}}$-approximation

$$0 \to Y \to X \to N \to 0,$$

where X is an S-finite S-projective module with $X/\mathfrak{m}_S X$ Δ-good, and Y good. As Y and N are good, X is also good. Hence, we have $X \in \omega_{G,S}^{\mathrm{pro}}$, and there is a tilting G-module V such that $X \cong S \otimes V$. As N is an A-module, we have a commutative diagram of (G, S)-modules

$$\begin{array}{ccccccccc} 0 & \to & I \otimes V & \to & S \otimes V & \to & A \otimes V & \to & 0 \\ & & \downarrow i & & \downarrow \cong & & \downarrow p & & \\ 0 & \to & Y & \to & X & \to & N & \to & 0. \end{array}$$

By the snake lemma, we have that p is surjective, i is injective, and $Z := \mathrm{Ker}\, p \cong \mathrm{Coker}\, i$.

We show that Z is good. Consider the push-out diagram

$$\begin{array}{ccccccccc} & & 0 & & 0 & & & & \\ & & \downarrow & & \downarrow & & & & \\ 0 & \to & I \otimes V & \xrightarrow{i} & Y & \to & Z & \to & 0 \\ & & \downarrow j' \otimes 1_V & & \downarrow & & \downarrow = & & \\ 0 & \to & J \otimes V & \to & Y'' & \to & Z & \to & 0. \\ & & \downarrow & & \downarrow & & & & \\ & & Y' \otimes V & \xrightarrow{=} & Y' \otimes V & & & & \\ & & \downarrow & & \downarrow & & & & \\ & & 0 & & 0 & & & & \end{array}$$

Then as Y' and V are good and V is R-finite projective, we have that $Y' \otimes V$ is good by Lemma 1.3.2. As Y is good, Y'' is also good. As $J \otimes V$ and Y'' are good, Z is also good.

Now consider the exact sequence of maximal Cohen–Macaulay (G, A)-modules

$$0 \to Z \to A \otimes V \xrightarrow{p} N \to 0.$$

Taking the canonical dual,

$$0 \to M \to K_A \otimes V^* \to \operatorname{Hom}_A(Z, K_A) \to 0$$

is exact. As we have $K_A \otimes V^* \in \omega_{G,A}$ and $\operatorname{Hom}_A(Z, K_A) \in \mathcal{X}_{G,A}$, it follows that $\omega_{G,A}$ is a cogenerator of $\mathcal{X}_{G,A}$. As we know that $\omega_{G,A}$ is $\mathcal{X}_{G,A}$-injective by Lemma 1.4.8, we have that $\omega_{G,A}$ is an injective cogenerator of $\mathcal{X}_{G,A}$. □

Combining the results obtained so far, we have:

Theorem 1.4.12 *$(\mathcal{X}_{G,A}, \mathcal{Y}_{G,A}, \omega_{G,A})$ is an Auslander–Buchweitz context of $\mathcal{A}_{G,A}$.*

2 An application to determinantal rings

Let R be a commutative noetherian ring.

2.1 Resolutions of determinantal rings — the problem and its background

Let m, n and t be integers such that $1 \le t \le m \le n$, and $V = R^m$ and $W = R^n$. The symmetric algebra $S := \operatorname{Sym}(V \otimes W)$ over R is isomorphic to the polynomial ring $R[x_{ij}]_{1 \le i \le m,\, 1 \le j \le n}$ in mn variables over R, as an R-algebra. We denote by $I_t := I_t(x_{ij})$ the ideal of S generated by all t-minors of the $m \times n$ matrix (x_{ij}). The ideal I_t is called the *determinantal ideal.* This ideal has been studied from various points of view and with different techniques by many mathematicians, see, e.g., [27].

In this section, we focus on the problem of free resolutions of S/I_t or I_t.

Problem 2.1.1 *Construct a finite S-free resolution*

(2.1.2) $$0 \to F_h \xrightarrow{\partial_h} F_{h-1} \to \cdots \to F_1 \xrightarrow{\partial_1} S \to S/I_t \to 0$$

S/I_t with "good properties."

Even if we restrict ourselves to consider this problem only, there is a long history about it.

(2.1.3) Projective dimension and Cohen–Macaulay property

Ignoring the chronological order, we review an important result due to M. Hochster and J. A. Eagon [84].

Theorem 2.1.4 (Hochster–Eagon) *The S-module S/I_t is perfect of codimension $(m-t+1)(n-t+1)$. If R is a normal domain, then so is S/I_t. Moreover, S/I_t is a free R-module.*

In particular, we must have $h \geq (m-t+1)(n-t+1)$ in (2.1.2), whatever the resolution is.

As S is locally Cohen–Macaulay over R, it follows that if R is Cohen–Macaulay, then so is S/I_t, by the theorem.

We will give a sketch of a proof of the main part of Theorem 2.1.4 in (2.2).

As S/I_t is an R-free module, if $\mathbb{F}$ is an S-free resolution of S/I_t, then for any commutative R-algebra R', the base change $R' \otimes_R \mathbb{F}$ is an S'-free resolution of S'/I'_t, where $S' = R' \otimes_R S \cong R'[x_{ij}]$ and $I'_t = I_t S'$. In particular, if we have a free resolution (2.1.2) over the ring of integers $\mathbb{Z}$, then we have a free resolution over arbitrary R by base change. However, the construction problem over $\mathbb{Z}$ is still a difficult problem, as we will see later.

(2.1.5) Graded minimal free resolution

Letting each variable x_{ij} be of degree one, S is a graded polynomial algebra over R, and I_t is a graded ideal of S generated by elements of degree t.

We say that the resolution (2.1.2) is *graded* if it is not only a finite free S-resolution, but also a complex in the abelian category of graded S-modules (that is, each F_i is graded and each ∂_i is degree-preserving).

We denote the ideal of S generated by all variables x_{ij} by $\mathfrak{M}_S$. In other words, $\mathfrak{M}_S = I_1$. We say that the resolution (2.1.2) is *minimal* if $\partial_i \otimes S/\mathfrak{M}_S = 0$ for $i \geq 1$. The following lemma explains the meaning of minimality.

Lemma 2.1.6 *Let R be a field. Then the following hold.*

1 *There is a graded minimal free resolution of S/I_t, uniquely up to isomorphisms of graded S-complexes.*

2 *If $\mathbb{G}$ is a graded S-free resolution of S/I_t, then $\mathbb{G}$ decomposes into a direct sum $\mathbb{F} \oplus \mathbb{G}'$, as a graded S-complex, where $\mathbb{F}$ is the graded minimal free resolution of S/I_t, and $\mathbb{G}'$ is an exact graded S-free complex.*

3 *If $\mathbb{F}$ is a graded minimal free resolution of S/I_t, then we have*

$$\operatorname{rank} F_i = \dim_R \operatorname{Tor}_i^S(S/I_t, S/\mathfrak{M}_S).$$

In particular, we have $\sup\{i \mid F_i \neq 0\} = (m-t+1)(n-t+1)$.

Note that the category of graded S-modules is equivalent to the category of (T, S)-modules, where $T = \mathbb{G}_m$ acts on S via $\deg x_{ij} = 1$. Hence, **1** and **2** are nothing but a special case of Corollary 1.3.18. **3** is a local version of Lemma I.2.8.3.

The number $\operatorname{rank} F_i = \dim_R \operatorname{Tor}_i^S(S/I_t, S/\mathfrak{M}_S)$ is called the ith *Betti number* of S/I_t (in the graded sense).

Returning to the case of a general noetherian base ring R, if $\mathbb{F}$ is a graded minimal free resolution of S/I_t over R, then for any commutative R-algebra R', the base change $R' \otimes_R \mathbb{F}$ is a graded minimal free resolution of S'/I'_t. So the construction over $\mathbb{Z}$ is important. However, the following negative result is known.

Theorem 2.1.7 ([72, 73, 132]) *If $R = \mathbb{Z}$, then there exists a graded minimal free resolution of S/I_t if and only if $t = 1$ or $m - t \leq 2$.*

(2.1.8) Equivariant resolution

Let G denote the split reductive R-group scheme $G = GL(V) \times GL(W)$. As $V \otimes W$ is a G-module in a natural way, the symmetric algebra $S = \operatorname{Sym}(V \otimes W)$ is a G-algebra. It is easy to see that I_t is a G-ideal.

A finite free S-resolution $\mathbb{F}$ of S/I_t is called *G-equivariant* if $\mathbb{F}$ is a (G, S)-module resolution. Any (G, S)-module has the grading corresponding to the morphism $\mathbb{G}_m \hookrightarrow Z(G)$ $(t \mapsto (t\mathrm{id}_V, \mathrm{id}_W))$ of R-group schemes, and this grading is the one given by $\deg x_{ij} = 1$. In particular, any G-equivariant finite free S-resolution is graded. As S is R-flat and positively graded, we may utilize the general theory in (1.2).

An explicit construction of a G-equivariant minimal free resolution over $R = \mathbb{Z}$ is known, in the following cases.

Example 2.1.9 The case $t = 1$. The Koszul complex (I.2.9) $\operatorname{Kos}(x_{ij}; S)$:

$$0 \to S \otimes \textstyle\bigwedge^{mn}(V \otimes W) \to \cdots \to S \otimes \bigwedge^1(V \otimes W) \to S \to S/I_1 \to 0$$

is a G-equivariant minimal S-free resolution of S/I_t.

Example 2.1.10 The case $t = m$. J. A. Eagon and D. G. Northcott [53] constructed a graded minimal S-free resolution of S/I_m, explicitly.

D. A. Buchsbaum [29] gave a description of the *Eagon–Northcott resolution* as a (G, S)-resolution, utilizing divided power representations. Buchsbaum's description shows that the Eagon–Northcott resolution has the structure of an $\mathcal{X}_{G,S}^{\mathrm{pro}}(\Gamma)$-resolution, where Γ is the t-closed subset of X_G^+ which is uniquely characterized by $C_G(\Gamma) = R[\operatorname{End}(V) \times \operatorname{End}(W)]$.

In the two extreme cases described above, the graded minimal S-free resolution is linear. The construction of D. Eisenbud and S. Goto [54] shows that if the graded minimal free resolution of S/I_t is linear over any field, then the graded minimal free resolution of S/I_t exists and is linear over $\mathbb{Z}$, and has a G-equivariant structure.

By Theorem 2.1.7, there is a graded minimal S-free resolution of S/I_t also for $t = m-1, m-2$. K. Akin, D. A. Buchsbaum and J. Weyman [3] gave an explicit description of the minimal free resolution for the case $t = m-1$, but it is not known that the minimal free resolution has a G-equivariant structure. The case $t = m-2$ is more difficult. The question whether the graded minimal S-free resolution of S/I_t has a G-equivariant structure in the case $t = m-1, m-2$, is open.

(2.1.11) Buchsbaum–Rim resolution

Almost at the same time as the construction of the Eagon–Northcott resolution (Example 2.1.10), D. A. Buchsbaum [28] independently constructed a finite free G-equivariant S-resolution

$$0 \to F_{n-m+1} \to \cdots \to F_1 \to S \to S/I_m \to 0,$$

of S/I_m, such that each F_i is of the form

$$\bigoplus_{1\le s(1),\ldots,s(i-1)\le m} S \otimes \textstyle\bigwedge^{s(1)} V \otimes \bigwedge^{s(2)} V \otimes \cdots \otimes \bigwedge^{s(i-1)} V \otimes \bigwedge^{m} V \otimes \bigwedge^{m+\Sigma_j s(j)} W$$

for $i \geq 1$.

Although this resolution is not minimal, it has a good property in the sense that each F_i is a direct sum of tensor products of exterior powers of V and W. Exterior powers have some good properties which divided powers do not have. For example, the universal quotient $\mathcal{Q}$ of the Grassmann variety enjoys the property that the higher cohomology of $\bigwedge^i \mathcal{Q}$ vanishes for $i \geq 0$ by Kempf's vanishing theorem. The resolution is known as a *generalized Koszul complex* or the *Buchsbaum–Rim resolution* [28, 31, 32].

(2.1.12) Lascoux–Pragacz–Weyman resolution

If $R = \mathbb{Q}$, then $G = GL(V) \times GL(W)$ is linearly reductive. It follows that the minimal free S-resolution (2.1.2) of S/I_t is uniquely G-equivariant. A. Lascoux [99] gave a complete description of the (G, S)-module F_i for any i for arbitrary m, n and t. He also gave a candidate for the boundary map ∂_i. P. Pragacz and J. Weyman [125] gave a complete description of the G-equivariant resolution later. So the minimal G-equivariant resolution of S/I_t is called the *Lascoux–Pragacz–Weyman resolution* (sometimes called *Lascoux's resolution*). K. Akin and J. Weyman study a larger action than that of G, namely the Lie super algebra $\mathfrak{gl}(m|n)$ action on the resolution, see [5]. Note that if $R \supset \mathbb{Q}$, then the base change of the Lascoux–Pragacz–Weyman resolution from $\mathbb{Q}$ is a G-equivariant minimal free resolution of S/I_t.

2.2 Buchsbaum–Rim type resolutions

Against the background reviewed in the last subsection, Buchsbaum and Weyman studied the following problem [30].

Problem 2.2.1 *Construct a finite free S-resolution*

$$\mathbb{F} : 0 \to F_h \to \cdots \to F_1 \to S \to S/I_t \to 0$$

of S/I_t explicitly, for arbitrary t, which satisfies the following conditions.

1 $\mathbb{F}$ *is G-equivariant.*

2 $h = \text{proj.dim}_S S/I_t = (m-t+1)(n-t+1)$.

3 *For any $i \geq 1$, $F_i \cong S \otimes T(i)$ as a (G,S)-module, where $T(i)$ is a direct sum of tensor products of G-modules of the form $\bigwedge^i V \otimes \bigwedge^j W$.*

The generalized Koszul complex by Buchsbaum for $t = m$ satisfies the conditions above. However, the Lascoux–Pragacz–Weyman resolution (2.1.12) in the case $R = \mathbb{Q}$ does not satisfy **3** in almost all cases. Instead of the condition **3**, we consider a weaker condition.

3′ For any $i \geq 1$, $F_i \cong S \otimes T(i)$ as a (G,S)-module, where $T(i)$ is a direct summand of a direct sum of tensor products of G-modules of the form $\bigwedge^i V \otimes \bigwedge^j W$.

Some good properties such as cohomology vanishing are inherited by a direct summand. Note that the LPW resolution satisfies **1**,**2**,**3′**.

Definition 2.2.2 We call a finite free S-resolution of S/I_t which satisfies the conditions **1**,**2**,**3′** a *resolution of Buchsbaum–Rim type*.

Related to this definition, we have the following, which is one of the main theorems in these notes.

Theorem 2.2.3 *Let R be a commutative ring. Then there is a Buchsbaum–Rim type resolution of S/I_t. If, moreover, R is a noetherian local ring, then there is a Buchsbaum–Rim type resolution $\mathbb{F}$ of S/I_t which satisfies the following condition, uniquely up to isomorphisms of (G,S)-complexes:*

4 *If $\mathbb{G}$ is an S-free resolution of S/I_t which satisfies the conditions* **1**,**3′**, *then there is a decomposition of a (G,S)-complex $\mathbb{G} \cong \mathbb{F} \oplus \mathbb{G}'$ such that $\mathbb{G}'$ is an exact sequence which satisfies* **1**,**3′**.

This theorem is merely an existence theorem, and it does not give any explicit construction. It does not give any information about the gap between conditions **3** and **3′**. The rest of this section is devoted to proving the theorem. The proof of the theorem is reduced to the following.

Theorem 2.2.4 *S, I_t, and $\mathrm{Ext}^h_S(S/I_t, S)$ are good as G-modules, where $h := (m-t+1)(n-t+1)$.*

The goodness of S and I_t is classical, and known as the *straightening formula*, whose complete proof is found in [4]. The problem is the goodness of $\mathrm{Ext}^h_S(S/I_t, S)$.

We prove that Theorem 2.2.4 really implies Theorem 2.2.3. First note that we may assume that R is noetherian, as a base change of a Buchsbaum–Rim type resolution is again of Buchsbaum–Rim type.

Set $\Gamma =: \Gamma_m \times \Gamma_n$, where Γ_i is the set of partitions whose lengths are at most i. Then Γ is a t-closed subset of X^+_G. In fact,

$$C_G(\Gamma) = C_{GL(V)}(\Gamma_m) \otimes C_{GL(V)}(\Gamma_n) \cong R[\mathrm{End}(V)] \otimes R[\mathrm{End}(W)]$$

is a subbialgebra of $R[G]$ (III.4.6).

Moreover, $S = \mathrm{Sym}(V \otimes W)$ belongs to Γ, since it is an $\mathrm{End}(V) \times \mathrm{End}(W)$-module. Hence, Γ is t-closed with respect to S. Hence, if the assumption of Theorem 2.2.4 holds, then the general assumptions in (1.3) are satisfied by S. In particular, we know that $(\mathcal{X}^{\mathrm{pro}}_{G,S}(\Gamma), \mathcal{Y}^{\mathrm{pro}}_{G,S}(\Gamma), \omega^{\mathrm{pro}}_{G,S}(\Gamma))$ is a weak Auslander–Buchweitz context of $\mathcal{A}_{G,S}(\Gamma)$ (Theorem 1.3.11). Assuming Theorem 2.2.4 is true, we have $\omega^{\mathrm{pro}}_{G,S}(\Gamma)$-resol.dim$(S/I_t) = h$ by Corollary 1.3.14 and Theorem 2.1.4. If, moreover, R is local, then by Corollary 1.3.18 there is an $\omega^{\mathrm{pro}}_{G,S}(\Gamma)$-resolution of S/I_t which is minimal with respect to direct summands. Hence, the following lemma completes the proof of the implication Theorem 2.2.4 $\Rightarrow$ Theorem 2.2.3.

Lemma 2.2.5 *Let T be an S-finite (G, S)-module. Then $T \in \omega^{\mathrm{pro}}_{G,S}(\Gamma)$ if and only if $T \cong S \otimes T_0$ with T_0 a direct summand of a finite direct sum of tensor products of G-modules of the form $\bigwedge^i V \otimes \bigwedge^j W$.*

Proof. By Proposition 1.3.9, we have

$$\omega^{\mathrm{pro}}_{G,S}(\Gamma) = S \otimes \omega^{\mathrm{pro}}_{C_G(\Gamma)}.$$

On the other hand, we have

$$\omega^{\mathrm{pro}}_{C_G(\Gamma)} = \mathrm{add}(\omega^{\mathrm{pro}}_{C_1} \otimes \omega^{\mathrm{pro}}_{C_2}),$$

where $C_1 := C_{GL(V)}(\Gamma_m)$ and $C_2 := C_{GL(W)}(\Gamma_n)$. Hence, the lemma follows from Proposition III.4.6.4. □

Now we know that Theorem 2.2.4 implies Theorem 2.2.3.

If R is a regular ring, then S and $A := S/I_t$ satisfy the general assumptions of (1.4) by Theorem 2.2.4. Hence, we have the following, by Theorem 1.4.12.

Corollary 2.2.6 *Set $A := S/I_t$. If R is a regular ring, then the triple $(\mathcal{X}_{G,A}, \mathcal{Y}_{G,A}, \omega_{G,A})$ is an Auslander–Buchweitz context.*

(2.2.7) Now we prove Theorem 2.2.4.

Lemma 2.2.8 *The construction of* $\mathrm{Ext}_S^h(S/I_t, S)$ *is compatible with a base change. Namely, for any morphism of noetherian commutative rings* $R \to R'$*, there is a* (G, S')*-isomorphism*

$$R' \otimes \mathrm{Ext}_S^h(S/I_t, S) \cong \mathrm{Ext}_{S'}^h(S'/I_t', S'),$$

where $(?)' = R' \otimes ?$*, and* $I_t' \to I_t S'$ *is an isomorphism.*

Proof. Obviously, we may assume $R = \mathbb{Z}$. The map $I_t' \to I_t S'$ is an isomorphism, since S/I_t is R-flat (Theorem 2.1.4). There is a G-equivariant finite free S-resolution $\mathbb{F}$ of S/I_t whose length is h by Lemma 1.2.13. Then $\mathrm{Hom}_S(\mathbb{F}, S)[h]$ is a (G, S)-resolution of $\mathrm{Ext}_S^h(S/I_t, S)$ by Theorem 2.1.4. As we have $\mathrm{Hom}_S(\mathbb{F}, S)[h]' \cong \mathrm{Hom}_{S'}(\mathbb{F}', S')[h]$ and $\mathbb{F}'$ is an S'-projective (G, S')-resolution of S'/I_t', we are done. □

Hence, to prove Theorem 2.2.4, it suffices to prove that $\mathrm{Ext}_S^h(S/I_t, S)$ is u-good, assuming that $R = \mathbb{Z}$. By the lemma again, we may assume that R is an algebraically closed field, since $\mathrm{Ext}_S^h(S/I_t, S)$ is a direct sum of $\mathbb{Z}$-finite $\mathbb{Z}$-free $G_{\mathbb{Z}}$-modules.

2.3 Kempf's construction

(2.3.1) In this subsection, we assume that $R = k$ is an algebraically closed field, and we prove

Theorem 2.3.2 $\mathrm{Ext}_S^h(S/I_t, S)$ *is good.*

This completes the proof of Theorem 2.2.4 (for general noetherian commutative R). We also prove Theorem 2.1.4 for the case that $R = k$ is an algebraically closed field. As S/I_t is an R-free module (see, e.g., [4]), the general case of Theorem 2.1.4 follows easily.

The following construction was given by G. Kempf, and used by Lascoux [99] and Roberts–Weyman [132] to study resolutions of determinantal rings effectively.

(2.3.3) Set $\mathbf{X} := \mathrm{Hom}(V, W^*) \cong \mathrm{Spec}\, S$, and let $\mathbf{Y}$ denote the closed subscheme $\mathrm{Spec}\, A = \mathrm{Spec}\, S/I$ of $\mathbf{X}$. We denote the Grassmann variety [96] of $(t-1)$-quotients of V by $\mathbf{G}$. There is a maximal parabolic subgroup P of $GL(V)$ such that $\mathbf{G} = GL(V)/P$. Letting $GL(W)$ act trivially, $\mathbf{G}$ is a k-smooth projective G-variety. Let

$$0 \to \mathcal{R} \to V \to \mathcal{Q} \to 0$$

be the tautological exact sequence of $\mathbf{G}$. Note that rank $\mathcal{Q} = t-1$, and this is an exact sequence of G-bundles.

We set $\mathbf{Z} := \mathcal{R} \otimes W^*$. This is also a G-bundle over $\mathbf{G}$. From the tautological exact sequence, we have an exact sequence

$$0 \to \mathbf{Z} \xrightarrow{i} \mathbf{X} \times \mathbf{G} \to \mathcal{Q} \otimes W^* \to 0$$

of G-bundles over $\mathbf{G}$. We denote the projection from $\mathbf{X} \times \mathbf{G}$ to $\mathbf{X}$ (resp. $\mathbf{G}$) by p_1 (resp. p_2).

The following is a variation of Kempf's vanishing theorem (I.4.7.6).

Proposition 2.3.4 *Let $M = M(V_1, V_2, V_3)$ be a universal module functor over k of type $(3, 0)$, and assume that M is a filtered inductive limit of objects of*

$$\operatorname{add} \mathcal{F}(\{L_\lambda(V_1) \otimes L_\mu(V_2) \otimes L_\nu(V_3) \mid \lambda, \mu \text{ and } \nu \text{ are partitions}\}).$$

Then we have $H^i(\mathbf{G}, M(V, \mathcal{Q}, W)) = 0$ $(i > 0)$, and $H^0(\mathbf{G}, M(V, \mathcal{Q}, W))$ is a good G-module. Moreover, the G-linear map

$$M(V, V, W) \cong H^0(\mathbf{G}, M(V, V, W)) \to H^0(\mathbf{G}, M(V, \mathcal{Q}, W))$$

induced by the canonical map $M(V, V, W) \to M(V, \mathcal{Q}, W)$ is surjective.

This proposition was proved by Roberts–Weyman [132], and used effectively.

Since p_2 is affine and $(p_2)_* \mathcal{O}_{\mathbf{Z}} = \operatorname{Sym}(\mathcal{Q} \otimes W)$ satisfies the condition in Proposition 2.3.4 by Akin–Buchsbaum–Weyman's straightening formula [4], it follows that $R^i \pi_* \mathcal{O}_{\mathbf{Z}} = 0$ $(i > 0)$, and $H^0(\mathbf{Z}, \mathcal{O}_Z)$ is good. Moreover, as we have $\bigwedge^t \mathcal{Q} = 0$, it follows that $I_t(\mathcal{Q}, W) = 0$, and hence $S(\mathcal{Q}, W) \cong S/I_t(\mathcal{Q}, W)$. More detailed investigations on the good filtration of S/I_t imply that the map

$$S/I_t = S/I_t(V, W) \to H^0(\mathbf{G}, S/I_t(\mathcal{Q}, W)) \cong H^0(\mathbf{G}, S(\mathcal{Q}, W))$$

is an isomorphism.

Proposition 2.3.5 (Kempf) *The composite morphism $p_1 i : \mathbf{Z} \to \mathbf{X}$ factors through $\mathbf{Y}$, and induces an $\mathbf{X}$-morphism $\pi : \mathbf{Z} \to \mathbf{Y}$. The morphism π is a resolution of singularities (i.e., a proper birational morphism with $\mathbf{Z}$ non-singular). Moreover, we have $R^i \pi_* \mathcal{O}_{\mathbf{Z}} = 0$ $(i > 0)$, and $\pi_* \mathcal{O}_{\mathbf{Z}} \cong \mathcal{O}_Y$. $\mathbf{Y}$ is a normal variety of dimension $mn - (m - t + 1)(n - t + 1)$.*

Proof. As we have $H^0(\mathbf{G}, S(\mathcal{Q}, W)) \cong S/I_t$, the morphism π is induced, and we have $\pi_* \mathcal{O}_{\mathbf{Z}} \cong \mathcal{O}_{\mathbf{Y}}$. As $\mathbf{Z}$ is a vector bundle over the k-smooth variety $\mathbf{G}$, it is a k-smooth variety. Hence, $\mathbf{Y}$ is a normal variety (see

Proposition I.2.11.13). We have already seen the vanishing $R^i\pi_*\mathcal{O}_{\mathbf{Z}} = 0$ $(i > 0)$. As $p_1 i$ is proper, π is also proper. We show that π is birational in order to prove that π is a resolution of singularities.

For any k-scheme $f : \mathbf{W} \to \operatorname{Spec} k$, we have $\mathbf{X}(\mathbf{W}) = \operatorname{Hom}(f^*V, f^*W^*)$. By the definition of Grassmann variety [96],

$$\mathbf{G}(\mathbf{W}) = \{\alpha : 0 \to \mathcal{R}' \xrightarrow{i'} f^*V \xrightarrow{p'} \mathcal{Q}' \to 0 \mid \operatorname{rank} \mathcal{Q}' = t-1\}/\sim,$$

where we say that $\alpha \sim \beta$ for two exact sequences

$$\alpha : 0 \to \mathcal{R}' \xrightarrow{i'} f^*V \xrightarrow{p'} \mathcal{Q}' \to 0$$

and

$$\beta : 0 \to \mathcal{R}_1 \xrightarrow{i_1} f^*V \xrightarrow{p_1} \mathcal{Q}_1 \to 0$$

if there is an isomorphism $\rho : \mathcal{R}' \cong \mathcal{R}_1$ such that $i_1 \circ \rho = i'$. In the sequel, we consider that $\alpha = \beta$ for simplicity, if $\alpha \sim \beta$. Hence, we have

$$\mathbf{Z}(\mathbf{W}) = \{(\alpha, \psi) \mid \alpha \in \mathbf{G}(\mathbf{W}),\ \psi : \mathcal{Q}' \to W^*\}.$$

The morphism $i : \mathbf{Z} \to \mathbf{X} \times \mathbf{G}$ is given by the natural transformation $(\alpha, \psi) \mapsto (\alpha, \psi p')$ via Yoneda's lemma (Lemma I.1.1.6). Hence, π is given by $(\alpha, \psi) \mapsto \psi p'$.

Next, we set $\mathbf{U} := \mathbf{Y} \setminus \operatorname{Spec} S/I_{t-1}$. Note that we have

$$\mathbf{U}(\mathbf{W}) = \{\varphi \in \mathbf{X}(\mathbf{W}) \mid \operatorname{rank} \varphi = t-1\}$$

(if $t = 1$, then we consider that $I_0 = S$ as a convention). Note also that $\mathbf{U}$ is an open subset of $\mathbf{Y}$, and it is non-empty, since there is an $m \times n$ matrix of rank $(t-1)$. The restriction $\pi : \pi^{-1}(\mathbf{U})(\mathbf{W}) \to \mathbf{U}(\mathbf{W})$ has the inverse $\varphi \mapsto (\alpha(\varphi), \psi(\varphi))$, where $\alpha(\varphi)$ is the exact sequence

$$0 \to \operatorname{Ker} \varphi \to f^*V \to \operatorname{Im} \varphi \to 0,$$

and $\psi(\varphi)$ is $\operatorname{Im} \varphi \hookrightarrow f^*W^*$. In fact, the image of a bundle map whose rank is $(t-1)$ everywhere is a quotient bundle of rank $(t-1)$. Conversely, the only rank-$(t-1)$ quotient through which a bundle map of rank $(t-1)$ factors is its image. By Yoneda's lemma, we have that $\pi : \pi^{-1}(\mathbf{U}) \to \mathbf{U}$ is an isomorphism. In particular, π is birational.

Finally, we prove that $\mathbf{Y}$ is $mn - (m-t+1)(n-t+1)$-dimensional. As is well-known, we have $\dim \mathbf{G} = (t-1)(m-t+1)$. As $\mathbf{Z}$ is a vector bundle of rank $(t-1)n$ over $\mathbf{G}$, we have $\dim \mathbf{Y} = \dim \mathbf{Z} = (t-1)(n+m-t+1) = mn - (m-t+1)(n-t+1)$. □

Proposition 2.3.6 *The canonical sheaf* $\omega_{\mathbf{Z}} := \bigwedge^{\mathrm{top}} \Omega_{\mathbf{Z}/k}$ *is isomorphic to*

$$(\textstyle\bigwedge^{\mathrm{top}} V)^{\otimes(t-1)} \otimes (\bigwedge^{\mathrm{top}} W)^{\otimes(t-1)} \otimes (\bigwedge^{\mathrm{top}} \mathcal{Q})^{\otimes(n-m)}$$

as a G*-bundle over* $\mathbf{Z}$.

Proof. As is well-known [90, p. 229], we have $\Omega_{\mathbf{G}/k} \cong \mathcal{Q}^* \otimes \mathcal{R}$. Hence, we have

$$\omega_{\mathbf{G}/k} \cong \textstyle\bigwedge^{\mathrm{top}}(\mathcal{Q}^* \otimes \mathcal{R}) \cong (\bigwedge^{\mathrm{top}} \mathcal{Q})^{\otimes(t-1-m)} \otimes (\bigwedge^{\mathrm{top}} \mathcal{R})^{\otimes(t-1)}$$
$$\cong (\textstyle\bigwedge^{\mathrm{top}} \mathcal{Q})^{\otimes(-m)} \otimes (\bigwedge^{\mathrm{top}} V)^{\otimes(t-1)}.$$

As we have $\Omega_{\mathbf{Z}/\mathbf{G}} \cong \mathbf{Z}^* \cong \mathcal{Q} \otimes W$, it follows that

$$\omega_{\mathbf{Z}/\mathbf{G}} \cong (\textstyle\bigwedge^{\mathrm{top}} \mathcal{Q})^{\otimes n} \otimes (\bigwedge^{\mathrm{top}} W)^{\otimes(t-1)}.$$

By Lemma I.2.11.12, we are done. □

Again by Proposition 2.3.4, we have $R^i\pi_*\omega_{\mathbf{Z}} = 0$ $(i > 0)$, and $\pi_*\omega_{\mathbf{Z}}$ is good. By Proposition I.2.11.13, $\mathbf{Y}$ is Cohen–Macaulay, and Theorem 2.1.4 for the case where R is an algebraically closed field follows. Moreover, there is an A-isomorphism (not necessarily a (G, A)-isomorphism)

$$\pi_*\omega_{\mathbf{Z}} \cong \omega_{\mathbf{Y}} = \operatorname{Ext}^h_S(S/I_t, K_S) \cong \operatorname{Ext}^h_S(S/I_t, S).$$

As $\pi_*\omega_{\mathbf{Z}}$ is good, $\operatorname{Ext}^h_S(S/I_t, K_S)$ is also good, by Lemma 1.1.11. Hence, Theorem 2.3.2 is proved.

Glossary

$\#X$	the cardinality of a set X, xv
$\mathbb{N}$	the set of positive integers, xv
$\mathbb{N}_0$	the set of non-negative integers, xv
$G^\times$	the group of invertible elements of a semigroup G, xv
$A^\times$	the unit group of a ring A, xv
${}_X\mathbb{M}$	the category of $\mathcal{O}_X$-modules, xv
$\mathrm{Qco}(X)$	the category of quasi-coherent $\mathcal{O}_X$-modules, xv
$\mathrm{Coh}(X)$	the category of coherent $\mathcal{O}_X$-modules, xv
$\kappa(\mathfrak{p})$	the field $A_\mathfrak{p}/\mathfrak{p}A_\mathfrak{p}$ for $\mathfrak{p} \in \operatorname{Spec} A$, xvi
$X(\mathfrak{p})$	the fiber of $\kappa(\mathfrak{p})$ in X, xvi
$\operatorname{Max} A$	the set of maximal ideals of A, xvi
$\operatorname{supp} M$	the support of M, xvi
$\operatorname{Min} M$	the set of minimal primes of M, xvi
$\tilde{M}$	the quasi-coherent sheaf associated to M, xvi
$\kappa(x)$	the residue field $\mathcal{O}_{X,x}/\mathfrak{m}_x$, xvi
$R[X]$	the coordinate ring of an affine R-scheme X, xvi
${}_A\mathbb{M}$	the category of (left) A-modules, xvi
$\underline{\mathrm{Set}}$	the category of (small) sets, 1
$\underline{\mathrm{Grp}}$	the category of groups, 1
$\underline{\mathrm{Ab}}$	the category of abelian groups, 1
$\mathrm{ob}(\mathcal{C})$	the set of objects of the category $\mathcal{C}$, 1
$\mathcal{C}(M, N)$	the set of morphisms from M to N in the category $\mathcal{C}$, 1
$\mathrm{Hom}_\mathcal{C}(M, N)$	the same as $\mathcal{C}(M, N)$, 1
$\mathcal{C}^{\mathrm{op}}$	the opposite category of $\mathcal{C}$, 1

$\mathrm{Func}(\mathcal{A},\mathcal{B})$	the category of functors from $\mathcal{A}$ to $\mathcal{B}$, 2
$\mathrm{Nat}(F,G)$	the set of natural transformations from F to G, 2
$\mathrm{coker}\, i$	the canonical map from the target of i to $\mathrm{Coker}\, i$, 6
$\ker p$	the canonical map from $\mathrm{Ker}\, p$ to the source of p, 6
$\mathrm{Sex}_R(\mathcal{A}^{\mathrm{op}}, {}_R\mathbb{M})$	the category of contravariant left exact R-functors from $\mathcal{A}$ to ${}_R\mathbb{M}$, 8
$C(\mathcal{A})$	the category of chain complexes in $\mathcal{A}$, 9
$C^+(\mathcal{A})$	the category of chain complexes bounded below in $\mathcal{A}$, 9
$C^-(\mathcal{A})$	the category of chain complexes bounded above in $\mathcal{A}$, 9
$C^b(\mathcal{A})$	the category of bounded chain complexes in $\mathcal{A}$, 9
$\mathbb{F}[n]$	the complex $\mathbb{F}$ shifted by n, 9
$\mathrm{Hom}^{\bullet}_{\mathcal{A}}(\mathbb{F},\mathbb{G})$	the Hom-complex from $\mathbb{F}$ to $\mathbb{G}$, 9
$K(\mathcal{A})$	the homotopy category of unbounded complexes in $\mathcal{A}$, 10
$E^?(\mathcal{A})$	the full subcategory of $K^?(\mathcal{A})$ consisting of exact sequences, 10
$D^?(\mathcal{A})$	the derived category of $\mathcal{A}$, 10
Ext^n_F	the nth extension group in the exact category determined by F, 15
$R^i_F Q$	the ith F-right derived functor of Q, 17
$\mathrm{sh}(T,\mathcal{C})$	the set of sheaves on T with values in $\mathcal{C}$, 21
$a(\mathcal{F})$	the sheafification of $\mathcal{F}$, 22
$\mathcal{D}_f$	the full subcategory of noetherian objects of $\mathcal{D}$, 24
$\mathrm{soc}\, A$	the socle of A, 27
$\mathrm{rad}\, A$	the radical of A, 27
$\mathrm{top}\, A$	the top of A, 27
$\mathrm{add}\, \mathcal{X}$	the smallest full subcategory containing $\mathcal{X}$ closed under finite direct sums and direct summands, 28
$\mathcal{F}(\mathcal{X})$	the smallest full subcategory containing $\mathcal{X}$ closed under extensions, 28
$\hat{\mathcal{X}}$	the full subcategory of objects with finite $\mathcal{X}$-resolutions, 28
$\mathcal{X}$-resol.dim A	the $\mathcal{X}$-resolution dimension of A, 28

$\check{\mathcal{X}}$	the full subcategory of objects with finite $\mathcal{X}$-coresolutions, 28
$\mathcal{X}$-cores.dim A	the $\mathcal{X}$-coresolution dimension of A, 28
$\mathcal{X}$-inj.dim A	the $\mathcal{X}$-injective dimension of A, 29
$\mathcal{X}^{\perp}$	the full subcategory of $\mathcal{X}$-injective objects, 29
$\mathcal{X}$-proj.dim A	the $\mathcal{X}$-projective dimension of A, 29
${}^{\perp}\mathcal{X}$	the full subcategory of $\mathcal{X}$-projective objects, 29
$\operatorname{rad} A$	the Jacobson radical of a ring A, 30
$\operatorname{flat.dim}_R M$	the R-flat dimension of M, 40
$W(M)$	the quasi-coherent faisceau induced by M, 44
N_a	stands for $W(\tilde{N})$, 44
$\operatorname{ann} M$	the annihilator of M, 46
$\operatorname{depth}_R(I, M)$	the I-depth of M, 46
$\operatorname{codim}_R M$	the codimension (the height of the annihilator) of M, 47
$\operatorname{ht} I$	the height of I, 47
$\operatorname{grade} M$	the grade of M, 47
$\operatorname{emb.dim} R$	the embedding dimension of R, 49
$\Gamma_Y(X, \mathcal{F})$	the group of sections of $\mathcal{F}$ over X with supports in Y, 51
$H^i_Y(X, ?)$	the ith local cohomology functor of X with supports in Y, 51
$\dim_{\varphi}(x)$	the relative dimension of φ at x, 52
$\beta^R_i(M)$	the ith Betti number of M, 56
$\mu^i_R(M)$	the ith Bass number of M, 56
$\operatorname{type} M$	the Cohen–Macaulay type of M, 57
I_Y	the dualizing complex of Y, 63
$\operatorname{Gr}_I M$	the graded module of M associated to I, 67
$\mathbb{M}^C$	the category of right C-comodules, 75
${}^C\mathbb{M}$	the category of left C-comodules, 75
$\operatorname{Hom}_C(M, M')$	the set of C-comodule maps from M to M', 75
$\operatorname{Cobar}_C(M)$	the cobar resolution of a C-comodule M, 79
$N \boxtimes^C L$	the cotensor product of N and L, 81

$\mathrm{Cotor}^i_C(M,N)$	the ith cotorsion module of M and N, 84
$\mathrm{Cobar}_C(M,N)$	the double cobar complex of M and N, 84
A°	the dual coalgebra of A, 90
$A\#U$	the smash product of A and U, 99
${}_B\mathbb{M}^H$	the category of (H,B)-Hopf modules, 100
Sch/X	the category of X-schemes, 101
${}_G\mathbb{M}$	the category of G-modules, 103
$\mathbb{G}_{m,X}$	stands for $\mathcal{O}_X^\times = GL(1,X)$, 105
G_u	the set of unipotent elements of G, 106
Σ_G	the set of roots of G, 107
$W(G)$	the Weyl group of G, 109
Δ_G	the base of the root system Σ_G of G, 109
Σ_G^+	the set of positive roots of G, 109
Σ_G^-	the set of negative roots of G, 109
$l(w)$	the length of w, 109
w_0	the longest element (of the current Weyl group), 109
X_G^+	the set of dominant weights of G, 110
λ^*	stands for $-w_0\lambda$, 110
R_λ	the rank-one R-free B-module whose restriction to T (resp. U) is λ (resp. trivial), 111
$\nabla_G(\lambda)$	the induced module of G of highest weight λ, 111
$\Delta_G(\lambda)$	the Weyl module of highest weight λ, 111
$L_G(\lambda)$	the simple G-module of highest weight λ, 112
$\mathrm{UMF}(r,s;X)$	the category of universal module functors of type (r,s) over X, 114
${}_{G,A}\mathbb{M}$	the category of (G,A)-modules, 127
$\varphi^\#$	the restriction with respect to φ, 131
$\varphi_\#$	the inflation with respect to φ, 131
Y^*	the image closure of the action $Y\times G\rightarrow X$, 139
x^*	the generic point of $\overline{\{x\}}^*$, 140
$\mathrm{Hy}\,G$	the hyperalgebra of G, 144

G-sch	the category of G-schemes, 146
${}_{G,X}\mathbb{M}$	the category of aff-$(G, \mathcal{O}_X)$-modules, 151
$\operatorname{ht} p$	the height of p, 158
$\operatorname{coht} p$	the coheight of p, 158
$\Delta_C(\lambda)$	the Weyl module of highest weight λ, 159
$\nabla_C(\lambda)$	the induced module of highest weight λ, 159
$\mathbb{M}^C(\pi)$	the category of C-comodules belonging to π, 160
$L_C(\lambda)$	the simple comodule of highest weight λ, 160
$V(\pi)$	the sum of all C-subcomodules of V which belong to $\mathbb{M}^C(\pi)$, 166
$Q_C(\lambda)$	the injective hull of $L_C(\lambda)$, 167
$[W : \nabla_C(\lambda)]$	the number of $\nabla_C(\lambda)$ in any good filtration of W, 176
$(V : L_C(\lambda))$	the Jordan–Hölder multiplicity of $L_C(\lambda)$ in V, 177
$R(\lambda)$	stands for $\operatorname{Hom}_C(\Delta_C(\lambda), \nabla_C(\lambda))$, 187
S_π	the Schur algebra with respect to π, 195
$\mathcal{A}_C$	stands for $\mathbb{M}^C_f$, 199
$\mathcal{Y}_C$	stands for $\mathcal{F}(\nabla)$, 199
$\mathcal{X}_C$	stands for $\mathcal{F}(\Delta)$, 199
ω_C	stands for $\mathcal{X}_C \cap \mathcal{Y}_C$, 199
$T_C(\lambda)$	the indecomposable tilting module of C of highest weight λ, 204
$C_G(\pi)$	the Donkin subcoalgebra of G with respect to π, 223
$S_G(\pi)$	the Schur algebra of G with respect to π, 223
$\tilde{\lambda}$	the transpose of a partition λ, 225
$\mathcal{X}^{\mathrm{pro}}_{G,A}(\Gamma)$	see the corresponding page, 240
$\mathcal{G}$	the category of good G-modules, 243
I_t	$= I_t(x_{ij})$, the determinantal ideal generated by all t-minors of (x_{ij}), 250

Bibliography

[1] E. Abe, *Hopf Algebras*, *Cambridge Tracts in Math.* **74**, Cambridge (1980).

[2] K. Akin and D. A. Buchsbaum, Characteristic-free representation theory of the general linear group, *Adv. Math.* **58** (1985), 149–200.

[3] K. Akin, D. A. Buchsbaum and J. Weyman, Resolutions of determinantal ideals: the submaximal minors, *Adv. Math.* **39** (1981), 1–30.

[4] K. Akin, D. A. Buchsbaum and J. Weyman, Schur functors and Schur complexes, *Adv. Math.* **44** (1982), 207–278.

[5] K. Akin and J. Weyman, Minimal free resolution of determinantal ideals and irreducible representations of the Lie superalgebra $\mathfrak{gl}(m|n)$, *J. Algebra* **197** (1997), 559–583.

[6] H. H. Andersen and J. C. Jantzen, Cohomology of induced representations for algebraic groups, *Math. Ann.* **269** (1984), 487–525.

[7] Y. Aoyama and S. Goto, On the type of graded Cohen–Macaulay rings, *J. Math. Kyoto Univ.* **15** (1975), 19–23.

[8] M. Artin, *Grothendieck Topology*, mimeographed notes, Harvard University (1962).

[9] M. Artin et al., *Théorie des Topos et Cohomologie Etale des Schémas, SGA 4*, *Lect. Notes Math.* **269, 270** (1972), **305** (1973), Springer-Verlag.

[10] M. Auslander and R. O. Buchweitz, The homological theory of maximal Cohen–Macaulay approximations, *Mém. Soc. Math. France (N.S.)* **38** (1989), 5–37.

[11] M. Auslander and I. Reiten, Applications of contravariantly finite subcategories, *Adv. Math.* **86** (1991), 111–152.

[12] M. Auslander, I. Reiten and S. O. Smalø, *Representation Theory of Artin Algebras*, Cambridge (1995).

[13] M. Auslander and S. O. Smalø, Almost split sequences in subcategories, *J. Algebra* **69** (1981), 426–454.

[14] L. L. Avramov, Infinite free resolutions, in J. Elias et al. (eds.), *Six Lectures on Commutative Algebra* (Bellaterra, 1996), *Progr. Math.* **166**, Birkhäuser (1998), pp. 1–118.

[15] L. L. Avramov, Locally complete intersection homomorphisms and a conjecture of Quillen on the vanishing of cotangent homology, *Ann. Math.* (2) **150** (1999), 455–487.

[16] L. L. Avramov and R. Achilles, Relations between properties of a ring and of its associated graded ring, in *Seminar Eisenbud/Singh/Vogel, Vol. 2*, Teubner, Leipzig (1982), pp. 5–29.

[17] L. L. Avramov and H.-B. Foxby, Gorenstein local homomorphisms, *Bull. Amer. Math. Soc.* (*N.S.*) **23** (1990), 145–150.

[18] L. L. Avramov and H.-B. Foxby, Ring homomorphisms and finite Gorenstein dimension, *Proc. London Math. Soc.* (3) **75** (1997), 241–270.

[19] L. L. Avramov and H.-B. Foxby, Cohen–Macaulay properties of ring homomorphisms, *Adv. Math.* **133** (1998), 54–95.

[20] L. L. Avramov, H.-B. Foxby and B. Herzog, Structure of local homomorphisms, *J. Algebra* **164** (1994), 124–145.

[21] L. L. Avramov, H.-B. Foxby and J. Lescot, Bass series of local ring homomorphisms of finite flat dimension, *Trans. Amer. Math. Soc.* **335** (1993), 497–523.

[22] J. Bernstein and V. Lunts, *Equivariant sheaves and functors*, *Lecture Notes in Mathematics* **1578**, Springer-Verlag (1994).

[23] G. Boffi, The universal form of the Littlewood–Richardson rule, *Adv. Math.* **68** (1988), 40–63.

[24] G. Boffi, On some plethysms, *Adv. Math.* **89** (1991), 107–126.

[25] M. Bökstedt and A. Neeman, Homotopy limits in triangulated categories, *Composito Math.* **86** (1993), 209–234.

[26] W. Bruns and J. Herzog, *Cohen–Macaulay rings*, first paperback edition, Cambridge (1998).

[27] W. Bruns and U. Vetter, *Determinantal rings*, *Lecture Notes in Mathematics* **1327**, Springer-Verlag (1988).

[28] D. A. Buchsbaum, A generalized Koszul complex, I, *Trans. Amer. Math. Soc.* **111** (1964), 183–196.

[29] D. A. Buchsbaum, A new construction of the Eagon–Northcott complex, *Adv. in Math.* **34** (1979), 58–76.

[30] D. A. Buchsbaum, oral communication (1996).

[31] D. A. Buchsbaum and D. S. Rim, A generalized Koszul complex, II, Depth and multiplicity, *Trans. Amer. Math. Soc.* **111** (1964), 197–224.

[32] D. A. Buchsbaum and D. S. Rim, A generalized Koszul complex, III, A remark on generic acyclicity, *Proc. Amer. Math. Soc.* **16** (1965), 555–558.

[33] L. Budach and R.-P. Holzapfel, *Localisations and Grothendieck Categories*, VEB Deutscher Verlag der Wissenschaften, Berlin (1975).

[34] M. P. Cavaliere and G. Niesi, On Serre's conditions in the form ring of an ideal, *J. Math. Kyoto Univ.* **21** (1981), 537–546.

[35] E. Cline, B. Parshall and L. Scott, Derived categories and Morita theory, *J. Algebra* **104** (1986), 397–409.

[36] E. Cline, B. Parshall and L. Scott, Algebraic stratification in representation categories, *J. Algebra* **117** (1988), 504–521.

[37] E. Cline, B. Parshall and L. Scott, Finite dimensional algebras and highest weight categories, *J. Reine Angew. Math.* **391** (1988), 85–99.

[38] E. Cline, B. Parshall and L. Scott, Integral and graded quasi-hereditary algebras, I, *J. Algebra* **131** (1990), 126–160.

[39] E. Cline, B. Parshall, L. Scott and W. van der Kallen, Rational and generic cohomology, *Invent. Math.* **39** (1977), 143–163.

[40] C. De Concini and C. Procesi, A characteristic free approach to invariant theory, *Adv. Math.* **21** (1976), 330–354.

[41] P. Deligne, Cohomologie à support propre et construction du foncteur $f^!$, in [69], pp. 404–421.

[42] M. Demazure, Schémas en groupes réductifs, *Bull. Soc. Math. France* **93** (1965), 369–413.

[43] M. Demazure and P. Gabriel, *Groupes Algébriques I*, Masson & Cie/North-Holland, Paris/Amsterdam (1970).

[44] V. Dlab and C. M. Ringel, Quasi-hereditary algebras, *Illinois J. Math.* **33** (1989), 280–291.

[45] V. Dlab and C. M. Ringel, The module theoretical approach to quasi-hereditary algebras, in H. Tachikawa and S. Brenner (eds.), *Representations of algebras and related topics* (Tsukuba, 1990), *London Mathematical Society Lecture Note Series* **168**, Cambridge (1992), pp. 200–224.

[46] S. Donkin, A filtration for rational modules, *Math. Z.* **177** (1981), 1–8.

[47] S. Donkin, *Rational representations of algebraic groups. Tensor products and filtration*, *Lecture Notes in Mathematics* **1140**, Springer-Verlag (1985).

[48] S. Donkin, On Schur algebras and related algebras, I, *J. Algebra* **104** (1986), 310–328.

[49] S. Donkin, Skew modules for reductive groups, *J. Algebra* **113** (1988), 465–479.

[50] S. Donkin, On tilting modules for algebraic groups, *Math. Z.* **212** (1993), 39–60.

[51] S. Donkin, On Schur algebras and related algebras, III. Integral representations, *Math. Proc. Cambridge Philos. Soc.* **116** (1994), 37–55.

[52] S. Donkin, *The q-Schur Algebra*, *London Mathematical Society Lecture Note Series* **253**, Cambridge (1998).

[53] J. A. Eagon and D. G. Northcott, Ideals defined by matrices and a certain complex associated with them, *Proc. Roy. Soc. Ser. A* **269** (1962), 188–204.

[54] D. Eisenbud and S. Goto, Linear free resolutions and minimal multiplicity, *J. Algebra* **88** (1984), 89–133.

[55] H.-B. Foxby, Isomorphisms between complexes with applications to the homological theory of modules, *Math. Scand.* **40** (1977), 5–19.

[56] P. Freyd, *Abelian categories. An introduction to the theory of functors*, Harper & Row (1964).

[57] E. Friedlander, A canonical filtration for certain rational modules, *Math. Z.* **188** (1985), 433–438.

[58] P. Gabriel, Des categories abéliennes, *Bull. Soc. Math. France* **90** (1962), 323–448.

[59] S. Goto and K.-i. Watanabe, On graded rings, I, *J. Math. Soc. Japan,* **30** (1978), 179–213.

[60] S. Goto and K.-i. Watanabe, On graded rings, II ($\mathbb{Z}^n$-graded rings), *Tokyo J. Math.* **1** (1978), 237–261.

[61] S. Greco and M. G. Marinari, Nagata's criterion and openness of loci for Gorenstein and complete intersection, *Math. Z.* **160** (1978), 207–216.

[62] J. A. Green, *Polynomial Representations of GL_n*, *Lecture Notes in Mathematics* **830**, Springer-Verlag (1980).

[63] A. Grothendieck, Sur quelques points d'algèbre homologique, *Tôhoku Math. J.* (2) **9** (1957), 119–221.

[64] A. Grothendieck, *Eléments de Géométrie Algébrique III*, *IHES Publ. Math.* **11** (1961), **17** (1963).

[65] A. Grothendieck, *Eléments de Géométrie Algébrique IV*, *IHES Publ. Math.* **20** (1964), **24** (1965), **28** (1966), **32** (1967).

[66] A. Grothendieck et al., *Revêtements Etales et Groupe Fondamental, SGA 1*, *Lecture Notes in Mathematics* **224**, Springer-Verlag (1971).

[67] D. Happel, On the derived category of a finite dimensional algebra, *Comm. Math. Helv.* **62** (1987), 339–389.

[68] D. Happel and C. M. Ringel, Tilted algebras, *Trans. Amer. Math. Soc.* **274** (1982), 399–443.

[69] R. Hartshorne, *Residues and Duality*, *Lecture Notes in Mathematics* **20**, Springer-Verlag (1966).

[70] R. Hartshorne, *Local Cohomology. A seminar given by A. Grothendieck, Harvard University, Fall, 1961*, *Lecture Notes in Mathematics* **41**, Springer-Verlag (1967).

[71] R. Hartshorne, *Algebraic Geometry*, *Graduate Texts in Math.* **52**, Springer-Verlag (1977).

[72] M. Hashimoto, Determinantal ideals without minimal free resolutions, *Nagoya Math. J.* **118** (1990), 203–216.

[73] M. Hashimoto, Resolutions of determinantal ideals: t-minors of $(t+2) \times n$ matrices, *J. Algebra* **142** (1991), 456–491.

[74] M. Hashimoto, *Auslander–Buchweitz Approximations of Equivariant Modules*, Tokyo Metropolitan Univ. Seminar Notes (1997), in Japanese.

[75] M. Hashimoto, Homological aspects of equivariant modules: Matijevic–Roberts and Buchsbaum–Rim, in D. Eisenbud (ed.), *Commutative Algebra, Algebraic Geometry, and Computational Methods* (Hanoi, 1996), Springer-Verlag (1999), pp. 259–302.

[76] M. Hashimoto, Good filtrations of symmetric algebras and strong F-regularity of invariant subrings, to appear in *Math. Z.*

[77] M. Hashimoto and K. Kurano, Resolutions of determinantal ideals, in *1989 Algebraic geometry symposium* (Kinosaki, 1989), (1990), pp. 116–133.

[78] A. Hattori, Semisimple algebras over a commutative ring, *J. Math. Soc. Japan* **15** (1963), 404–419.

[79] A. Heller, Homological algebra in abelian categories, *Ann. Math.* (*2*) **68** (1958), 484–525.

[80] J. Herzog and E. Kunz, *Der kanonische Modul eines Cohen–Macaulay-Rings*, *Lecture Notes in Mathematics* **238**, Springer-Verlag (1971).

[81] W. Hesselink, Depth and normal flatness, two examples, *Math. Nachr.* **79** (1977), 189–191.

[82] G. Hochschild, Note on relative homological dimension, *Nagoya Math. J.* **13** (1958), 89–94.

[83] M. Hochster, *Topics in the homological theory of modules over commutative rings*, *AMS CBMS* **24**, AMS (1975).

[84] M. Hochster and J. A. Eagon, Cohen–Macaulay rings, invariant theory, and the generic perfection of determinantal loci, *Amer. J. Math.* **93** (1971), 1020–1058.

[85] M. Hochster and L. J. Ratliff, Jr., Five theorems on Macaulay rings, *Pacific J. Math.* **44** (1973), 147–172.

[86] M. Hochster and J. Roberts, Rings of invariants of reductive groups acting on regular rings are Cohen–Macaulay, *Adv. Math.* **13** (1974), 115–175.

[87] J. E. Humphreys, *Linear Algebraic Groups*, *GTM* **21**, Springer (1975).

[88] J. E. Humphreys, *Introduction to Lie Algebras and Representation Theory*, *GTM* **9**, seventh corrected printing, Springer (1997).

[89] S. Iitaka, *Algebraic Geometry*, *GTM* **76**, Springer (1982).

[90] J. C. Jantzen, *Representations of algebraic groups*, Academic Press (1987).

[91] M. Kaneda and S. Takashima, Tilting modules in algebraic groups, in *Proceedings of the 6th symposium on representation theory of algebras*, Tateyama (1997), pp. 57–81.

[92] I. Kaplansky, Projective modules, *Ann. Math.* (2) **68** (1958), 372–377.

[93] C. Kassel, *Quantum Groups*, *GTM* **155**, Springer (1995).

[94] T. Kawasaki, On Macaulayfication of Noetherian schemes, *Trans. Amer. Math. Soc.* **352** (2000), 2517–2552.

[95] G. Kempf, F. Knudsen, D. Mumford and B. Saint-Donat, *Toroidal Embeddings, I*, *Lecture Notes in Mathematics* **339**, Springer-Verlag (1973).

[96] S. L. Kleiman, Geometry on Grassmannians and applications to splitting bundles and smoothing cycles, *IHES Publ. Math.* **36** (1969), 281–297.

[97] K. Kurano, On relations on minors of generic symmetric matrices, *J. Algebra* **124** (1989), 388–413.

[98] K. Kurano, Relations on pfaffians, I. Plethysm formulas, *J. Math. Kyoto Univ.* **31** (1991), 713–731.

[99] A. Lascoux, Syzygies des variétés déterminantales, *Adv. Math.* **30** (1978), 202–237.

[100] D. Lazard, Autour de la platitude, *Bull. Soc. Math. France* **97** (1969), 81–128.

[101] S. Lichtenbaum, On the vanishing of Tor in regular local rings, *Illinois J. Math.* **10** (1966), 220–226.

[102] J. Lipman, *Notes on Derived Categories and Derived Functors*, preprint 1997, available at `http://www.math.purdue.edu/~lipman`

[103] W. Lütkebohmert, On compactification of schemes, *Manuscripta Math.* **80** (1993), 95–111.

[104] I. G. Macdonald, *Symmetric Functions and Hall Polynomials*, second edition, Oxford University Press (1995).

[105] S. Mac Lane, *Homology*, reprint of the 1975 edition, Springer (1995).

[106] S. Mac Lane, *Categories for the Working Mathematician*, second edition, *GTM* **5**, Springer-Verlag (1998).

[107] O. Mathieu, Filtrations of G-modules, *Ann. Sci. École Norm. Sup.* (4) **23** (1990), 625–644.

[108] J. Matijevic, Three local conditions on a graded ring, *Trans. Amer. Math. Soc.* **205** (1975), 275–284.

[109] J. Matijevic and P. Roberts, A conjecture of Nagata on graded Cohen–Macaulay rings, *J. Math. Kyoto Univ.* **14** (1974), 125–128.

[110] H. Matsumura, *Commutative Ring Theory*, first paperback edition, Cambridge (1989).

[111] J. S. Milne, *Étale Cohomology*, Princeton (1980).

[112] J.-i. Miyachi, Duality for derived categories and cotilting bimodules, *J. Algebra* **185** (1996), 583–603.

[113] Y. Miyashita, Tilting modules of finite projective dimension, *Math. Z.* **193** (1986), 113–146.

[114] S. Montgomery, *Hopf Algebras and their Actions on Rings*, *AMS CBMS* **82**, AMS (1993).

[115] D. Mumford, J. Fogarty and F. Kirwan, *Geometric Invariant Theory*, third edition, Springer (1994).

[116] J. P. Murre, *Lectures on an Introduction to Grothendieck's Theory of the Fundamental Group*, Tata Institute, Bombay (1967).

[117] M. Nagata, Complete reducibility of rational representations of a matrix group, *J. Math. Kyoto Univ.* **1** (1961), 87–99.

[118] M. Nagata, A generalization of the imbedding problem of an abstract variety in a complete variety, *J. Math. Kyoto Univ.* **3** (1963), 89–102.

[119] M. Nagata, Some questions on Cohen–Macaulay rings, *J. Math. Kyoto Univ.* **13** (1973), 123–128.

[120] A. Neeman, The derived category of an exact category, *J. Algebra* **135** (1990), 388–394.

[121] A. Neeman, The Grothendieck duality theorem via Bousfield's techniques and Brown representability, *J. Amer. Math. Soc.* **9** (1996), 205–236.

[122] T. Ogoma, Existence of dualizing complexes, *J. Math. Kyoto Univ.* **24** (1984), 27–48.

[123] R. S. Pierce, *Associative Algebras*, *GTM* **88**, Springer (1982).

[124] N. Popescu, *Abelian Categories with Applications to Rings and Modules*, Academic Press (1973).

[125] P. Pragacz and J. Weyman, Complexes associated with trace and evaluation. Another approach to Lascoux's resolution, *Adv. Math.* **57** (1985), 163–207.

[126] D. Quillen, Higher algebraic K-theory, I, in *Higher K-theories* (Seattle, 1972), *Lecture Notes in Mathematics* **341**, Springer-Verlag (1973), pp. 85–147.

[127] A. Ragusa, On openness of H_n-locus and semicontinuity of nth deviation, *Proc. Amer. Math. Soc.* **80** (1980), 201–209.

[128] M. Raynaud, Flat modules in algebraic geometry, *Composito Math.* **24** (1972), 11–31.

[129] M. Raynaud and L. Gruson, Critères de platitude et de projectivité. Techniques de "platification" d'un module, *Invent. Math.* **13** (1971), 1–89.

[130] J. Rickard, Derived categories and stable equivalence, *J. Pure Appl. Algebra* **61** (1989), 303–317.

[131] C. M. Ringel, The category of modules with good filtrations over a quasi-hereditary algebra has almost split sequences, *Math. Z.* **208** (1991), 209–223.

[132] J. Roberts and J. Weyman, A short proof of a theorem of M. Hashimoto, *J. Algebra* **134** (1990), 144–156.

[133] P. Roberts, Le théorème d'intersection, *C. R. Acad. Sci. Paris Sér. I Math.* **304** (1987), 177–180.

[134] L. L. Scott, Simulating algebraic geometry with algebra, I. The algebraic theory of derived categories, in *The Arcata Conference on Representations of Finite Groups* (Arcata, 1986), *Proc. Symp. Pure Math.* **47** (1987), pp. 271–281.

[135] C. S. Seshadri, Geometric reductivity over arbitrary base, *Adv. Math.* **26** (1977), 225–274.

[136] R. Y. Sharp, Finitely generated modules of finite injective dimension over certain Cohen–Macaulay rings, *Proc. London Math. Soc.* (3) **25** (1972), 303–328.

[137] N. Spaltenstein, Resolutions of unbounded complexes, *Composito Math.* **65** (1988), 121–154.

[138] T. A. Springer, Linear algebraic groups, in A. N. Parshin and I. R. Shafarevich (eds.), *Algebraic Geometry IV*, EMS **55**, Springer (1994).

[139] T. Svanes, Coherent cohomology on Schubert subschemes of flag schemes and applications, *Adv. Math.* **14** (1974), 369–453.

[140] M. E. Sweedler, *Hopf Algebras*, Benjamin (1969).

[141] R. W. Thomason and T. Trobaugh, Higher algebraic K-theory of schemes and of derived categories, in *The Grothendieck Festschrift, Vol. III*, *Progr. Math.* **88**, Birkhäuser (1990), pp. 247–435.

[142] J.-L. Verdier, Base change for twisted inverse image of coherent sheaves, in *Algebraic Geometry* (*Internat. Colloq.*), (Bombay, 1968), Oxford Univ. Press (1969), pp. 393–408.

[143] J.-L. Verdier, Catégories dérivées, quelques résultats (état 0), in *Cohomologie étale, SGA* $4\frac{1}{2}$, *Lecture Notes in Mathematics* **569**, Springer-Verlag (1977), pp. 262–311.

[144] C. A. Weibel, *An Introduction to Homological Algebra*, Cambridge (1994).

[145] D. Woodcock, Schur algebras and global bases: new proofs of old vanishing theorems, *J. Algebra* **191** (1997), 331–370.

[146] J. Xu, *Flat Covers of Modules, Lecture Notes in Mathematics* **1634**, Springer-Verlag (1996).

[147] Y. Yoshino, *Cohen–Macaulay Modules over Cohen–Macaulay Rings, London Mathematical Society Lecture Note Series* **146**, Cambridge (1990).

[148] Y. Yoshino, Cohen–Macaulay approximation, *Proceedings of the 4th Symposium on Representation Theory of Algebras*, Shimoda (1993), pp. 119–138, in Japanese.

Index